Baobab

John Rashford

Baobab

The Hadza of Tanzania and the Baobab as Humanity's Tree of Life

Springer

John Rashford
Department of Sociology and Anthropology
College of Charleston
Charleston, SC, USA

ISBN 978-3-031-26469-6 ISBN 978-3-031-26470-2 (eBook)
https://doi.org/10.1007/978-3-031-26470-2

This Springer imprint is published by the registered company Springer Nature Switzerland AG
The registered company address is: Gewerbestrasse 11, 6330 Cham, Switzerland

Introduction

Modern humans are the only living representative of a founding population that diverged from chimpanzees some four to eight million years ago, giving rise to a branching group of bipedal African apes called hominins. The baobab tree (*Adansonia digitata* L.) is likely to have been an important partner in the story of hominin evolution; it is not only the most conspicuous tree of Africa's woodlands and grasslands but also widely recognized as one of the most useful. Understandably, the theme of this study is the baobab as the exemplary multipurpose tree of the mosaic savanna of hominin evolution.

"Tree of life" identifies a tree whose exceptional nature is that it is life manifesting, life giving, and life representing, and it is noteworthy that the baobab is already widely referred to as *a* tree of life, *the* tree of life, and Africa's tree of life. There is a growing awareness of the baobab because it is often featured in nature documentaries as emblematic of Africa and of the African savanna. The tree and its image and tree-of-life name are often incorporated into the logos of diverse organizations; they are referenced in children's books and academic publications, and they are a growing part of popular culture generally. They are, for example, associated with theme parks, films, and other forms of commercial entertainment. And now the tree, image, and name are also linked to the increasing international trade in the fruit pulp and seeds, which are promoted for their health benefits in food, pharmaceutical, and cosmetic products. What is particularly striking, however, is that the baobab is not only known as a tree of life and the tree of life; it has also been identified as humanity's ancestral tree of life. Speculations that the baobab was important in human evolution are not new. Hints of this idea appear in Owen's (1968:36) characterization of the baobab as the "silent chronicler of events down the centuries," in Blurton Jones's (2016:17) reflection on the long history between the Hadza and their baobab trees, and in various cultural cosmologies in which human origin is linked to the baobab. This idea occurs most famously in the title of Peter Matthiessen's classic 1972 book, *The Tree Where Man Was Born*.

This book makes the case for identifying the baobab as the tree of life in the hunting and gathering adaptation that was the economic foundation of hominin evolution until as recently as some twelve thousand years ago when the cultivation of plants and the keeping of livestock developed. The specific purpose of this study is to provide support for accepting as a reasonable assumption the theory of the baobab as humanity's ancestral tree of life. This support involves two premises. First, the overall argument presented is based on the importance of the baobab as an exceptional resource environment for Hadza foragers of northeastern Tanzania; Hadza savanna foraging is an historically contextualized instance of early hominin foraging in the same environment. Second, the baobab is the Hadza exemplary multipurpose tree, and the idea of an exemplary multipurpose tree is fundamental to the definition of a tree of life. The baobab's extraordinary appearance, longevity, diverse ecological associates, and numerous uses – especially uses associated with its abundant fruiting, long harvest period, and nutritious fruit pulp and seeds – have made it the outstanding resource tree of the Hadza; it is their tree of life, so to speak. These premises provide support for the theory of the baobab as the tree of life in the forager adaptation of early hominins.

This book stems from an interest in the baobab's importance for Africans, especially foragers like the Hadza and San. I first saw baobab trees in East Africa when, as an undergraduate, I spent almost a year traveling in Uganda, Kenya, Tanzania, and Ethiopia. I saw them again in 1971 when, as a group leader for Crossroads Africa, I traveled with my fellow Crossroaders north from Wenchi, Ghana, to Ouagadougou, the capital of Burkina Faso, and from there south by train through the length of the Ivory Coast to Abidjan, the country's capital. Although I was duly impressed with the magnificence of the baobab, as I am sure most people are when they see a mature baobab for the first time, I initially had no interest in learning more about the tree. That interest developed when I began encountering old baobab in the Caribbean while researching the cultural significance of the cotton tree (*Ceiba pentandra L.*) in the region (Rashford 1985). My incidental discovery of the species in Antigua was a surprise that made me realize how little was known about the ecology, history, and cultural significance of the baobab in the Caribbean, and in the Americas generally (e.g. Rashford 1987a, b, 1991, 1992, 1994, 1995, 1997b, 2022). In time, my interest in the introduction of the baobab to the Americas made me curious to learn more about its out-of-Africa dispersal, especially to India and to other parts of the Indian Ocean (Rashford 1993, 2019). It was not long, however, before my interest would lead to a curiosity about the baobab in human evolutionary history, given the presence of the tree in the landscape of the African savanna.

The book is divided into seven sections. The first chapter of Part I identifies the distinguishing features of the baobab that allow us to characterize it as a tree of life, and the second introduces both the Hadza as successful foragers of the landscape of human origins and Hadza studies that document the use of the tree.

The two chapters of Part II present the development of hominin gendered foraging supported by the correlated evolution of handiness and bipedality as the general theoretical framework for identifying the baobab as the Hadza and Humanity's tree of life.

The three chapters of Part III begin the discussion of the importance of material culture in hominin evolution and the importance of the baobab in Hadza material culture. The baobab as Africa's premier fiber tree is highlighted in Chap. 5; Chapter

This book is dedicated to the Hadza who are as tenacious as their baobab tree.

Preface

Modern humans, descendants of a founding population that separated from chimpanzees theorized to be anywhere from four to eight million years ago, are today the only living representative of a branching group of African apes called hominins. Because of its extraordinary size and shape, the baobab (*Adansonia digitata* L.) has long been identified as the most striking tree of Africa's mosaic savanna, the landscape generally regarded as the environment of hominin evolution. This book makes the case for identifying the baobab as a "tree of life"—the exceptional life-manifesting, life-giving, and life-representing tree—in the development of the hunter-gatherer adaptation that became the economic foundation of hominin evolution. The argument is based on the significance of the baobab a resource-rich environment for the Hadza of northeastern Tanzania, who continue to be successful hunter-gatherers of the African savanna. The Hadza are known for having one of the best baobab cultures in the world, though it is yet to be documented in a systematic way. They are also widely recognized by researchers from various disciplines as among the most important hunting and gathering cultures for theorizing hominin evolution in association with Africa's mosaic savanna (Hawkes 2016). In May 1991, the *Deseret News* published an article describing an exhibition of forty-five Hadza photographs at the Utah Museum of Natural History. They were taken by James F. O'Connell, an archaeologist at the University of Utah well known for his Hadza research, and the exhibition was aptly titled "Children of the baobab: Growing up Hadza." As foragers in the landscape of human evolution, the Hadza model for us the possible practical and inspirational significance of the baobab for prehistoric hominins, especially species of the genera *Australopithecus* and *Homo*. From this perspective, we are all in a sense "children of the baobab."

In his influential article on the medical, social, and cultural significance of the baobab in African communities, the anthropologist John Owen (1968:36) referred to the baobab as a "silent chronicler of events down the centuries," noting that archaeologists, physical anthropologists, cultural anthropologists, ecologists, historians, and folklorists "can learn a great deal from studying this very unusual tree." Blurton Jones (2016:17), another major contributor to the development of Hadza studies, wrote: "Given that baobab trees have been found by radiocarbon dating to live up to about 1000 years (Patrut et al. 2007), we can suspect that the Hadza have not been without these and other of their staple food plants at any time in the last thousand years." There is indeed a lot to be learned from studying the baobab/human association. But if we are to think of ourselves as "children of the baobab," we will need a much more expanded time frame than suggested by "centuries," or even millennia. It is in this context of the long view that the baobab is identified in this book as humanity's tree of life. The rich baobab culture of the Hadza makes them an appropriate group for exploring the idea of the baobab as humanity's tree of life because many Hadza remain viable foragers in the ancestral landscape of human evolution.

Charleston, SC, USA John Rashford

6 discusses the importance of containers in hominin evolution and looks at the many ways in which the baobab is linked to Hadza container use; and Chap. 7 discusses the significance of fire in hominin evolution and the baobab's many contributions to the Hadza use of fire.

The two chapters of Part IV discuss Hadza baobab use in relation to environmental considerations. The baobab and the Hadza use of water is the focus of Chap. 8. Chapter 9 discusses the seasonal value of the baobab for the Hadza. This seasonal importance as a resource-rich environment is relevant to many aspects of the argument for identifying the baobab as a tree of life for the Hadza and the tree of life in hominin evolution.

The eight chapters of Part V assess the baobab's practical value for the Hadza, includes fruit, beverage, honey, birds, other animal foods, hunger-time tree of life, health, and exchange. The discussion is based on comparing and contrasting the Hadza with early hominins. The Hadza are theorized to be similar to early hominins with respect to foraging on the African savanna. From the point of view of similarities, if the traditional Hadza use of the baobab does typify the use of the tree in hominin evolution, then the Hadza can be viewed as a window on the importance of the baobab as a resource-rich environment in hominin evolution; from this perspective, the baobab can justly be described as humanity's tree of life. However, the Hadza are also clearly different from early hominins who were foragers of the African savanna before the Broad-Spectrum Revolution, the origin of agriculture, and the development of the world market. In terms of the contrast between the Hadza and early hominins, the Hadza have long been subjected to the influences of increasing resource competition and environmental and social circumscription involving their traditional food-producing agricultural, pastoral, and agropastoral neighbors, and especially the powerful influence of local, national, and global commerce and politics (Bird-David 1992).

The baobab's inspirational value for human beings has been largely overlooked in the literature. This aspect of the tree's significance for the Hadza and other Africans is discussed in the five chapters of Part VI. This discussion includes the baobab's widespread links to various kinds of spiritual beings, its status as a fertility tree, its contribution to aesthetics, oral traditons, recreation, and education, and its many associations with danger and death.

The three chapters of Part VII discuss the Hadza influence on baobab regeneration. The first chapter describes Hadza central-place residential camps and discusses the baobab as a preferred camping site for the Hadza. The second considers the ways in which Hadza camping may be influencing the regeneration of the baobab in their environment; it looks at human dispersal of the baobab from the vantage point of the coevolution of humans and baobab. The third considers the baobabs landscape of the Hadza their ultimate retreat.

This is a good time to write a book on the baobab as our ancestral tree of life since it is already widely identified in the popular imagination as *a* tree of life and as *the* tree of life. It is also identified today as the new "cinderella tree," the new "super fruit," and "the world's newest super food" (Starin 2009). As first described by Roger Leakey of The International Center for Research in Agroforestry (ICRAF), "cinderella trees" are those that until recently have largely been overlooked by science, industry, and the media, even though they clearly have the potential to alleviate poverty and contribute to food security by providing a variety of nutritious foods that are capable of long-term storage. The global commercialization of baobab fruit pulp and seeds is proceeding at a rapid pace and is full of promise. The justification for this commodification is that it will promote local development through international trade aimed at improving rural life, especially for women and children, and that it will provide income-producing work associated with the harvesting, gathering, processing, packaging, and transporting of baobab products. There is, however, cause for concern regarding the conservation of the species and the rich biological community it supports, and the task of ensuring that local communities continue to have access to the fruit and other baobab resources.

At issue is the fact that at this stage of market development, baobab fruits are harvested from trees in traditional landscapes that benefit local populations rather than from trees that have been systematically planted to meet the growing national and international market demand. In addition, some traditional harvesting techniques used to supply the commercial market, such as breaking or chopping off branches to harvest the fruit and leaves, are destructive. These concerns have led Starin (2009) to title her op-ed article in *The New York Times* "What Will Happen When the Baobab Goes Global?"

> The baobab was approved for European markets last year, and the Food and Drug Administration is expected to follow suit soon. The fruit's dry pulp will be sold as an ingredient in smoothies and cereal bars. Already, a small jar of African baobab jam made in England sells for around $11. According to the National Resources Institute in Britain, an international baobab industry could bring in about $1 billion a year and provide jobs for 2.5 million African families.

The aim of Starin's article was to warn that "there's another side to the picture," and this is a warning that has also been voiced by others (Buchmann et al. 2010). In making this point, Starin would have been more accurate in titling her article "What Will Happen Now That the Baobab Has Gone Global?" even though we are still at the early stages of this development. "Fair Partnership" agreements suggest fruit-growers earn more from selling whole fruit than from selling fully transformed products on local markets, and women in particular who process baobab in rural communities in Benin earn much more than they would from other activities. This fact raises the prospect of market competition for the fruit that will be to the detriment of local fruit consumption, and it highlights the immediate need for the commercial production of the baobab based on selective breeding of the most productive varieties, a development that also raises important issues regarding the protection of baobab biodiversity (Raebild et al. 2011; Jensen 2011).

The ideal of commercialization sounds positive, but what about all the people who have traditionally depended on local markets and on largely naturally occurring trees for their baobab products? The Hadza are on the front line of this process, and the fate of their baobabs will be the fate of the Hadza themselves. What will happen to the Hadza way of life when they face stiff competition for the fruit of the baobab in the wake of its global commercialization? Consider, for example, the following observation of Blurton Jones (2016:21):

> In 2000, Gudo Mahiya and I were distressed to see near Yaeda, that "Swahilis" were coming with sacks and donkeys to collect and take away baobab fruit, presumably to transport for sale in towns and villages. The developing "commercialization" of baobab (Kamatou et al. 2011) may become the next way in which the Hadza will be deprived of their heritage.

The baobab is not yet heavily commercialized in Hadza territory. Its current economic value is associated chiefly with honey and fruit, and with "safari tourism " and recreational activities such as baobab sightseeing, tree-climbing, and fruit tasting.

It is hoped that this study of the relationship between humanity and this most extraordinary of trees will be of interest not only to researchers, teachers, and students, but also to policymakers whose decisions could determine the fate of the Hadza, their baobab trees, and their rights to the annual baobab harvest of their home environment (Prance 2007). It is also hoped that this study will enrich not only the perspective of those who live with the baobab, including everyone who enjoys it in their garden or as a potted plant or bonsai specimen, but also those who have seen it in their travels, and even those who know it in the form of jewelry, paintings, carvings, photographs, or the subject of religious and literary works.

I thank my family, friends, colleagues, and students; an extended work such as this is never successfully completed without their sustained support. I am also indebted to all Hadza researchers whose studies have made this assessment of Hadza baobab use possible. Completing this book would have taken much longer without the encouraging comments from colleagues on my baobab-related papers presented at the annual meeting of the Society for Economic Botany. A special thanks to my colleagues of the College of Charleston's Department of Sociology and Anthropology. Hector Qirko, in particular, provided helpful comments on an early draft of this book, and Allison Foley for her helpful discussion and comments regarding hominin evolution. A special thanks also to Larry Lepionka, Steve Hoffius, and Bruce Tomaso for their expert editorial assistance. I am very appreciative of the assistance I received from the College of Charleston's Marlene and Nathan Addlestone Library, especially the library's interlibrary loan service. This book has definitely benefited from the many valuable comments received from reviewers.

Fig. 1 Baobab Flower (1), examples of rounded and elongated baobab fruit shape (2 and 3), with elongated fruit showing the arrangement of the fruit pulp (3), and baobab seeds (4)

Fig. 2 Hadza hunters headed in the direction of the baobab as a resource-rich environment

Fig. 3 The Hadza use a peg ladder to scale the massive trunk of the baobab for a variety of resources including honey, fruit, water, and bark fiber

Fig. 4 Hadza woman pounding baobab fruit pulp to make baobab powder. (Photograph by Alyssa Crittenden)

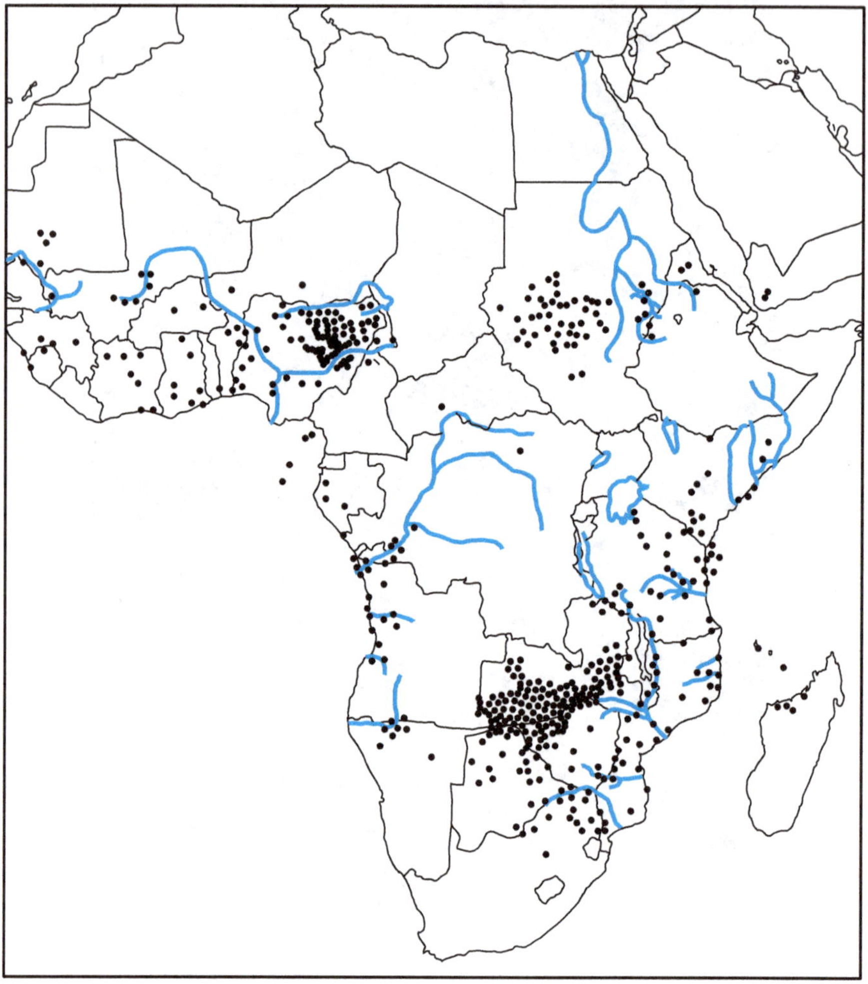

Fig. 5 Map of Africa. The general study area (including Lake Eyasi) is indicated in red and is shown in Google Maps on the opposite page. (The map is copied from "One Stop Map [online]: Vector Map Africa continent shaded relief")

Fig. 6 Google satellite map of northern Tanzania with Lake Eyasi to the southwest of the volcanic highlands

Contents

Part I
The Baobab

Chapter 1
The Distinguishing Features of the Tree of Life and the Baobab

In popular culture, the baobab is identified not only as a symbol of the Africa continent and its diverse peoples but also as Africa's tree of life (Layser 2001: 152). A number of trees have been identified as trees of life in scholarly works, general publications, and online sources, and with the notable exception of evergreens, the baobab shares the defining tree-of-life features of all of them. However, the baobab differs from all other trees of life because we can theorize that as a part of Africa's mosaic savanna, it has had an enduring association with the evolution of our species. That is, during the longest and most consequential period of hominin history, the baobab would likely have been among the trees earliest recognized in the human imagination as life-manifesting, life-giving, and life-representing. Kaare and Woodburn (1999) described the Hadza landscape as "dominated by the fan acacias and baobab trees which are so familiar to viewers of television wildlife documentaries." It is probably true that most readers of this book are now familiar with the baobab's extraordinary appearance and many uses. The same cannot be said of its likely role in human history, which deserves to be better known. If the baobab is indeed the tree of life in the landscape of our evolutionary history, then it can be identified as humanity's ancestral tree of life. This chapter, which is in three parts, discusses the distinguishing features of trees of life and highlights the baobab as the exemplary representative. The first describes the baobab; the second discusses its evolution; and the third identifies the kinds of trees of life and the criteria for regarding the baobab as the exemplary multipurpose tree of life.

Baobab Description

Its imposing presence has made the baobab the most celebrated member of a small group of Old-World tropical trees in the Malvaceae family. In addition to the African baobab, there are six species endemic to Madagascar, the center of the genus's diversity, and one to Australia (Baum 1995a, b; Baum et al. 1998; Wickens and

J. Rashford, *Baobab*, https://doi.org/10.1007/978-3-031-26470-2_1

Lowe 2008). Pettigrew et al. (2012) claim to have identified a second species *Adansonia* that is native to Africa, but a recent study suggests this is not the case (Cron et al. 2016). Linnaeus named the genus *Adansonia* in honor of the celebrated French naturalist Michel Adanson, who encountered the African baobab while exploring and collecting in Senegal from 1749 to 1754. Adanson was the first to give a technical description of the tree and to provide a variety of accurate illustrations (Adanson 1757, 1771). Bernard de Jussieu was Adanson's teacher, and it was his report of Adanson's findings that led Linnaeus to mention the tree in his *Species Plantarum*, published in 1753. The species epithet, *digitata*, is in reference to the tree's alternate compound leaves, with their long leaf stalks and five oval-shaped leaflets radiating from the top like fingers from a hand. The smooth bark, which is black, grayish, or pinkish in color, contains an inner layer composed of tough longitudinal fibers.

The African baobab is readily distinguished by its immense trunk that seems strangely disproportionate to its thick, rapidly tapering branches (Pakenham 2004; Watson 2007). Young trees develop a conical shape by the time they begin flowering. From that point on, the ballooning shape of the baobab as it ages assumes different forms which generally have been characterized as cylindrical, barrel-like, bottle-shaped, and multistem. The baobab's extraordinary adaptation to the dry conditions of the African savanna – an adaptation that makes it the biggest tree of the continent – is evident in its wide-spreading shallow roots, moisture-laden wood, photosynthesizing bark, water-conserving leaves, deciduous habit, and well-protected seeds. The unique configuration of an obese trunk or trunks, a dome-shaped canopy of drooping branches, and many spreading surface roots are the essential characteristics of old baobabs, especially of ancient baobabs, which are only found in Africa (Rashford 2022).

The baobab sheds its leaves after the autumn rains and is leafless in the winter dry season. Through the annual rains, it produces large glossy white flowers 12–15 cm in diameter. These flowers, which first appear with the new leaves of spring, are abundant in summer. They are suspended upside down at the end of long flexible stalks and are night-blooming, odiferous, nectar-rich, and bat-pollinated. From the flowers develop large, woody, gourd-like pods that are oblong in shape, up to 24 × 12 cm in size, and covered by what feels like velvet. These indehiscent fruits mature through summer and early autumn and ripen and fall in winter and spring, sometimes lasting into the summer. Fruits vary widely in size with a thick shell and fuzzy exterior; their variable shape can be broadly categorized as rounded, oblong, oval, and elongated (Gurashi and Kordofani 2014; De Smedt et al. 2011; Sanchez et al. 2011). Baobab fruits may contain from 30 to more than 200 kidney-shaped seeds embedded in a white acidic pulp that is laced together by a mass of tough stringy fibers.

The baobab does not generally occur above 1500 meters. In eastern Africa, it is common in coastal environments from Kenya to Mozambique. Africa is a vast plateau whose coastal lowlands form a small part that never extends very far from the sea. The highest point of the African plateau is in the eastern part of the continent and includes such famous volcanic peaks as Kilimanjaro and Mount Kenya, which are geologically part of the great Rift Valley that stretches from the Zambezi River in Mozambique to the Jordan Valley. The baobab is also common in dry low-lying areas of the Rift Valley and of interior river valley systems (Pielou 1952). The tree is thought not to occur naturally in Uganda (Gebauer et al. 2016a, b) and is considered a human introduction into the highlands of Kenya and Tanzania (Wickens 1982: 181). The baobab is generally recognized as one of Africa's best-known multipurpose trees. In Tanzania, where over 30 uses have been recorded, it is one of the country's most important indigenous trees. According to Buchmann et al. (2010: 145), "Three hundred traditional uses of the baobab were documented in Benin, Mali, and Senegal across 11 ethnic groups and 4 agroecological zones."

The Tree of Life and the Kinds of Trees of Life

The prominence of the tree-of-life concept in human consciousness is evident in its association with specific trees, particular kinds of trees, trees in general, tree-like things, and things that are not even trees or tree-like; in its worldwide occurrence as a central motif of religious beliefs and practices; in its antiquity in diverse cultural traditions as documented by anthropologists, archaeologists, art historians, historians of religion, and folklorists; and especially in its contribution to theorizing biological evolution as the phylogenetic tree of life. It is surprising that recent books on the relationship between human beings and trees have not made use of the tree-of-life concept (Rival 1998; Anderson 2003; Jones and Cloke 2002). The concept merits consideration in this work, because baobab is already popularly identified not only as a tree of life, but also as the tree of life – the true, real, actual, or genuine tree of life. The baobab's tree-of-life image and name have been popularized in recent years because of their incorporation into organizational logos; their identification with theme parks, films, and other forms of entertainment; and their association with the growing international trade in baobab fruit pulp and seeds.

Because the tree-of-life concept is used to frame the discussion this book presents, it is necessary to specify exactly what kind of tree of life the baobab represents, and this means distinguishing the baobab from other kinds of trees of life. The popular identification of the baobab with the tree of life is only a start, however, since it is impossible to argue persuasively that the baobab is indeed the Hadza's and humanity's tree of life without considering the religious, philosophical, and scientific uses of the tree-of-life concept. To make the case for the baobab as the Hadza's and humanity's tree of life, this chapter identifies the ways in which the baobab compares to other trees and to other things that have also been identified as trees of life.

As familiar as the tree-of-life concept is, rarely has an attempt been made to systematically distinguish the kinds of trees of life. The six forms recognized in this book are the botanic, imaginative, generic, figurative, metaphorical, and symbolic trees of life, and they are based on identifying tree-of-life attributes, dictionary definitions, and popular usage. The three attributes that define a tree (or some other thing) as being a tree of life are that it is life-manifesting, life-giving, or life-representing or some combination of the three. The baobab is an exemplary tree of life because it manifests all three attributes. Dictionaries are an authoritative introduction to the meaning of the "tree of life" because they provide not only definitions, but also etymology, current usage, and illustrative examples. The second, unabridged *Webster's New International Dictionary of the English Language* (WNIDEL 1949) provides a convenient way to approach the tree-of-life concept because this dictionary gives the most complete characterization of the conventional view, containing as it does a reference to all the kinds of trees of life mentioned in other dictionaries. In addition to lexical definitions of tree-of-life attributes, the popular usage of the concept has contributed to the identification of the six kinds of trees of life recognized in this work.

Biblical Trees of Life

The Edenic Tree of Life

Dictionaries recognize the Biblical, representational, and botanical trees of life, and these three are incorporated into six forms of the tree of life discussed in this work. The most important of the Biblical trees of life are the Garden of Eden and Heavenly Jerusalem trees; both are imagined trees of life – the first mythical and the second envisioned. In the Jewish, Christian, and Islamic traditions, the mythical fruit-immortalizing tree of life was located "eastward in Eden" at the center of the beautiful, well-watered, mountaintop orchard "planted" by God as the original home of humanity. "And out of the ground made the LORD God to grow every tree that is pleasant to the sight, and good for food; the tree of life also in the midst of the garden, and the tree of the knowledge of good and evil" (Genesis 2: 12). Adam and Eve were to "till and keep" the garden and were free to eat all the fruits it produced, including the life-prolonging fruit of the tree of life, with one exception: they were forbidden to eat the fruit of the tree that held the knowledge of good and evil. Adam and Eve made the life-and-death choice to disobey God and eat the forbidden fruit. So, religious tradition holds, began the realities of human life as we know them today. It is noteworthy that the Garden of Eden's tree of life is the only meaning given to "tree of life" in both the published and online versions of several dictionaries (e.g., EWED 1999; WNRUD 1984; OEDO 2005).

The Heavenly Jerusalem Tree of Life

The Heavenly Jerusalem tree of life is mentioned by the Apostle John in his vision of "a new heaven and a new earth" (Revelations 21: 1). Recounting his vision he said (Revelation 22: 2–5):

> … the angel showed me the river of the water of life, as clear as crystal, flowing from the throne of God and of the Lamb down the middle of the great street of the city. On each side of the river stood the tree of life, bearing twelve crops of fruit, yielding its fruit every month. And the leaves of the tree are for the healing of the nations. No longer will there be any curse. The throne of God and of the Lamb will be in the city, and his servants will serve him. They will see his face, and his name will be on their foreheads. There will be no more night. They will not need the light of a lamp or the light of the sun, for the Lord God will give them light. And they will reign for ever and ever.

The Edenic and Heavenly Jerusalem trees are the most often mentioned in lexical definitions of "tree of life," and in several cases, they are the only trees identified as trees of life (e.g., RHDEL 1969; RHWCD 2000).

Representational Trees of Life

The third form of the Biblical trees of life identified in several dictionaries is the representational tree of life, which may be metaphorical, figurative, or symbolic. Metaphorical trees of life are non-tree things portrayed as if they were life-giving trees. The Biblical metaphorical trees of life include "wisdom" (Proverbs 3: 16–18), the "fruit of the righteous" (Proverbs 11: 30), "realized hope" (Proverbs 13: 12), and "a wholesome tongue" (Proverbs 15: 4).

The figurative tree of life is one of the most important expressions of the tree-of-life concept; it specifically represents anything called a tree of life that is also graphically tree-like in the sense of having a branching structure. Significant expressions of the figurative tree of life include such varied things as traditional cosmologies, cross, menorah, genealogy, phylogeny, human body, and brain. The figurative tree of life is noteworthy because it includes one of the oldest meanings of "tree of life" in the form of the tree-like cosmos and also because it is one of the most recent expressions of the concept with respect to biology and the phylogenetic or evolutionary tree of life (Cracraft and Donoghue 2004; Baum and Smith 2013).

The third representational form of the tree of life identified in Webster's dictionary (WNIDEL 1949) is the symbolic, and it includes the Oriental or archaeological tree of life, described as the "sacred life-preserving tree represented in the art of the ancient Near East, whence it probably was taken over by Aegean countries." Characterized as "a highly conventionalized and often ornate representation of a tree or vine used as a decorative motif," this Near Eastern tree of life has been much discussed in the literature, and it is identified with architecture, sculpture, carpets, tapestries, and paintings. The Near East symbolic tree of life is the only definition of "tree of life" given in the *Merriam-Webster Online* dictionary (MWO 2005).

The symbolic tree of life remains an important representational form of the tree of life, and its meaning is widely associated with strength, tenacity, longevity, continuity, fruitfulness, fertility, protection, and biodiversity. In popular culture, it is a highly stylized generic tree with ties to religion and aesthetics. In the latter case, it is especially familiar in the form of personal adornments, such as jewelry and tattoos, and in association with clothing, photographs, prints, paintings, stickers, wall hangings, and logos.

The category of the generic tree of life is added here because of its association with tree-of-life symbols. Generic trees of life, like imagined trees of life, are simply those presented in narratives and graphic illustrations that are not botanically specifiable. It is the "tree" in general that is generic tree of life. This is evident, for example, in popular tree-of-life symbols. Some represent actual trees, but many are examples of the generic tree of life.

Botanic Trees of Life

Evergreens

The tree of life is more widely known today as a representational tree with ties to aesthetics, religious, and scientific thought, rather than as a botanically specifiable tree whose practical and inspirational value identifies it as life-giving, life-manifesting, or life-representing. Practically, both fruitfulness and multipurpose use are attributes of the botanical tree of life as a symbol of abundance and, by extension, of fertility. Manifest trees of life convey the idea of vigor not only in their fruitfulness and multipurpose use but also in their size, shape, longevity, or evergreenness. The botanic tree of life is particularly important in this book, which is specifically concerned with actual trees of life, rather than simply with imagined, generic, or representational trees of life.

After the Biblical trees, the botanical is the second meaning of "tree of life" provided by Webster's dictionary (WNIDEL 1949), and the three trees so identified, in the order in which they are presented, are the arbor vitae of North America, Bhutan cypress of India, and date palm of the Near East. Exemplary botanical trees of life are often both life-manifesting and life-giving; with respect to these characteristics, dictionaries indicate that evergreenness and medicinal use are the distinguishing features that characterize arbor vitae as a tree of life. In contrast, multipurpose use characterizes the palm as a tree of life; it is the only multipurpose tree of life mentioned in dictionaries. A Dictionary of American English on Historical Principles (ADAEHP 1944) gives 1712 as the earliest reference to arbor vitae as a tree of life. In Webster's Dictionary (WD 1828), "tree of life" is identified only as an "evergreen tree of the genus *Thuja*." The same is true of other dictionaries (e.g., TFD 2004). Arbor vitae is a leaf-healing evergreen tree of life that is by far the botanical tree of life most frequently mentioned in dictionaries; it is noteworthy that some dictionaries regard "evergreen" as the original and most important meaning of the tree of life (e.g., Jobes 1962: 1595; Utley 2005). In temperate latitudes in particular, trees that shed their leaves in autumn and replace them in spring are generally thought to symbolize transience. But they can also symbolize permanence, with the loss and replacement of leaves conceived of as renewal, resurrection, rejuvenation, restoration, revival, or regeneration.

Evergreen trees like arbor vitae and other conifers represent permanence in contrast to the transitory nature of things expressed by deciduous trees. In botany, an evergreen, generally speaking, is a plant with green leaves throughout the year. However, evergreen specifically identifies the families, genera, and species of the order Coniferales; it includes some of the best-known trees of the families Pinaceae, Cupressaceae, and Taxodiaceae, such as pine, yew, cedar, and larch. Within the dualistic symbolic framework of human thinking, evergreen trees, particularly conifers, retain their green leaves in the face of the shortening, darkening, and cooling days of autumn, unlike deciduous trees, which mark the end of the year's growing seasons when their leaves turn brilliant colors, fall, and fade. Evergreens, not manifesting this seasonal death, exhibit instead an abundance of life's vigor; they have become the preeminent symbol of what endures, especially for the peoples of temperate Asia and Europe. Evergreens are ritually, aesthetically, and recreationally used to represent what is alive, fresh, thriving, and ultimately inspiring. They express the deep human desire to be free of death – forever living. Because of arbor vitae's name, all the dictionaries identify it as a tree of life; only *Webster's* (WNIDEL 1949) specified the Bhutan cypress as a second life-manifesting evergreen trees of life. The Bhutan cypress is the national tree of Bhutan. Tradition holds that it came from the walking stick of Guru Rinpoche, the second Buddha, who is said to have brought Buddhism to Bhutan.

Palms

A growing awareness of the multipurpose value of palms (Brondizio 2008) has led to their widespread designation as trees of life. With over 100 genera comprising some 2,800 species found throughout the tropics and subtropics, palms constitute one of the oldest and most diverse plant families in the world (Britannica 2006). After grasses and legumes, palms have been identified as the most important family of plants for human beings worldwide (Harrison et al. 1969: 194). They are considered trees of life primarily for their distinctive form and many uses (Mogea et al. 1991; Johnson 2012); this is what distinguishes palms as trees of life from the Biblical trees of life, from the arbor vitae, and from the Bhutan cypress. Palm fruits are highly valued as a source of food and oil, and the terminal bud and sap of several species are eaten. The large leaves are resistant to decay and are widely used for thatch and cordage and for weaving such things as containers and mats. And palm trunks are used in construction.

In addition to the evergreens, *Webster's Dictionary* (WNIDEL 1949) identified the date palm (*Phoenix dactylifera*) as a tree of life, and so did Nixon (1951). Like Jobes (1962: 1595), Utley (2005) regarded the tree of life as "a variety of evergreen," which he thought was "originally perhaps the Babylonian date palm." This multipurpose palm has an erect, branchless, tapering trunk up to 3 meters tall. Its distinctive trunk, its spherical crown of evergreen feather-like leaves, and its abundant clusters of fruit set it apart from all other trees of the warm temperate latitudes of North Africa and Western Asia, where it is widely dispersed; these characteristics have made it one of the most consistently identified trees of life and an important cultural influence in the development of the West. We have only to think of its practical and inspirational significance in the religious traditions of Ancient Egyptians, Jews, Christians, and Moslems.

The fourth edition of *The American Heritage Dictionary of the English Language* (TAHDEL 2000) is the only dictionary that identifies the moriche palm (*Mauritia flexuosa*) as a tree of life, and it does so without identifying any other tree as a tree of life. This palm is a native of northern South America with large fan-shaped leaves and diverse uses. Referred to in the dictionary as the Mauritia palm and now identified by the common name moriche palm, it is one of the most well known Amazonia multipurpose palms, easily recognized by its fan-shaped leaves, which are among the largest in the world. It is especially appreciated for its fruit, which is eaten fresh and used to flavor drinks and desserts. In addition to the date palm and coconut, the Forestry Department of the United Nation's Food and Agriculture Organization (FAO 2012) noted that a "number of other palms could similarly be represented as 'trees of life'" including the African oil palm (*Elaeis guineensis*), palmyra palm (*Borassus flabellifer*), babassu palm (*Attalea speciosa*), and pejibaye palm (*Bactris gasipaes*). Similarly, a 1949 report titled "The Society's Activities" published in the journal of the *Royal African Society* noted that for a joint meeting, "the United Africa Company was good enough to loan a film, titled *The Tree of Life* depicting the development of palm oil in the Belgian Congo" (Anonymous 1949: 160). According to Sowunmi (1985: 127):

> The [African] oil palm is one of the most important tree crops in the forest zone of West Africa today (e.g., Harris 1976: 326). Whilst its primary importance lies in its value as a source of cooking oils (obtained from both the pericarp and the kernel), practically every other part of the tree is used as well: the sap (later fermented) collected at the base of the inflorescences for palm wine, the shell and residual fibre of the pericarp for fuel, leaves for thatching, fine fibres of young leaflets for fishing lines, the trunk for rafters, fences, and bridges, and fibres from the base of the stem for cordage and fish traps (Dalziel 1937).

Given the many practical and inspirational uses of palms, it is easy to understand why Michael Balick (1988) titled his book, *The Palm – Tree of Life: Biology, Utilization and Conservation* and Ohler (1984) titled his book *Coconut: Tree of Life.*

Other Botanical Trees of Life

The full awareness of the multipurpose value of palms is the result of European oceanic expansion in the fifteenth century; it laid the foundation for the making of our present global economy, a process in which the palm became emblematic of the tropics. Preeminent among the tree-of-life palms is the now globally dispersed coconut (Persley 1992). "According to the South Sea proverb," wrote Lehner and Lehner (1962: 70), "'He who plants a coconut tree plants food and drink, vessels and clothing, a habitation for himself and a heritage for his children'." The multipurpose coconut is also an important crop in local economies and international trade, the latter involving copra, coconut oil, copra meal, and desiccated coconut. This palm is one of the most recognizable trees in the world because of its identification with the white-sand beaches and blue-green waters that are emblematic of global tourism in the tropics.

No account of the botanically specifiable trees of life would be adequate without mentioning the ceiba, oak, breadfruit, and fig, even though they are not identified as trees of life in dictionaries. These are multipurpose trees. Ceiba (*Ceiba pentandra* (L.) Gaertn.), a native of tropical America and West Africa, is preeminently a landmark, shade tree, gathering place, and inspirational source. Like the baobab, it belongs to the family Bombacaceae. Ceiba is especially a tree of life in the religious traditions of the peoples of Central America and of members of African religious traditions in the Americas, particularly in Cuba, Haiti, and Jamaica (Rashford 1985). The size of oaks and their durable wood and multipurpose use have made them a tree of life for the people of temperate latitudes. In the Pacific, however, the breadfruit (*Artocarpus altilis* (Parkinson) Fosberg) is the tree of life for many people. It too has many uses but is chiefly valued for its large edible fruit produced in great abundance for several months of the year.

With some 900 tropical and subtropical species comprised mostly of trees, *Ficus* is one of the largest and most celebrated genera of the diverse Moraceae family (Britannica 2023). The genus is well represented in human environments worldwide including temperate latitudes, where several genera and a number of species have become popular potted plants (Compton et al. 1996). Many figs are valued as multipurpose trees, and they are often associated with religion, especially as cosmic trees, altars, altar sites, and sanctuaries (Rashford 2012a, 2022). Some version of the cosmic-tree motif appears in religious cosmologies worldwide, and it is often a fig. Some of the best-known examples are *Ficus natalensis* (Hochst.) and *Ficus sycomorus* (L) of Africa; *Ficus benghalensis* (L) and *Ficus religiosa* (L) of India; *Ficus rumphii* (Blume) of India and Southeast Asia; Ficus *carica* (L) of Western Asia and the Mediterranean; and, the amate figs of Aztec and Mayan papermakers (Von Hagen 1943).

The best-known fig trees are remarkable for their aerial roots, which are essential to their epiphytic beginning and strangler habit; they give the fig a striking appearance. The ability of many figs to start life in the canopy – along with their enclosed inflorescence, coevolved wasp pollinators, abundant fruiting, and diverse seed dispersers – helped them to compete successfully with other trees for light and space. *Ficus* is a major promoter of biodiversity, with its many keystone species in tropical rainforest ecosystems, and it is an evolutionary triumph in this regard. In addition to their extraordinary growth habit, appearance, and reproductive strategy, other factors that place figs as the tree of life at the center of cultural traditions worldwide are their intimate association with rivers, lakes, and groundwater, their ability to establish themselves and grow quickly from seeds and cuttings, and their multipurpose use for such things as food, medicine, wood, fuel, fiber, fencing, fodder, and shade.

The Baobab as the Exemplary Botanical Tree of Life

While dictionaries and popular usage identify the palm as the classic form of the multipurpose tree of life, the same can be said of the fig. Importantly, however, in the landscape of the Hadza that is the focus of this book, the palm and the fig, while present, are less significant than the baobab as the Hadza's multipurpose tree of life. And coniferous evergreens are not significant at all. The baobab is unique on the African savanna as both a manifest and multipurpose tree of life. It is noteworthy that with the exception of evergreenness, the baobab is a tree of life in all ways a tree can be so described. Its distinguishing characteristics as a tree of life are its association with the landscape of human evolution, its life-manifesting and life-giving qualities, and its widespread adoption as a representational tree of life.

Although mentioned in most dictionaries, the baobab is never identified as a tree of life. However, it is significant that the tree-of-life label has been applied to the baobab by many other sources. The characterization is being popularized around the world through the websites of bloggers, storytellers, businesses, associations, religious organizations, and other interest groups. Merchants sell baobab jewelry, wall hangings, posters, paintings, quilts, batiks, sculptures, bronze figures, and ostrich eggs celebrating the baobab as the tree of life. The reasons for this identification often are not made clear, however. Still, just as dictionaries provide a basis for distinguishing the broad forms of the tree of life, the Internet provides a framework for determining in what sense the baobab today is popularly "known as" or "called" the tree of life. The reasons given overwhelmingly involve the baobab's status as life-manifesting, especially regarding its ecological support of diverse species, and its status as life-giving, especially its fruitfulness and multipurpose use. There are also sites that attribute the baobab's tree-of-life status to African legends.

Manifest Tree of Life

The baobab is distinguished as a life-manifesting tree of life by its size, shape, and longevity, the features that make it a natural landmark and savanna icon. It is also distinguished by its tenacity of life, evident in its ability to survive drought, grazing, toppling, fire, and ring-barking; its productivity evident in its fruitfulness and its many-seeded fruit; and its contribution to biodiversity evident in diverse ecological associations. To further the analysis this book presents, however, it is useful to separate the baobab's manifest vigor from its ecology and utility.

Ecological Tree of Life

The baobab is an ecological tree of life because its diverse microenvironments support a great variety of species. The areas in, on, under, and around the baobab benefit many resident and visiting insects and other arthropods, as well as several reptiles and a number of birds and mammals that rely on the tree for such things as shelter, refuge, drink, food, or breeding place. The baobab also provides habitats for microbes, fungi, and plants. It is therefore not surprising that Elizabeth Marshall Thomas and Stephanie Thomas (1976), in their review of the documentary film *Baobab: Portrait of a Tree*, produced by Alan and Joan Root (1973), described the baobab as "teeming with life," and *The Baobab: Home for Everyone* is how Dellatola (1983) titled her article on the interdependence of life in and around a baobab. According to Smith (1882: 37), a baobab "is said to be more like a forest than a single tree," and the tree has also been characterized as an "island ecosystem" (Wickens and Lowe 2008: vii).

The features of the baobab that contribute to its status as a biodiverse ecosystem include the shade of its massive trunk and thick spreading branches, even when leafless; its bark, wood, sap, leaves, flowers, and fruit; and its stored water, both in its wood and hollow trunk, a critical resource in the dry savanna. The numerous holes that develop in its trunk and branches are one of the baobab's most important attributes, essential to its status as an ecological tree of life and a multipurpose tree of life. The holes range in size from small cavities to the tree's spacious hollow trunk, befitting the flowering tree with the greatest spread in the world. Most trees develop holes from breaches to their trunk or branches. The initiation of these tree holes has been attributed to such things as lightning and fires; self-pruning in response to shading; the fused development of multistem trees; weathering, especially the effects of wind and water; the influence of bacteria and fungi; and especially the activity of birds and mammals for whom baobab holes provide shelter, protection, and a breeding place. The humid nature of some baobab holes provides an ideal habit for reptiles, especially snakes, as well as for the growth of fungi and bacteria. As we will see throughout this work, the tree's hollow trunk has been put to a wide variety of uses throughout its range, including reservoir, shelter, recreation, gathering place, dwelling, workshop, storage, refuge, hideout, tomb, and shrine. As a symbolic space, the hollow baobab is significant in religious beliefs and practices, especially where the tree serves as an altar or altar site.

Many authors mention the baobab's association with animals, especially birds and a wide variety of mammals, including bats, baboons, and elephants, but with the exception of the tamarind and fig (Rashford 1994a, 1997a; Watson 2007), they rarely mention its association with other plants. The tamarind and different figs are important components of the baobab's flora; they are part of the microenvironments it creates and the variety of species it supports (Wickens and Lowe 2008; Imoro et al. 2015). The baobab/tamarind relation appears to be a mutualistic association involving a coevolved mimicry of taste that enhances the dispersal of both species (Rashford 1994a, 1997a, 2019). The colonizing fig is the only tree that kills the baobab. It is also worth noting that the fig's many visitors, especially fruit-eating bats, bring fig seeds to the baobab, where they germinate and begin their early life as epiphytes. Watson (2007) has provided remarkable photographs of figs in various stages of colonizing baobab trees.

It is not just tamarind and figs, however; the baobab's wrinkled and folded bark and the crevices of branches are all places where sediment collects, giving rise to the baobab's own herbaceous and woody flora that Sweeney (1974: 58) described as "miniature gardens of lichens, grasses, orchids, *Sansevieria* and many others." The complex nature of the baobab's community of life is only one of many factors that make it an essential resource for a variety of species, including our hominin ancestors (Sweeney 1974). We know the Hadza eat the fruit of one species of fig and of the tamarind, and the latter is also traded. And *Sansevieria* they value as a source of fiber.

The baobab contributes to enhanced growing conditions for plants and animals under and around the tree. It supports biodiversity by reducing soil temperatures and water loss due to evaporation and transpiration. The baobab's abscised branches, leaves, and floral parts, and the droppings, food waste, and decomposing bodies of the many baobab-supported species, enrich the soil around the tree with the organic nutrients they provide (Belsky et al. 1989; Du Puy 1996; Imoro et al. 2015).

The baobab is the hub of a biologically diverse community, and this makes it a resource destination for humans and many other species. It is a worthy candidate for being humanity's ancestral tree of life because in the final analysis, the baobab as a resource-rich environment would hardly have been overlooked by early hominins. One of the important reasons for considering the baobab a central-place tree of life is its role as a gathering place for humans for the same reasons it is a living environment (or part one) for a variety of other species.

The baobab was featured as an ecological tree of life in Barbara Bash's award-winning children's book *Tree of life: The World of the African Baobab* (1989). The success of this book inspired the students of Highwood Hills Elementary School in St. Paul, Minnesota, to perform an original musical titled *Baobab: Tree of Life* in June 2005. According to the brief description of the musical on the school's website (HHES 2005):

> "Baobab: Tree of Life" follows the events surrounding an ancient Baobab Tree on the African savannah. A wise old impala tells of a secret that the Baobab Tree has been keeping for centuries, and a variety of African animals come to the tree, searching for the secret. Only at the end of the musical do the animals discover that the secret is that we must protect our resources and protect the Earth for all animals and people. We need to work together. We all need each other to survive.

Today, a common reason for identifying the baobab as an ecological tree of life is its role in conservation, given the biodiversity the tree supports. The baobab is bat-pollinated, and this coevolved association has long been known and studied. And in publications and on Internet sites dealing with bats and bat conservation, the baobab is often identified as an ecological tree of life.

Multipurpose Tree of Life

In general, all trees in human environments worldwide can be thought of as having multipurpose value. They provide not only pleasure, landmark, shade, and gathering place but also environmental services, such as the protection of watersheds, soil and biodiversity, and carbon sequestration. But trees typically labeled as having multipurpose value are those that provide such things as food, fuel, fencing, construction material, and health products. Woodburn (1970: 40) reported that the baobab tree is "important in Hadza economy in a number of different ways," and it is this fact, along with the baobab's deep inspirational significance for the Hadza, that makes it their multipurpose tree of life. As earlier noted, popular identification of the baobab as the tree of life is not always accompanied by an explanation of why this should be so. When explanations are offered, however, they are usually along the lines of the baobab's vigor, ecology, and utility. The most important utility explanation is the baobab's association with such varied aspects of life as a central place, food, water, shelter, fertility, health, and cosmic modeling. The baobab is the Hadza multipurpose tree of life because it is both a manifest tree of life and an ecological tree of life that can equally well be described as their resource-rich tree of life identified by Pakenham (2004: 16) as "The Great Provider" and characterized throughout its range as "mother."

Chapters 10, 11, 12, and 13 (which cover the baobab and food, including fruit, beverage, honey, birds, other animal foods, and hunger-time tree of life) discuss the baobab as a multisource food tree, and a provider of life-sustaining staples, given its long fruiting season and its nutritious leaves, fruit pulp, and seeds. One of the baobab's most important practical values for many Africans is that, as a diverse food source, it is a well-known hunger-time tree of life (e.g., Kranjac-Berisavljevic et al. 2009). During the annual period of seasonal hunger and in times of famine, the baobab provides not only staples in the form of leaves, fruit pulp, seeds, and honey but also edible root, bark, buds, flowers, and seed sprouts (Rashford 1987a, 2002). Chapter 16 discusses the health uses of the baobab, and Chap. 17 discusses its exchange uses. The baobab is indeed the Hadza multipurpose tree of life, and one use of the tree that has largely been overlooked is its value as a source of trade goods, including honey and beeswax, for which the Hadza receive agricultural crops, chiefly grains and sweet potato, and money, which they use to acquire alcohol and other store-bought resources.

Central-Place Tree of Life

The baobab is identified as a central-place tree of life because it is a sheltered place where people gather. The heart of the Hadza landscape is their hilly baobab retreat, and the baobab is their preferred camping place. The idea of the baobab as a central-place tree has been especially emphasized by storytellers for whom the tree is also linked to fertility, generation, genealogy, tradition, and culture.

Inspirational Tree of Life

The baobab as an inspirational tree of life is covered in Part V of this work, and the discussion includes the baobab as both a fertility tree of life and a cosmic tree of life. The connection between the baobab and fertility is understandable. In the oppositionally unified symbolism of African cultures, the baobab is most often placed on the female side. In addition to its many links to water and to the darkness of night, the baobab's female associations are also related to its fat trunk, pregnant shape, womb-like hollow trunk, breast-shaped fruits, milk-like beverages, upside-down nocturnal flowers, and nighttime floral associates.

The baobab is a cosmic tree of life because it is used to model the universe. As such, it provides the framework for culturally structured individual and social behavior. The cosmic tree of life is so named because it is the universe conceived as a vast life-manifesting, life-giving, and life-representing tree whose furthest upward branches constitute the sky and whose deepest earthly roots form the earth. In encompassing all existence, the intersection of its trunk and the earth's surface is the unifying central axis that is the spiritual crossroads of the cosmos. The concept of the cosmic tree of life expresses the oneness of the universe, and in idea and iconography, it is at the center of human religious imagination. It is in association with the cosmic tree that people seek contact with the spiritual dimension in order to bring about the outcomes they desire. The cosmic tree-of-life sanctuary is one of the earliest forms of the shrine, the forerunner of temple, church, and mosque.

The Baobab as Humanity's Ancestral Tree of Life

Recognition that the Hadza live on the African savanna in an area rich in hominin fossils; that they are viable contemporary foragers in the ancestral landscape of human evolution; and that they have been extensively studied, especially from an evolutionary perspective, makes the Hadza's use of the baobab an ideal perspective for exploring the significance of the baobab as the tree of life in human evolution (Mabulla 2003, 2007; Kusimba 2003; Marlowe 2010). The qualities that make the baobab so remarkable are not only its manifest vigor, rich ecology, and diverse uses, but also the fact that it grows best in dry places where its many products, especially shade, leaves, fruit, honey, and water, are greatly appreciated. It is a reasonable assumption that human beings are largely responsible for the tree's near-universal distribution throughout the dry lowlands of tropical Africa and its presence in wet areas or at high elevations where it would not normally be found (Wickens 1982; Kusimba 2003: 30). It is against this background that this book provides a basis for the baobab as the icon of the African savanna and as worthy of being regarded as the Hadza's tree of life and humanity's ancestral tree of life. As such, the baobab can also be thought of as humanity's primordial and archetypical tree of life, since it represents best what it means to be an actual tree of life and has been with us over the course of our evolutionary history. The rest of this book presents the assembled documentation for identifying the baobab as the Hadza's and humanity's tree of life.

The Evolutionary Origin of the Baobab

The theoretical approach to determining the place of origin for cultivated crops by identifying the geographical center of diversity of their wild relatives dates back to the distinguished nineteenth-century French botanist Augustin Pyramus de Candolle (2011). He was keenly interested in plant geography, especially as it related to the origin of cultivated crops, and he pursued this question from an interdisciplinary perspective that included evidence from botany, archaeology, history, and linguistics. Because Madagascar has six endemic species of the eight species of *Adansonia*, the island has been viewed as the baobab's center of diversity and place of origin (Daniel Aubréville 1975; Armstrong 1977, 1979, 1983; Wickens 1982; Bigham 1994; Wickens and Lowe 2008). From Madagascar, the African baobab (*Adansonia digitata*) spread west throughout Africa prior to hominin evolution, and the Australian baobab (*Adansonia gregorii*) spread east to northern Australia before the arrival of humans, or with the arrival of humans (McGregor 2019). Although the Russian geneticist Nikolai Vavilov (1926) was critical of de Candolle's conception of the relationship between cultivated crops and wild relatives, he adopted de Candolle's concepts of center of diversity and center of origin, to which he added the approach of a field researcher and plant breeder. Rindos (1984: 83) noted, however, that although "Vavilov recognized that diversity might appear as a secondary development in centers far from the original one, he gave the matter little attention." Yet, this very question is fundamental for understanding the relationship of the baobab's center of diversity in Madagascar with its place of origin (Tsy et al. 2009).

According to Tsy et al. (2009: 1713), of "the 23 genera in the Bombacoideae recognized by Kubitzki and Bayer (2002), 21 are found in the New World and only two (*Adansonia* and *Bombax*) are observed exclusively in the Old World, mainly in African and in the surrounding islands." There is now a growing consensus that the African baobab's place of origin is West Africa with an ancestry that traces back to the Neotropics. In the early twentieth century, Chevalier (1906) theorized, based on his study of morphological variation and palaeogeography, that the African baobab was likely native to the Atlantic coastal regions of Africa with progenitors in the Neotropics, just as has been argued for the related silk cotton tree, *Ceiba pentandra* (Dick et al. 2007). In his construction of the *Adansonia* phylogenetic tree, Baum (2003) indicated that the origin of the African baobab occurred before the differentiation of the six Madagascan species, and Baum and colleagues (Baum et al. 1998, 2004) have come to regard Africa as the place of origin of *Adansonia digitata* subspecific diversity and as the progenitor of the Madagascan and Australian species. Using genetic analysis involving "chloroplast DNA fragments obtained from 344 individuals of *Adansonia digitata*, collected from 74 populations covering the entire extant distribution range of the species," Tsy et al. (2009: 1709) identified five distinct haploid groups. They wrote:

> Most populations were fixed for one haplotype and only two contained two different haplotypes. H1 was observed in 38.66% individuals, exclusively in West Africa and in La Reunion and Mauritius islands. H2 was found in the 10.17% individuals growing exclusively in West Africa and in La Reunion and Mauritius islands. H4 was the predominant haplotype and was observed in 50.29% individuals, all distributed in southern and eastern Africa, in the Arabic Peninsula and in the Indian Ocean islands of Zanzibar, Comoros, Mayotte, Madagascar and La Reunion. H3 and H5 were found in 0.2% and 0.6% individuals, respectively, and were restricted to one site in Congo and to a single site in Madagascar (Moramba bay) respectively.

Based on this evidence, they have argued that from its place of origin in West Africa, the baobab spread throughout the savanna of Africa and "the divergence between *A. digitata* and the six endemic *Adansonia* taxa of Madagascar, which form a monophyletic group, occurred approximately 10 million years ago."

This means the baobab was part of the African savanna at the time when chimpanzees diverged from a common ancestry with gorillas, some 3–5 million years before the hominin lineage began. The baobab is dispersed in a variety of ways, including the influence of gravity and rainfall-runoff on slopes. However, given its food bribe dispersal strategy it is especially dispersed by large primates, most of all humans, as well as by other large mammals.

Our current understanding of the baobab is the result of a growing body of literature that includes not only academic publications covering a wide range of topics, but also traveler accounts and newspapers and magazine articles. It was the trade in the baobab fruit pulp in Egyptian markets that brought the tree to the attention of Europeans. The Italian physician and herbalist Prospero Alpino described the medicinal properties of the fruit pulp using the name *bahobab*. This name is generally attributed to the Arabic word *bu hibab*, meaning "many-seeded fruit" (Nicolas and Nicolas 1955; Wickens 1982); according to Baum (1995b: 440), the "description and accompanying illustrations published in Alpino [in 1592] constituted the main information about the plant for over 150 years (e.g., Clusius 1605; Veslingius 1638; Ray 1693; Lippi 1704)."

One of the earliest traveler's description of the baobab was that of the fourteenth-century Arab scholar Ibn Battuta (1969). He encountered the tree in the Sahel and was impressed by its size and age, its generous shade, and its hollow trunk, which stored water, harbored honey bees, and, in one instance, served as the workplace of a weaver. Historical references to the baobab from the nineteenth century onward became common in the reports, diaries, and correspondence of the explorers, administrators, and missionaries who were part of Europe's expanding colonial frontier in Africa. As Sweeney (1974: 2) noted, "the writing of inveterate travelers such as Richard Burton, Georg Schweinfurth, Heinrich Barth, Samuel Baker, David Livingstone and Henry Stanley contain references to baobabs they came across during their travels in the southern Sudan and East and Central Africa."

The earliest European reports were by Portuguese explorers who discovered the tree in West Africa as they made their way down Africa's Atlantic coast in search of an oceanic route to the Indian Ocean and the wealth of Asia (Cadamosto 1507: 70; Thevet 1558: 47). Wickens (1982: 175) wrote: "Recognizable descriptions were given by Portuguese navigators such as Gomes Eannes de Azurara for trees seen on Bisiguiche (Bisiguienne), off Guinea in 1447–48, by Bento Banha Cordoso in 1622, as well as from the mouth of the Senegal by Venetian Aloysius de Cada Mosto in 1454." The fifteenth-century Portuguese who came to West Africa in the wake of the navigators knew the baobab as the calabash tree (*Calabaceira*).

It was Adanson (1757), however, who realized that the fruit of the baobab tree he had seen in Senegal was the same as the fruit described by Alpino in 1592. Adanson named the tree the baobab, which remains the common name for *Adansonia digitata* as well as other species of the genus *Adansonia*. Today, the most important botanical studies of the African baobab (*Adansonia digitata*) and of the genus *Adansonia* have been made by Baum (1995a, b, 2003) and by Wickens and Lowe (2008).

Early baobab reviews such as Gerber (1895), Chevalier (1906), and Adam (1962) that dealt with the botany, ecology, and use of the baobab were largely regional in nature. In a more comprehensive treatment that was an extension of the Baobab Map Project initiated by Lucas (1971), Wickens (1982: 174) presented "a summary" of all that was known at the time "about the baobab through its range." He also expressed concern for baobab conservation in the face of reckless development, highlighted gaps in our knowledge of the tree, and encouraged further fieldwork. In 2009, Wickens and Lowe (2008) produced what is today the most comprehensive botanical and ethnobotanical treatment of the genus *Adansonia*.

While travel accounts, early floristic studies, and the works of Wickens (1982) and of Wickens and Lowe (2008) dealt with the botany, ecology, and worldwide use of the baobab, scientific publications of the past 30 years have covered a wide range of topics influenced not by the curious nature of the tree as reported in the earlier travel accounts, or the need for floristic documentation, but by the international commercialization of the fruit pulp, seed oil, and bark fiber. A number of important development-oriented organizations have been involved in raising awareness about the economic and social value of the international trade in baobab products (Gruenwald and Galizia 2005). Bioversity International (previously the International Plant Genetic Resources Institute) identified the baobab as one of the top ten agroforestry trees deserving of conservation and domestication. The International Centre for Underutilized Crops has made it a priority to promote research and development of the baobab (Sidibé et al. 2002). And the aim of BAOFRUIT "is the exploration of sustainable use and commercialization" of baobab products to "improve food and nutrition security and to combat rural poverty in East Africa" (North et al. 2014). BAOFRUIT was a joint project from September 2013 to July 2014 funded by the German Federal Ministry of Education and Research and representing cooperation between research institutions, non-governmental organizations, and the private sector in Kenya, Sudan, Malawi, the United Kingdom, and Germany.

Recent ethnobotanical studies of the baobab in Africa have been especially focused on the baobab's cultural importance, and on cultural comparisons of its practical use, mostly with respect to food and medicine (Gebauer et al. 2002; Diop et al. 2006; Assogbadjo et al. 2008; Chadare et al. 2008, 2009, 2013; De Caluwé et al. 2010; Kaboré et al. 2011). But baobab commodification has resulted in growing concerns about the impact of the world market on the sustainable use of the species. This has led to increasingly sophisticated studies focused on baobab demography, especially studies dealing with the problem of baobab regeneration, and on baobab genetics (Kyndt et al. 2009; Assogbadjo et al. 2010), agronomy, and domestication (Jensen et al. 2011, Raebild et al. 2011). The destructive nature of some of the most important traditional harvesting techniques, which require the removal of bark and the cutting of branches to gather leaves and fruits, is now part of the production process for the international market. Romero et al. (2014) and Luckert et al. (2014) have provided two recent studies of the commercialization of baobab bark, which they argue is economically important, but likely unsustainable. In fact, the justification for the Romero et al. (2014) study was "to provide baseline information for the development of management guidelines for this important … [non-traditional forest product] for the benefit of the resource base and the inhabitants of the Save-Odzi Valley [in Zimbabwe]." Much contemporary baobab research has been fact-finding in response to the growing international trade in baobab products, which has raised concerns for continued local access, the influence of harvesting techniques on sustainable use, and the long-term success of baobab regeneration.

Following Wicken's (1982) influential summary, which was based on the curious nature of the baobab, and which, even then, raised questions about the conservation of the species, came the comprehensive review of Sidibé et al. (2002). This review continued the concern with baobab conservation, but it reflected a growing interest in the tree's global commercial potential and the community benefits that were said to justify this commodification. Sidibé et al. (2002) noted that the purpose of their book, *Baobab, Adansonia Digitata L.*, was "to assemble information on production, processing, marketing and utilization of baobab … in order to identify research constraints and highlight the importance of the species for nutrition and poverty alleviation." Recent studies, however, have been greatly concerned with the sustainability of current commercial practices. According to Schumann et al. (2010: 2036), "There is virtually no information about the tolerance of baobabs to harvesting," especially the impact of traditional harvesting techniques on fruit production. "In addition," they wrote, "no studies have assessed the impact of harvesting the different parts of the tree and if the response to harvesting varies with tree size and/or with land-use types with different disturbances such as fire, logging or grazing."

One of the most interesting and important areas of contemporary baobab research with implications for how we understand the Hadza/baobab relation is the impact of different land-use areas on their baobab populations. A few studies have attempted to assess this impact, particularly with respect to sustainable harvesting and baobab regeneration. Assogbadjo et al. (2005) presented the results of a comparative field study of three baobab populations in Benin across the wet-to-dry climatic gradient of the study, aimed at evaluating the ecologically based diversity of Benin's baobab populations, which included a quantitative assessment of relative productivity with respect to fruit pulp, seeds, and seed kernel. And in a later study, Assogbadjo (2008) documented the cultural understanding of baobab variation in Benin, particularly perceptions of morphological differences in fruit pulp, seed, leaves, and bark, as well as differences in taste preferences with respect to the fruit pulp, seeds, and leaves. Using a comparison of natural baobab stands in a national park with those in fallow fields, croplands, and villages identified as communal lands, Schumann et al. (2010) assessed the influence of these land-use types on baobab populations in Burkina Faso, paying particular attention to density, size-class distribution, harvesting tolerance, sustainable use, and conservation. In their study comparing baobab productivity across five land-use types in South Africa, Venter and Witkowski (2010) reported that fruit production was slightly better in villages and fields than in nature reserves, plains, and rocky outcrops. Similar studies of the influence of land-use types on baobab populations have not been done for the landscape of the Hadza and their food-producing neighbors.

The Baobab Tree in the Landscape of the Hadza

Cooper (1949: 8) described the Kidero Hills, which "may be said, roughly, to define the limits of the country of the" Hadza (including the regions of the Tli'ika to the south, the Han!abi at the center, and Mangola to the north), as "mostly covered with fairly open forest of acacia and baobab." So did Woodburn (1968a, b), in his account of Hadza ecology, and Mabulla (2003: 48), who noted that "Baobab trees dominate the Hadzabe landscape, with dense patches in the Kidero horst and inland areas that fringe the Yaeda Endanyawishi valley." According to Blurton Jones (2016: 106), "each region [of Hadzabeland] includes patches of the major resources, soft berry patches, *Grewia* fields, baobab forest, tuber hillsides, and water," and he described the distribution of the baobab as "being especially abundant in Tliika but also plentiful in Han!abe, parts of Siponga, the Endanawish area, and the southern slopes of the Balai Valley [which is part of the Mangola region]." In her article on Hadza plant foods of the Mangola region, Vincent (1985a: 134) said the area of her fieldwork comprised the floodplains of the Barai River and the northern part of the Kidero Hills to the east of Lake Eyasi, which she referred to as baobab woodlands. In the Mangola region, the baobab is also described as plentiful in Endamagha.

It is surprising indeed that the overall importance of the baobab in Hadza life has not been fully appreciated and the attention it has received has focused largely on its value as a multisource food tree appreciated for its nutritious fruit pulp, seeds, and leaves and for the honey gathered from hives in the hollow trunk and branches of old trees. But the baobab is not only the Hadza's multisource food tree; it is, more importantly, their multipurpose tree of life. The baobab is a dominant feature of its savanna landscape where there are often few other features to rival its imposing presence. Its enormous size and peculiar shape make it an ideal landmark useful in orientation and as a place marker. The tree provides shelter in the form of shade and rain cover. It is an ideal resource destination that, in addition to providing such things as fruit, seeds, leaves, root, honey, water, and fiber, is valued as a rest spot or campsite. One of the most important yet least explored uses of the baobab are in the material culture of Africans as both material means and material ends. The baobab enters into all areas of Hadza life because it is the exemplary ecologically based multipurpose tree of life that is a rich source of Hadza material culture.

Chapter 2
The Hadza and the Studies That Document Their Use of the Baobab

The 1300 or so Hadza are traditionally highly mobile foraging people whose homeland includes the grassy flood plains, gallery forests, and baobab-clad rocky outcrops and uplands that surround Lake Eyasi in the Eastern Rift Valley of northern Tanzania. What sets them apart from their neighbors is their central-place hunting and gathering way of life; the development of this mode of adaptation became the economic foundation of hominin evolution for most of the past four to eight million years, lasting worldwide until the conditions for the success of foraging as an economic way of life began to change with the advent of agriculture and herding a mere 12,000 years ago. The Hadza are now among a small number of extant foragers whose way of making a living is an expression of humanity's ancestral mode of adaptation. Because of this, they have become one of the most intensely studied kin-ordered societies in the world. By looking at their multipurpose use of the baobab in the context of their central-place hunter-gatherer way of life, this book lays the ground for identifying the baobab as the Hadza's tree of life and, by extension, humanity's ancestral tree of life. This chapter introduces the Hadza and the studies that document their use of the baobab.

Evidence of the intimacy of the Hadza/baobab relationship can be found from the earliest accounts of the Hadza right down to the present (Baumann 1897; Obst 1912; Bleek 1931; Cooper 1949; Woodburn 1968a, 1970; Marlowe 2010; Blurton Jones 2016; Crittenden 2016). The Hadza first attracted serious academic interest in the 1930s, but it was Woodburn's pioneering ethnographic research in the 1950s that brought them to prominence in the scientific literature (Woodburn 1968a, b). His interest in forager ecology, material culture, social organization, egalitarian life, and political challenges has had a significant impact on theorizing hunter-gatherer prehistory and the origin of agriculture (Cohen 1977). Ethnobotany, however, was not a major concern of Woodburn's Hadza research. Nevertheless, the materialist and ecological orientation presented in his many publications, an ethnographic film, and a museum exhibition documented the fundamental importance of the baobab in the largely plant-based way of life of the Hadza (Woodburn 1970). In the years following Woodburn's studies, evolutionary scholars cited throughout this book have

J. Rashford, *Baobab*, https://doi.org/10.1007/978-3-031-26470-2_2

provided excellent quantitative data that touch on the baobab's importance for the Hadza. These studies provide insight into the Hadza/baobab relation while exploring such specific topics as Hadza hunting, meat consumption, bone assemblages, male provisioning, demography, life history, time allocation, and honey collecting. In contrast to the many publications on Hadza hunting, only three studies have dealt with Hadza plant foods in a significant way, including a study of the "sources" of Hadza foods (Tomita (1966); the use of tubers, the year-round mainstay of Hadza diet (Vincent 1985a, b); and the nutritional composition of the most important wild plant foods and kinds of honey used by the Hadza (Murray's et al. 2001). Although none of these studies focused on the baobab, they are important contributions to our understanding of the significance of the baobab in Hadza life.

Scholarly publications are the foundation of this work, but relevant information on Hadza baobab use has also been gleaned from newspapers, magazines, official reports, newsletters, trade publications, reliable websites, and documentary films. In recent years, the Hadza have been subject to the many influences of mass tourism, and the result of this contact increasingly shows up in books, articles, and Internet postings that discuss exotic vacations, adventure travel, recreational trekking, pilgrimages, and journeys of self-discovery (Stephenson 2000). Some of these accounts have brought to light aspects of the baobab's importance for the Hadza and are included in this book.

Species of the genus *Adansonia* occur in Africa, Madagascar, and Australia. In accounting for this disjunct distribution, Armstrong (1983:156) thought that the African baobab's religious significance was an important factor in its dispersal in the Indian Ocean. In this context also, he wrote of the Hadza's intimate relationship with the baobab as a thing of the past: "Woodburn documents the manner in which the economy and ecology of the Hadza … in northern Tanzania, until about 1965, were intimately associated with the baobab. They lived amongst the trees, used them as a source of food and material, and climbed into them seeking for honey." Armstrong was convinced that Hadza traditional life had ended because of forced governmental and missionary-sponsored settlement schemes centered on farming, but he was wrong. Hawkes et al. (1997:552) wrote, "None of these [settlement] projects have been successful, and in every case, most of the Hadza involved have returned to the bush, usually within a few months. In each instance, some Hadza have managed to avoid settlement and continued to live as hunters" (see also McDowell 1981a, b, 1982; Ndagala 1988). In the view of Lee and Hitchcock (2001:262), the Hadza "are foragers who despite interactions with their neighbors, have maintained a fierce independence, owning few livestock, not cultivating crops, and remaining in remote areas (Woodburn 1964, 1970; Blurton Jones et al. 1989; Blurton Jones and Marlowe n.d.; O'Connell et al. 1988a, b, 1990)." One important component of their continued independence is their association with the baobab as their tree of life.

The two chapters of Part II of this book introduce the theoretical framework for understanding the importance of the baobab in the context of the division of labor and in relation to explanations for the origin of hominin bipedality.

Part II
Theoretical Framework: Societal Specialization and Bipedality

Chapter 3
Hominin Adaptation as the Development of a Gendered Forager Division of Labor

"Division of labor," generally defined, is the efficiency of dividing the overall process of producing something into specialized tasks, each performed by a different individual or group. Many anthropologists have regarded the gendered forager division of labor (with men hunting and women gathering) as a fundamental feature of hominin development (Washburn and Lancaster 1968; Murdock and Provost 1973; Isaac 1978a, b; Panter-Brick 2002; Nakahashi and Feldman 2014). This is understandable; the ethnographic literature documents the importance of this mode of life for contemporary and historical foragers (Marlowe 2007; Blurton Jones 2016). This documentation has led many to speculate that gender-specialized foraging has been a defining feature of human adaptation from early on in the evolution of the hominin lineage. How early, however, is a matter of significant debate. This chapter on the division of labor (including exchange) and the next on bipedalism introduce the evolutionary framework for looking at the baobab as the Hadza and humanity's tree of life.

Division of Labor and Societal Specialization

There are many ambiguities in the use of the division of labor concept, including the seminal works of Adam Smith (1776) and Durkheim (1893). Jolly and Plog (1987: 234) provided an example of these ambiguities when they wrote "all human societies have some kind of *division of economic labor* [italics in the original]." If "labor" is understood to mean "subsistence," which is to say economic activity, then the idea of "economic labor" is redundant. But if "labor" is used in ambiguous or arbitrary ways, then it is necessary to specify that one means "labor" in the strict economic sense. And yet, in most anthropological works, "labor" is synonymous with economic activity. This suggests that the concept of "division of economic labor" is an unnecessary redundancy. Narrowly defined in economic terms, the division of labor concept overlooks the specialization involved in other kinds of practical activities.

J. Rashford, *Baobab*, https://doi.org/10.1007/978-3-031-26470-2_3

It also overlooks the specialization of inspirational activities associated with religion, oral traditions, arts, and recreation.

Halperin (1980) attempted to "broaden" the meaning of "division of labor," arguing that we had to "qualitatively determine" the "features of [foragers'] production process." This required a "revision" of the concept that expanded its meaning beyond gender-based food procurement (as merely male hunting and female gathering) to include other aspects of "material livelihood" such as ecologically based movements, making clothing, obtaining food-processing materials (e.g., "firewood" and "water"), and food preparation for storage and consumption. But Halperin's view is still limited with respect to the range of practical and inspirational activities encompassed by the division of labor concept. For example, he did not systematically correlate reproduction specialization with gendered food procurement, and these with the wider concept of societal specialization. The concept of societal specialization identifies a people's overall division of practical and inspirational activities based on sex, age, interest, and talent, and the most important practical activities are reproduction and production.

Since the publication of Brown's (1970) influential article on "the sexual division of labor," there has been a much better appreciation of the important relationship between reproductive and economic specialization. Before that, explanations for the development of the "sexual division of labor" were based on ideas of innate male/female differences (especially in physical strength), reproductive biology, and aspects of consciousness, such as psychology and thinking (Murdock 1937, 1949; Murdock and Provost 1973; Brightman 1996). Brown (1970: 72) argued, instead, that it was specifically childcare compatibility that determined what women did economically in contrast to men; the birth and nursing of infants and providing primary caring for weanlings limit women's activities to those that were safe, interruptible, and located in or near their residential environment. Brown's childcare compatibility theory has become one of the most often cited explanations for the gendered division of labor (e.g., Williams 1971; Burton et al. 1977; Mackey and Coney 2000; Marlowe 2007).

An alternate expression of the reproductive/productive perspective is the idea that men and women are assigned dangerous and safe activities, respectively, not because of their relative strength or their mobility as child-free versus child-restricted (Murdock and Provost 1973: 203), but because women, by virtue of their unique reproductive value in bearing and nursing infants, are not as reproductively "expendable" as men (Ember and Melvin 1998: 245). Lower reproductive value and relative child-free status provide a basis for men to orient themselves to risky activities far from camp, where they target energy-dense foods that are difficult to acquire (Hawkes 1990, 1991; Apicella et al. 2017). This especially includes the much-loved meat and honey, not only to provision their mates, offspring, and other kin but also for camp-wide distribution. From the perspective of forager "sexual division of labor" as merely gender-specialized economic activity, reproductive specialization seems an unrelated concern. From the perspective of societal specialization, however, reproductive specialization is just as important as production specialization.

Societal specialization is a natural feature of the hominin lineage since the evolution of the lineage would necessarily have involved reproductive specialization in association with forager-generalized or gender-specialized production.

Societal Specialization and the Hadza/Baobab Relation

The most important reason for considering societal specialization in this work is that it frames our understanding of the Hadza/baobab relation and our efforts to theorize the significance of the baobab in human evolution. For example, if we only look at the baobab in terms of Hadza hunter-gatherer "division of labor," we learn that baobab fruit, unlike any other in-camp food, is gathered by both men and women, even though women bring more baobab back to camp than men. Nevertheless, it is significant that men bring back almost as much baobab fruit to camp as women. If we go beyond gendered food production to the broader view of societal specialization, we get a more complete view of the baobab's importance. The tree is not only a source of key resources like fruit pulp and seeds and indirectly a source of honey; it is also an important factor in the acquisition of other resources given its influence on their seasonal availability, spatial distribution, and density. A more complete understanding of the baobab's significance for the Hadza must also include such things as shade, water, fiber, containers, and fire equipment. We see, for example, that the baobab is significant as both a fertility tree and the source of one of the Hadza's most important weaning foods. It is only when we consider the baobab's importance in Hadza societal specialization, including children's activities and inspirational life, that we get a full view of the tree's contribution to Hadza foraging and its likely significance in hominin evolution.

When Did the Gendered Forager Division of Labor Develop in Hominin Evolution

The centrality of gendered economic specialization in the lives of known foragers has led many to theorize that it has been a defining feature of human adaptation from early on in the history of the hominin lineage (Lancaster and Landcaster 1980; Lancaster and Whitten 1979, 1980; Ember et al. 2019). How early, however, is much debated in anthropology (Panter-Brick 2002; Nakahashi and Feldman 2014). According to Leacock and Lee (1982: 5), gendered foraging was "the system of production that prevailed during virtually 99 per cent of human history." Many researchers have argued that it originated in the late Pliocene in association with the genus *Homo* and most of all with *Homo erectus* (e.g., Isaac 1978a, b; Lovejoy 1981; Estalrrich 2018). Leaky and Lewin (1992: 46–47) wrote, for example:

> Stone tools for the first time gave the impression of standardization, the imposition of a mental template; fire was harnessed for the first time; for the first time hominids expanded beyond the African continent. And surely the rudiments of language – perhaps even consciousness – were produced in a dramatically expanding brain. Yes, *Homo erectus* marks a clear move from an apelike past to a humanlike future. If we are to understand the origin of humanity, we have to understand *Homo erectus*, its anatomy, its biology, its behavior.

In their study of the division of labor in relation to Neanderthals and modern humans in Eurasia, Kuhn and Stiner (2006: 953) argued that gender-specialized foraging was associated only with modern humans, but recent studies have suggested otherwise (Estalrrich and Rosas 2015; Estalrrich 2018).

Gendered foraging is here theorized to have been the mode of adaptation that developed out of the context of generalized production, which would have been the earliest production system associated with the evolution of the branching hominin tree. As Lancaster and Whitten (1979: 48) expressed it:

> The elementary human adaptation, the one upon which all else is built, depends on a simple but unique change in ecological and social relations. This change was the shift from the individual foraging and feeding pattern of other primates to a system of sharing and cooperation, in which adults feed infants and juveniles. Associated with this assumption of responsibility to feed the young was a new economic interdependence between the sexes, one in which females gathered and males hunted. This ancient division of labor, permitting the flexible, joint dependence on plant and animal food, probably accounts for the early success of the first hominids.

It is likely that the forager adaptation of the Australopithecines was primarily in the form of generalized production with only a rudimentary development of the gendered division of labor (Zihlman 1991). From this perspective, the evolution of *Homo erectus* and of modern humans are merely more intensive modes of the hunter-gatherer way of life, not its origin. The fundamental assumption of this book is that hominins evolved for an increasingly efficient hunter-gatherer division of labor and its success was the foundation for the rise of the genus *Homo* some 2.5 million years ago, with *Homo erectus* as its most successful representative. The forager way of life was perfected with the highly variable gendered economic specialization typical of hunter-gatherers described in the ethnographic literature (Lee and DeVore 1968; Dahlberg 1981; Blurton Jones 1996; Panter-Brick 2002; Robson and Kaplan 2003; Marlowe 2007; Bird and Bird 2008; Bird and Codding 2015).

Exchange in Hominin Evolution

Exchange, generally defined as reciprocal giving and receiving, is commonly identified as a universal feature of human life of fundamental importance from the earliest development of hominin hunter-gatherer specialization to the world market of today. "It should be no surprise," wrote Barnard and Spencer (1996: 218), "that some argue that exchange is a key to social life." As a basic feature of gender-specialized foraging, exchange merits attention in this work because it is essential to Hadza use of the baobab tree not only among themselves and in relation to their traditional neighbors but also in relation to the more recent influx of commercial producers, merchants, government officials, missionaries, and tourists.

Resource Transfer and Exchange

"Transfer" is the most general concept for the process by which resources are circulated and distributed in any society; broadly speaking, it can involve theft, begging, gifting, and exchange. Of these, exchange in its various forms has probably long been the most important form of resource transfer in hominin evolution; begging, theft, and gifting are best explained in the context of the evolved hominin capacity for exchange. For example, gifting, strictly defined, is not exchange since exchange is specifically reciprocal giving and receiving. To give a gift is to transfer a resource to another without any thought, requirement, or expectation of return, and this means the idea of a "free gift" or "pure gift" is a problematic expression. Ideally, gift is a free transfer of a resource without the obligation of an immediate or delayed return. In its ideal form, delayed exchange appears to be like gifting. That is, delayed exchange often clothes itself in the ideology of gifting. Unlike gifting, however, delayed exchange always has the hidden fact of an obligation to reciprocate; a return is definitely expected.

Exchange has always been a topic of important discussion in anthropology. Significant contributions to exploring the concept were made in the 1920s with Malinowski's (1922) ethnographic study of Trobriand traders and Mauss's (1925) comparative study of gift-giving; these studies were later incorporated in the influential publications of Polanyi (1957, 1959) and Sahlins (1972). Mauss dismissed the idea of a gift that is completely free; instead, gift-giving as the "archaic" form of exchange that reflected the deep inner structuring of the human mind was the moral obligation derived from the collective conscience to give, receive, and repay. Mauss developed his argument in the ethnographic context, not of foragers, but in association with the horticultural way of life in New Guinea.

Exchange in Foraging Societies

Anthropological studies of exchange have long been presented within a polarizing framework. Various permutations of the opposites savage/civilized, primitive/modern, and traditional/contemporary have been routinely correlated with the dichotomies of exchange, such as gifting and market exchange, gift-giving and trade, and reciprocity and market exchange. For Mauss, gifting was presented as the characteristic mode of exchange in kin-ordered societies which contrasted with commercial exchange of "the civilized world." The exchange system of foragers has been widely characterized as sharing or gifting (Mauss 1925; Kaplan and Hill 1985), as reciprocity or reciprocal exchange (Polanyi 1957, 1959), as reciprocal altruism (Trivers 1971), as generalized reciprocity (Sahlins 1972), and as cooperation (Jaeggi and Gurven 2013; Kramer 2018). Recent textbooks also identify forager exchange as generalized or delayed reciprocity; both concepts are consistent with what is identified in this work as delayed exchange. However, in addition to delayed exchange in the specific form of delayed reciprocal gifting associated with family, relatives, and friends, there is also delayed exchange in the form of delayed reciprocal entitlement characterized by behavioral ecologists as "tolerated theft" or "tolerated scrounging" (Kaplan and Hill 1985).

In contrast to the popular characterization of forager resource transfer as sharing, behavioral ecologists like Winterhalder (2001: 31) have claimed that their approach "shows that sharing may be much more heterogeneous than it has seemed from standard ethnographic accounts." In his review of the subject, Winterhalder (2001: 31) concluded that in contrast to the seemingly simple ideas of sharing, gifting, or reciprocity, "intra-group food transfers [among foragers] are likely to result from some combination of scrounging, reciprocity, risk minimization, exchange, showing off, and costly signaling." He continued by noting that this "is an embarrassment of difficult-to-separate causal possibilities. Although there have been many fine ethnographic studies of food sharing, most did not record the detailed, quantitative information needed to assess the relative importance of these overlapping hypotheses." Behavioral ecologists have indeed challenged the sharing conception of forager exchange, and on the basis of field research, they have presented a number of exchange models. However, they have not offered a compelling synthesis of the forms of exchange they model, nor have they explained how these forms are systematically related to the overall anthropological literature and to theorizing hominin evolution.

From the theoretical approach presented in this chapter, the five forms of exchange that are discussed include sharing, delayed exchange, barter, redistribution, and market exchange. All societies are regarded as dependent on a system of exchange involving a combination of three or more of these five forms. Gender-specialized foraging has three interlocking forms of exchange that include sharing, delayed exchange, and barter. And while all three are associated with resource transfer involving household, kin, and camp, barter is especially associated with intersocietal trade. This system of sharing, delayed exchange, and barter is characteristic not of foragers in general, but of highly mobile immediate-return foragers like the Hadza. It stands in contrast to foragers who develop descent systems involving the control of key resources and who operate with a system of redistributive exchange because of intraspecific competition. Intensive foragers of the Mesolithic as well as intensive cultivators and herders also have redistributive systems, as well as sharing, delayed exchange, and barter. Market exchange, however, is particularly associated with money as the medium of exchange and is linked to the rise of feudal states, merchants, and long-distance trade – a historical process that ultimately led to the triumph of merchants and capitalist producers in association with the development of the world market (Polanyi 1944).

The Importance of Exchange

Among the "functions" attributed to exchange is the idea that it facilitates specialization as a more efficient form of resource production; exchange allows the results of specialization to be converted into the diversity of foods and other resources required for daily life. And exchange makes the best use of an abundant resource that would otherwise be lost to spoilage. It ensures a steady flow in the distribution of food and other resources throughout a community. Among foragers in particular, it evens out the flow of resources acquired through the differential success of hunting and gathering individuals and groups. Social solidarity has also been identified as an important reason for the development of exchange.

Summary

Societal specialization and exchange were discussed in this chapter as important aspects of the evolutionary framework for looking at the baobab's importance in Hadza life and its theorized importance in hominin evolution. The other important aspect of this framework, which is discussed in the next chapter, is the evolution of bipedality as a marker for the development of the forager way of life.

Chapter 4
Bipedality as the Outcome of the Multidimensional Selective Pressure of the Developing Forager Way of Life

Fossil indicators of bipedality dating back more than three million years provide supporting evidence of the theorized divergence of hominins from chimpanzees four to eight million years ago (Sarich and Wilson 1967; Gibbons 2012; Langergraber et al. 2012). This makes bipedality one of the most important fossil markers for a suite of evolutionary changes that together distinguish hominin ancestry from that of all other primates. Bipedality, viewed as a "consequence," is widely thought to have resulted from a fundamental change in the environment or in mode of food procurement. Viewed as a "cause," it is considered the linchpin of hominin evolution; it set in motion the evolutionary trends that ultimately gave rise to modern humans. Whether from the perspective of consequence or cause, bipedality is a key part of the suite of traits that include a divergence from the situational upright posture and bipedal walking associated with arboriterrestrial non-human apes, especially the knuckle-walking quadrupedal gorillas and chimpanzees.

Russell Tuttle (1981) presented his own summary of labeled bipedality explanations, and others have built on this in producing their own. According to Niemitz (2010:241), for example, there are "approximately 30 hypotheses." Given extensive overlap, these 30 explanations are reduced to 9 scenarios including a change in the method of food procurement, handiness, savanna adaptation, upright advantage, feeding behaviors, physiological adaptations, the use of tools and weapons, and carrying behavior. A significant limitation of these scenarios is that they have largely been single-factor or minimally multifactor approaches, and even multifactor approaches have often been reduced to single-factor labels. Nevertheless, bipedality explanations are valuable because they highlight a number of important hominin-associated traits with a long history of discussion in anthropology. This chapter argues for a unifying perspective on these hominin traits, which are identified in this work as the dynamic suite of reciprocally determining selective advantages that together exerted the overall selective pressure for the correlated evolution of handiness and bipedality. Simply stated, early hominins became increasingly bipedal in conjunction with the multiple selective pressures of gender-specialized foraging, a process that also involved the intelligent use of the hands in conjunction with material culture and a growing capacity for thought and articulate speech. This review of bipedality explanations and the unifying approach advocated provides an appropriate evolutionary framework for identifying the baobab as the Hadza tree of life and for inferring that the baobab was the tree of life in hominin evolution.

J. Rashford, *Baobab*, https://doi.org/10.1007/978-3-031-26470-2_4

A Change in the Method of Food Procurement

Although rarely presented as a labeled category, a change in the way food is acquired is part of most bipedalism explanations, and it can involve a change in such things as resource environment, feeding posture, material culture, dietary options or preferences, or mode of life. The review of bipedality scenarios presented here suggests that the development of gender-specialized foraging was based on these kinds of changes and this makes it the likely underlying explanation for hominin bipedality.

Darwin and the Correlated Evolution of Handiness and Bipedality

Darwin's account of bipedality is much broader in scope than the simple idea of the use of "tools and weapons" to which it is often reduced. For Darwin, the preconditions that made the selection for bipedality possible were the evolving primate brain and hand, and the initiating cause was a change in environment or in resource procurement. These changes did not result in an abrupt move from living exclusively in trees to living exclusively on the ground. They resulted, instead, in an arboriterrestrial ape that was distinguished from other African arboriterrestrial apes by spending more time on the ground and less time in trees. This increase in terrestrial adaptation meant the outcome "would have had to become either more strictly quadrupedal or bipedal" (Darwin 1871, vol. 1, 140–141). There is no doubt that the true hero of human evolution for Darwin is the intelligent use of the hands, also identified in this work as hand versatility, intelligent handiness, or simply handiness. He noted that it was "the free use of the arms and hands … [as] partly the cause and partly the result of … [humanity's] erect position" that "appears to have led in an indirect manner to other modifications of structure" (1871:280). As Darwin expressed it, "Man could not have attained his present dominant position in the world without the use of his hands, which are so admirably adapted to act in obedience to his will" (1871:279). The "freeing of the hands" is one of the most widely referenced bipedality explanations that is often linked to scenarios that include terrestrial living, material culture, carrying behavior, and food-procurement strategy.

Savanna Adaptation

Many environmental stimuli have been posited in explanations for the evolution of bipedality including climate, geomorphology, vegetation, and waterscapes. Among these, the "savanna perspective" has long been the most versatile. It came to the fore in the twentieth century, especially with the discovery of hominin fossils in Africa and the development of hunter-gatherer studies (Shreeve 1996; Potts 1998; Bender et al. 2012). The minimal definition of "savanna" in dictionaries is "flat grassland," "grassy plain," or "grass-covered area." Current bipedality explanations emphasize forests, woodlands, and grasslands, the latter with varying combinations of trees and shrubs. But the landscape of human evolution cannot be adequately conceptualized as merely forests or woodlands in contrast to wooded or pure grasslands. The versatile nature of the developing forager sexual division of labor would have made it possible for early hominins to adapt to a variety of environments including African mosaic savanna (Kingston 2007; Wood et al. 2021).

Desiccating Climate Change

The two major savanna explanations of bipedalism are desiccating global climate change and desiccating geomorphologically based regional climate change. Both processes are identified with transforming forests into woodlands and grasslands in eastern Africa, and it was the "retreat of forests" and the expansion of grasslands that triggered bipedality. The assumptions are that either people were left stranded on the expanding grassland or they opted to colonize the savanna rather than continue with arboreal living in the retreating forests.

Drier conditions produced by global climate change might well have been of decisive importance at the point of the chimp-human divergence some four to eight million years ago (Vrba 1993, 1995; Vrba et al. 1995). However, the more immediate fact of geomorphologically based climate change in eastern Africa in combination with global climate change is a more reasonable scenario (Coppens et al. 1976). Significant volcanic activity and massive faulting and uplifting were in full swing in eastern Africa in the middle of the Tertiary period during the Miocene epoch, when extraordinary mountain building was occurring worldwide. These changes produced the Rift Valley. Over 5000 miles long, it is one of the earth's major geological features, extending from Lebanon south to Mozambique. The dramatic changes in the physical landscape, including the creation of uplands such as volcanic mountains, plateaus, escarpments, and hills, together with geomorphologically determined climate change, produced a highly diverse environment of springs, rivers, and lakes; forested and wooded mountains, hills, and riversides; and grasslands that were treeless, bushy, or tree-studded. This varied landscape (Harris 1980a, 1980b; Leakey and Lewin 1992; Winder et al. 2013, 2014) is identified in this book as a mosaic savanna, and it is regarded as the most likely environment of human evolution; it is also the landscape of the Hadza.

Why Did the Descendants of an Arboriterrestrial Ape End Up on the Savanna?

In the wake of a drying climate, early hominins were "forced" to leave the disappearing forests and adapt to the savanna. Linked to this environmental trigger is the idea that the savanna is a resource-poor environment because of widely scattered food sources requiring efficient long-distance travel and the ability to cope with direct exposure to the sun at the hottest time of the day. The idea of a climatic change as the motivation for adapting to the savanna has been preferred by many researchers, compared with the idea of hominin choice. The alternatives to an external environmental trigger are explanations in which early hominins were attracted to living on the savanna because it is a resource-rich environment, an idea that is often linked to bipedality explanations based on handiness, material culture, carrying behavior, and various hunting and meat-eating scenarios. As life-manifesting, life-giving, and life-representing, the baobab would have been a powerful attraction of the mosaic savanna.

The Upright Advantages of Bipedalism

Jean-Baptiste Lamarck and Raymond Dart

de Lamarck (2006 [1809]) offered one of the earliest explanations for the evolution of bipedalism, emphasizing the transition from living in trees to living on the ground, but he did not explain the reason for the transition, referring in passing only to "force of circumstances or some other cause." But once on the savanna, he theorized that the earliest hominins would have become erect because they were "impelled by the desire to command a large and distant view." It was Raymond Dart (1925), however, discoverer of the first hominin ancestor in Africa (*Australopithecus africanus*), who brought the savanna perspective into the mainstream of anthropology. Reviews of bipedality explanations typically begin with Dart's (1925) idea that "upright vigilance" (in the sense of predator awareness) was necessary in the "dangerous" environment of the savanna. Following his initial emphasis on upright vigilance, Dart and Craig (1959:195) would later argue that the persistent use of defensive and hunting weapons on the southern African grassland savanna was "the only habit capable of shifting the weight of apes from their knuckles and buttock bones onto their feet." This was the development supposedly responsible for Dart's view of the evolution of humanity as an armed killer ape (Dart 1953), a view made popular by a number of authors (Ardrey 1961, 1966, 1971; Lorenz 1966, 1973; Morris 1967, 1969; Tiger and Fox 1969). According to Dart, our ancestral hominins left the ease of the arboreal fruit-eating life of forest apes to become "carnivorous creatures, that seized living quarries by violence, battered them to death, tore apart their broken bodies, dismembered them limb from limb, slaking their ravenous thirst with the hot blood of victims and greedily devouring livid writhing flesh" (cited in Reader 1988:140).

The idea of upright advantages suggested by Lamarck and Dart has been subject to a great deal of elaboration in the literature including arboreal feeding posture; the need for spatial orientation; the practical value of landscape scanning involving resource assessment and alertness to the presence of others; and the importance of social awareness, group communication, and behavioral displays. The "threat posture model" was proposed by Livingstone (1962), and Wescott (1967a, b) characterized it as the "exhibitionistic origin of human bipedalism." Livingstone (1962:302) reported that the gorilla (as an analogy of early hominins) was often bipedal during "chest thumping" and charging display. Livingstone (1962) and Jordania and Jordania (2011) viewed the charging display as a threat behavior, and Jablonski and Chaplin (1993) described it as a mechanism of social control. Michael Dale (2018) recognized the link between bipedal displays and hierarchy-dominance displays, but argued instead for bipedality as a fitness marker in the process of sexual selection.

Feeding Behaviors Leading to Obligate Bipedalism

Postural Feeding

The "postural feeding hypothesis" (Hunt 1996) includes a number of ideas. Jolly (1970), for example, identified eating grassland seeds (as well as corms and tubers) from a squat position as pivotal in the transition to bipedality, but Kingdon (2003) theorized it was reaching up for the fruits of trees and shrubs of coastal seasonal forests which gave rise to the bipedal gait.

Arboreal Bipedality to Terrestrial Bipedality

Some researchers have theorized that the earliest hominins were already pre-adapted for habitual terrestrial bipedality; they point to feeding strategies in trees involving both upright posture and bipedality and vertical climbing (Hunt 1992, 1994). The origin of hominin bipedality as an adaptation to living in trees has received increasing attention in recent years (Rose 1976; Thorpe et al. 2007, 2014; O'Higgins and Elton 2007; Hart and Sussman 2009; Crompton et al. 2010; Kimura 2019). In arguing against the widely held view of a knuckle-walking phase preceding hominin terrestrial bipedality (summarized by Richmond et al. 2002), Thorpe et al. (2014) theorized instead that arboreal bipedality led directly to terrestrial bipedality and that this transition was initiated by environmental change. Based on this view, Hart and Sussman (2009:178) suggested that the conventional single-factor explanations were not the causes of bipedality, but its consequences. They wrote:

> If we are correct, rather than looking for the factor or factors that caused our ancestors to make those initial bipedal steps, we can simply accept that it was a combination of many factors that likely made bipedal locomotion advantageous in the particular habitat in which the first hominids lived. Such things as carrying food, tools, or weapons, sitting upright while eating, feeding on tall bushes, hunting, gaining better vision over the tall grass, or thermoregulatory needs did not cause other edge- or arid savanna-living primates to abandon their quadrupedal postures. Bipedal locomotion was a given; the successes or added advantages were simply a by-product. These advantages, plus the ability exploit both arboreal and terrestrial habitats, added to the success of these versatile creatures, our ancient ancestors.

While Hart and Sussman (2009) viewed arboreal bipedality as leading directly to terrestrial bipedality, the ultimate cause of terrestrial bipedality in their scenario was the environmental change that made terrestrial living necessary. But if hominins are a branch of African apes that were arboriterrestrial, and if we share this common ancestry with gorillas and chimps, then why did they not become bipedal as hominins did? The multidimensional nature of the selective pressures for bipedality cannot simply be explained away as "a by-product" of a transition from bipedal tree living to bipedal ground living.

Physiological Adaptions and the Savanna Origin of Bipedalism

Recent subsets of the savanna scenario identify several physiological adaptations for grassland living as the basis for the evolution of bipedality. For Wheeler (1984, 1991, 1992), bipedality provided a thermoregulatory advantage compared with the quadrupedal locomotion of extant African apes. It meant exposure to cooler air, sweating as an evaporative cooling system, and a minimization of exposure to direct solar radiation on the body, especially at noon. Along similar lines, Falk (2007) argued in his "radiator hypothesis" that the selective advantage of terrestrial savanna living was the brain-cooling effect of erect walking. Arguments for the importance of bipedality in energetically efficient long-distance travel (Rodman and McHenry 1980) and long-distance scavenging (Shipman 1984) have also been made.

Bipedality and Tool Use, Weapons, and Other Kinds of Equipment

According to Lovejoy (1981), "The most commonly cited distinction between man and apes is the former's reliance on material culture. The belief that tools were pivotal to the divergence of hominids was initiated by Darwin (1871) and has remained the most popular view (Bartholomew and Birdsell 1953; Mann 1972; Washburn 1960, 1963, 1968; DeVore 1965; Tobias 1971; Washburn and Ciochon 1974)." Two common reasons for rejecting the tool-use explanation are the relative timing of the fossil evidence for bipedalism and for brain enlargement and the significance of chimp material culture in theorizing the origin of hominin bipedality. First, it is said that the fossil evidence for bipedality dates back more than three million years. This significantly predates the archaeological evidence of fabricated stone tools, which only dates back some 2.5 million years (Sassaman 1992). As many have pointed out, the major weakness of this critique of the tool-use perspective is that it overlooks or minimizes the central importance of unmodified stones and of plants and other perishable materials in the making, using, and transporting of tools. The argument for a time gap of several million years between bipedality and tool use undervalues the importance of the use of "tools" by apes and by primates generally (McGrew 1987, 2004, 2013; McGrew et al. 1979, 2003; Biro et al. 2003). It underestimates the importance of naturally occurring stones used as "tools," which would not be readily identifiable in the archaeological record; and, in identifying "tool use" with the appearance of fabricated stone tools, it presupposes that tool use was a relatively sudden occurrence in human evolution, rather than a process of development over a long period of time (de la Torre 2011; Panger et al. 2002), especially in association with plants.

The extent of chimp material culture is the second reason for rejecting the tool-use explanation. The tool-use scenario was dealt with a severe blow when Jane Goodall's (1986) field research beginning in the early 1960s documented that "tool use" occurred among chimps; it was therefore not exclusively human as had previously been argued in the idea of "man the tool-maker" (e.g., Oakley 1959). It is now clear that humans are not the only tool-using species; there is ample documentation of tool use by many arthropods, fish, birds, and mammals (Alcock 1972; Pierce 1986; Boinski et al. 2001; Koops et al. 2010; Carvalho and McGrew 2012). Chimps use a variety of things including plants as protective cover; tree canopy for nighttime sleeping security; bedding from interlaced branches and vines; sticks for probing, picking, leveraging, reaping, and digging and for catching termites and ants; hammers and anvils (also of stones) especially for cracking nuts and hard-shell fruits like the baobab (Ambrose 2001:1749; Marchant and McGrew 2005); brushes for extracting honey; natural containers such as leaves and fruit pods (Sousa et al. 2009); leaf sponges for drinking water and leaf wipes for hygiene; and weapons. But chimps have not evolved to be handy bipeds (Campbell and Loy 1996:266; Langdon 2005:121; Scupin and DeCorse 2008). Equipment use similar to chimps and other non-human primates (and likely similar to early hominins) should not be expected to select for bipedality, however, since this is a single-factor perspective.

But hominin material culture (including equipment) as a critical component of the multidimensional selective pressures of gender-specialized foraging offers an entirely different perspective. We human beings and our hominin ancestors are the preeminent tool-using species in nature, as evidenced by the emerging global civilization that tool use has now made possible (Sayers and Owen Lovejoy 2008). This is so because tools are an obligate aspect of human evolution that is of decisive importance in the success of practical activities and their associated inspirational framework. Yet, even though we now recognize that plant and stone-based technology among primates includes chimpanzees, capuchins, and macaques, and, even though, as Ambrose (2001:1748) pointed out, chimps "have diverse and regionally varied repertoires of tool-using, tool-making, and other 'cultural' behaviors [McGrew 1992, Whiten et al. 1999, and Van Schaik et al. 1999]," what is genuinely significant is that humans have long used tools in myriad ways that go well beyond the ability of chimpanzees, other primates, and, indeed, all other species.

Carrying Things as the Selective Advantage of Bipedality

Food Carrying

Routine bipedal food carrying, especially food carrying in the form of "provisioning," is unique to humans and has long been recognized as having profound implications for social life, particularly with respect to family and community-based systems of resource exchange. Foremost among the carrying scenarios highlighted in the 1960s was the food-carrying explanation of Gordon Hewes (1961), a major critic of the tool-use perspective. He argued that the need for efficient food carrying was the selective advantage for the evolution of bipedalism, and he regarded the short-distance, situational, bipedal food carrying of monkeys and apes as support for this perspective. Hewes theorized that food carry was motivated by the need for a safe eating place for maximizing the consumption of meat, tendons, and bone marrow away from the dangers of predators associated with a scavenging site. Along similar lines, others have reported situational short-distance food carrying among monkeys and apes not for safety, but for monopolized feeding (Carvalho et al. 2012). Hewes's explanation is inherently ambiguous, however, since food transport was the immediate reason for the evolution of bipedality, but climatic change was presented as the ultimate cause of the need to transport food. As Gruber (1962:605) pointed out, the scenario Hewes presented was one in which the carrying of food "in a new manner depends upon new environmental conditions (i.e., living in the open)." This meant transporting food would seem "to be more of an effect rather than a cause of bipedalism" (Gruber 1962:605). Although Niemitz's (2010:247) assessment was that "wild great apes rather avoid bipedal locomotion when transporting food," Niemitz thought, nevertheless, that the food-carrying scenario could "perhaps remain a possible contributing factor to the evolution of habitual upright bipedalism." For this reason, some critics have accepted the importance of food-carrying in the evolution of bipedality, but they have not accepted Hewes's specific food-carrying explanation.

Carrying and the Things Carried

Understandably, the efficient long-distance carrying of things, which is one of the key modalities of handiness, is widely regarded in the literature as a significant factor in the evolution of hominin bipedality. In light of this, it is surprising that there are no bipedality explanations based on the general need to carry things, rather than the particular things that are carried. We find instead that carrying explanations are mostly single-factor scenarios that highlight the transport of tools and weapons (Bartholomew and Birdsell 1953; Washburn 1960; Marzke 1986), infants (Etkin 1954; Lancaster 1978; Tanner 1981; Cann and Wilson 1982; Sinclair et al. 1986), food (Hewes 1961, 1964; Brace 1962), male mate and offspring provisioning (Morris 1967; Lovejoy 1981, 1982, 1988; Stanford 2003), female provisioning of offspring (e.g., Linton 1971; Zihlman, 1978a, b; Zihlman and Tanner 1978c; Fisher 1982; Tanner 1981), resources for central-place exchange (Johansson 1999), and the pyrophilic primate hypothesis and the gathering of fuel (Burton 2009; White et al. 2009; Medler 2011:11; Parker 2015; Parker et al. 2016). There is no reason to restrict our attention to only a few selected features, however. It is probably the case that handiness and the carrying behavior, in conjunction with the overall development of material culture, were important aspects of the multidimensional nature of forager selective pressures that resulted in the correlated evolution of bipedality.

Gender-Specialized Foraging as a Unifying Explanation of Bipedality

Early accounts of bipedalism were often single-factor explanations (e.g., Napier 1963; Relethford 1990:332–5), but trends toward a unifying approach have become common (Isaac 1978a; Lancaster and Whitten 1979; Day 1986; Foley 1995; Scupin and DeCorse 2008; Hart and Sussman 2009:178; Niemitz 2010; Ko 2015). The multifactor approach is especially associated with gender-specialized foraging as the overall selective pressure for bipedalism, which is not a new idea. One precursor of this unifying approach is the idea of bipedality as a marker of gender-specialized foraging that Isaac (1978a) identified as "a novel adaptive complex" and Lancaster and Whitten as the "sharing ape."

According to Isaac (1978a), the new adaptive complex of "interconnected behavioral elements that may well have been characteristic of Plio-Pleistocene hominids" was an "inseparable set of distinctive behaviors which reinforced each other, and which were jointly intensified through the action of natural selection without any one factor being an isolated prime mover." But Isaac's (1978a:320) multifactor scenario highlights the difficulty of avoiding one-sided explanations. We learn that "Since no one element is held to be the prime mover it is difficult to assign a name to this model; perhaps it can be designated as a food exchange, or reciprocity model, to distinguish it from the technological, the hunting or the gathering models." Identifying his synthesizing explanation as a "food exchange" or "reciprocity model" reduces it by name back to the status of a prime-mover perspective and in so doing deemphasizes the importance of his unifying approach. To serve as a reasonable account of hominin bipedality, the "novel adaptive complex" explanation should be a synthesis of the positive content of all other explanations. That is, they should be unifying explanation.

Summary

The overview of bipedality explanations presented in this chapter shows that they involve a number of hominin traits, and it suggests these traits represent a dynamic combination of reinforcing selective advantages of the gender-specialized foraging that exerted the overall selective pressure for the correlated evolution of handiness and bipedality. This development provides the appropriate evolutionary framework for identifying the baobab as the Hadza tree of life and for inferring that the baobab was the tree of life in hominin evolution.

Part III
Material Culture and Technology

Chapter 5
Africa's Premier Fiber Tree

Although the baobab can be described as a multipurpose tree wherever it grows, it is also true that different groups throughout its range emphasize some uses over others in keeping with their way of life. In the wetter parts of Africa to which the baobab has been introduced, other plants provide preferred alternatives to some of its many products, and the baobab retains its natural form consisting of one or more massive smooth trunks, with a dome-like canopy. The tree also retains its natural form where appreciated for its shade, its symbolic and religious value, and its much-loved fruit. However, in the dryness of the savanna where alternative resources are scarce, the baobab bears the conspicuous marks of its diverse uses. For foragers like the Hadza who climb the baobab to gather honey, water, fiber, birds, and other resources, the tree is dotted with holes and small sticks used as climbing pegs. For many who value baobab leaves for food and fodder, the baobab is a pollarded tree with a swollen trunk and stubby branches, a sight especially common in West Africa. For those interested in the baobab's strong versatile bark fiber, which is often obtained from the lower part of the trunk, the telltale scars of ring-barking are clearly evident. Many marvel at the baobab's remarkable ability to survive ring-barking, which highlights its status as a manifest tree of life. The value of plant fibers, especially for binding, stringing, bundling, fastening, hoisting, hanging, lashing, packaging, wrapping, sowing, weaving, anchoring, and mooring, is often overlooked or poorly treated in theoretical discussions of early hominin material culture (Lancaster 1968; Ambrose 2001). But as Plotkin (1988) notes, "Fibre plants are second only to food plants in terms of their usefulness to humans and their influence on the advancement of civilization" (Cited in Cotton 1996: 190).

"The Baobab as a Fiber Tree" is the title of an anonymous 1876 article in Scientific America that reported:

> Twisted into string and rope, [baobab fiber] … is used for all sorts of purposes, and in untwisted strips it serves to secure loads and to bind together the poles employed in making their huts. Finer pieces are pulled out so as to resemble coarse netting; and the edges being

J. Rashford, *Baobab*, https://doi.org/10.1007/978-3-031-26470-2_5

> sewn together, they make handy bags for cotton, gum, grain, and the like. Coffee and ground nuts are brought down from the interior to the coast in very strong bags woven from thin strips of the bark.

According to Palmer and Pitman (1961: 231), "The bark [compared to the fruit and wood] is of far more use; it is, in fact, the most useful part of the tree." Riley and Brokensha (1988: 197) also identified fiber and fruit as the "chief utilization" of the baobab for the Mbiri of Kenya. In listing the most important plants of the six major regions of the world, Harlan (1992: 71–72) included the baobab among the "fruit and nuts" and "fiber plants" of Africa. The use of baobab fiber is so widespread the tree can justifiably be described as one of Africa's best-known fiber plants. This chapter introduces the baobab as Africa's premier fiber tree and discusses the significance of this aspect of the tree in Hadza life.

The use of plant fibers, especially cordage in the form of thread, string, cord, and rope, is a development that would have enhanced the effectiveness of the forager way of life. These four kinds of lines have many links to carrying devices such as baskets and bags. They would have been essential for making composite tools requiring the combination of stick with stone, bone, horn, shell, and feathers; for fastening projectile points and arrow fletchings; for making the handles of fruit pod containers; and for making bows and musical instruments. Rope-making is most often highlighted in discussions of the use of baobab bark fiber (Lely 1925: 9; Mabberly 1990: 8). It is also important to those who tend livestock, where it is used to tether animals and to make harness straps. For Akamba beekeepers of Kenya, baobab rope is used to hoist artificial hives up to the protective branches of tall baobab trees. The Dogon of Mali use baobab rope to raise the dead who are buried in caves in the cliff face of the Bandiagara Mountain. Baobab string and cord have also been used to make hunting nets and snares. For fishing populations like the Tonga of southern Africa, as well as coastal populations of East Africa and western India, the baobab provide not only place markers, shade, food, and inspiration – as it does for people across Africa – but also fishing equipment that included bailers, fishnets, fishnet floats, fishing lines, ropes for mooring and anchoring, and lashing cords for boat rigging (Pickering 1879; Mittre 1981). Palmer and Pitman (1961: 231) also identified clothing as one of the main uses of baobab fiber.

Today, handicraft producers in southern Africa use baobab fiber to make not only containers but also hats and mats including table mats, door mats, floor mats, and sleeping mats. These are sold to tourists in urban markets and internationally. This has meant a thriving market for the sale of baobab fiber and a growing concern for the sustainability of intensive harvesting (Romero et al. 2014; Luckert et al. 2014). The baobab and *Sansevieria* are two of the most useful fiber plants of the African savanna, and both are important to the Hadza. In light of this, it is the declining availability of *Sansevieria pearsonii* resulting from overharvesting for the handcraft commerce that has led to the destructive intensification of the harvesting of baobab bark fiber (Romero et al. 2014: 109). Similarly, the cultivation of cotton and sisal are important fiber crops of world trade, and it is interesting to note that their cultivation in Africa has resulted in the destruction of local baobab populations because the baobab is thought to host insects injurious to these crops (Wickens 1982: 194–195).

Hadza Use of Baobab Bark Fiber

Cordage

Cordage is one of the Hadza's practical concerns. Peterson et al. (2013: 120) noted that they make rope and string from the "inner bark of many trees and shrubs," but the trees and shrubs were not identified. The same for Marlowe (2010: 77–78). He reported the Hadza use of "bark twine" for the "twine noose snare" they sometimes make; for tying string to carry meat and baobab fruit; for the handles of gourds used to carry water and honey; for binding the long stalks of ripe baobab fruits into a bunch in order to carry a larger quantity; and for making the "gourd with handle and rope," which Marlowe reported is used to "get water in tree." Peterson et al. (2013: 122) provided an illustration of the "gourd with handle," identified in the Hadza language as *Hlok'oko* and described as a baobab pod "used as a drinking vessel for water drawn out of a natural baobab cistern." They also discussed a string instrument the Hadza occasionally make (2013: 145).

Hadza Bowstrings

Published accounts indicate that for some specific uses, fiber taken from the canopy branches of a "comparatively" young baobab (Dalziel 1937: 14) or from the roots or branches of a mature baobab is preferred (Watt 1885: 106; Wickens and Lowe 2008: 116). Watson (2007: 132, 134) presented a photograph of a Hadza man harvesting fiber from the canopy of a baobab and wrote, "Bark is usually harvested from the trunk, but for bow strings, this Hadzabe hunter in Tanzania preferred to strip it from young branches at the top of the tree." According to Matthiessen (2010: 175), "Ordinarily, Hadza bow strings are of zebra tendon." Citing Woodburn (1970), Marlowe (2010) wrote "The [bow] string [of the Hadza] is made from strips of nuchal ligament of eland, buffalo, or zebra or the sinew of giraffe." Bow string is certainly an important component of Hadza hunting equipment, and it is clear that several kinds of material have been used. Bleek (1931: 274) wrote, for example, "The [bow] string is usually sinew, but sansevieria cord may be substituted." And now, Peterson et al. (2013: 40) have reported that "With fewer large animal kills nowadays and the introduction of nylon string, bowstrings are often made of nylon."

Hadza Bead Stringing

The traditional importance of beads is evident in centuries of African bead-making and in the acquisition of exotic beads through trade with Europe, the Near East, India, and Asia. According to Dubin (1987: 122), the earliest examples of locally manufactured African beads – disk-shaped beads made from ostrich eggshells dating back some 12,000 years – are from archaeological sites in Libya and Sudan. It is noteworthy that ostrich-shell beads have long been worn by the San. Stone beads have also been recovered from several sites in Africa. The earliest accounts of the Hadza, published toward the end of the nineteenth century and through the first half of the twentieth century, indicate that they traded with their traditional neighbors for beads; in the 1950s, Kohl-Larsen (1958) mentioned the Hadza trading for beads with the Isanzu. According to Marlowe (2010: 95), "It is from the Masai and other neighboring people that Hadza used to get the beads, but today they get them from us researchers," a fact which points to the importance of beads as currency. Bleek's 1930s photographs (see Peterson et al. 2013: 22–23) show men and women dressed in skins and wearing beaded headbands, necklaces, and bracelets, with only women wearing beaded waistbands. Judging from the comment of a 90-year-old Hadza named Gudo, women wearing beaded waistbands is a practice that is now disappearing, probably because of the colorful cloth they have adopted (Peterson et al. 2013: 173):

> Many of the old ways are no longer with us, clothing has changed and the girls no longer wear beads around the waist, men don't wear the goyogoda (genet loin skin), the large mammal kills that bring joy as people gather are fewer, new Swahili (outsider) behavior and things are now common.

The Hadza are known for wearing body attachments of string, leather, metal, and strung objects, and strung objects include headbands, chokers, necklaces, shoulder bands, armbands, bracelets, rings, waistbands, leg bands, and anklets. Strung attachments are mostly made by women, but worn by women and men of all ages. According to Berbesque et al. (2012), "The Hadza frequently use their teeth as tools. Men use their teeth to work arrows …, tighten bow strings, and strip bark from branches to make twine. Women use their teeth to cut beading string, and they chew leather hides and bow strings to make them supple." Matthiessen (1972: 221) reported seeing the young wife of a Hadza hunter "threading beads in long strands of fiber from the baobab." However, because the Hadza make cordage from "many trees and shrubs," it is impossible to say at this point the extent baobab fiber is important in the stringing of objects. Stringing beads for use in a variety of ways is popular now, judging from the many photographs available in print and online (e.g., Peterson et al. 2013). And while beads of vibrant colors are the most widely used material today, strung attachments are traditionally made from a variety of objects. As Marlowe (2010: 94) noted, "Most all the jewelry the Hadza make out of glass beads they used to make out of organic materials, and one can still see plenty of this traditional organic jewelry worn. They use bones, often finger and toe bones of baboons, little hollow sticks, pods of various sizes, the quills of porcupines, shells, the scales of pangolins (*Smutsia temminckii*), teeth, and occasionally feathers."

The strung body attachments the Hadza make are often identified as "jewelry," and they are usually discussed under the heading of beadwork for personal ornamentation (e.g., Fernández and Jöris 2008; Anderson 1990). As Marlowe (2010: 94) puts it, "The main form of [Hadza] art is body adornment," and he speculated that "It seems likely this would have been the first form of art." In thinking about the stringing of objects, it is helpful to adopt a broader framework with more precise descriptive categories. The subject heading adopted here is body communication. It consists of behavior that includes facial expressions, stances, gestures, and vocalizations; body modification including hairstyling, scarification, painting, and tattooing; and body attachments, including piercings and rings, as well as the use of string and strung objects. And while body modification and attachments are clearly significant aspects of personal ornamentation, they are also done for other reasons.

Strung objects are also used to decorate such things as containers, clothing, and dolls. They are associated with religion, including objects of spiritual significance; with wellness, where the concerns include health, protection, and luck; with social marking, including identity, sexual attraction, kinship affiliation, status, rank, and accomplishment; and with currency, including tradable goods, maintaining economic reserves, manifest wealth, and researchers' payments. The baboon, for example, is a featured animal in Hadza religious narratives (Peterson et al. 2013: 3). They describe the baboon as their "greatest competitors for food" and one of their favorite meats. They also wear headbands, armbands, and bracelets of baboon hair, and they use it to decorate their bows. It is not surprising that they string baboon bones, not just because they are readily available and considered ornamental but probably also because of the baboon's deep cultural significance in their lives. It is likely, therefore, that there are practical, religious, and wellness concerns associated with the baboon-incorporating necklace.

The Hadza make three kinds of wellness necklaces. One is the *kelaguko*, described by Agova (2015: 4) as having a "medical-protective function for children, in particular (J. Woodburn 20th March 2013 pers. Comm.; 1970, 53)." Peterson et al. (2013: 31, 139–140, 164) published several photographs of children as well as adults wearing this necklace (pages 109, 156–159), and they also show children and adults wearing both ornamental and wellness necklaces at the same time. The second wellness necklace discussed by Agova (2015) is *ukurukakako*, which is the Hadza name for both pangolin scales (*Manis temminckii*) and for necklaces made of pangolin scales. She noted that pangolin necklaces were used as "medicine against chest pains and spitting of blood (Woodburn 1970: 53; see Heath 1992)." The last of the wellness necklaces is called *kirurube*, "from the seed (*kirurube*) of a small bush to suck on when your heart starts racing or is irregular" (Peterson et al. 2013: 119).

With respect to the attachments' function as social marking, it is noteworthy that young men and women are particularly decked out with beads during initiation rites marking the transition from childhood to social maturity (Peterson et al. 2013: 56, 150–151).

It could be argued that the use of strung beads is now mostly related to personal ornamentation and currency. But if Hadza body attachments, especially strung beads, were essentially for personal adornment, then young Hadza women would be expected to wear more beadwork than older women. But as Marlowe (2010: 95) reported, older women wear the most beadwork, often concealed under clothing, because of the value of beadwork as currency with respect to tradable goods, maintaining economic reserves, and portable wealth, rather than only as a social marker in the form of personal ornamentation associated with sexual attraction or with status, such as manifest wealth, rank, and accomplishment.

The Use of Animal Skins Versus Baobab Fiber

The Hadza have traditionally depended on the skins of animals they hunted for cordage, containers, sleeping mats, and clothing. And while the baobab is a source of cordage and containers, there are no reports of the Hadza using baobab mats and clothing as reported for the San in the early twentieth century. "The [San]," wrote Stow (1905), "convert the fibrous bark into a kind of matting which is sometimes used in lieu of a blanket or kaross," and he noted that these products "look as if made of a material like coir." He noted also that "Bags, ropes, etc., are also made of it" (1905: 56). With the development of fishing, cultivation, and rearing livestock, there would not have been the abundance of skins from hunting. Baobab bark fiber may have substituted for various kinds of skins used for cordage, containers, cloth, and mats.Given that there is as yet no significant baobab handicraft industry in the Hadza environment, ;that they use baobab baobab bark fiber from the baobab's tree's upper branches, and that their traditional use of skin for clothing and household items such as sleeping mats, it is not surprising that the many photographs of baobab trees in the Hadza landscap do not show the scars of intense harvesting of bark fiber from the trunk.

Summary

The baobab is indeed one of Africa's most valuable fiber trees as is evident when we look at this use of the tree throughout its range. The focus of this chapter has been the Hadza use of baobab fiber in the context of their forager adaptation to the mosaic savanna bearing in mind that the use of plant fibers significantly differentiates contemporary human beings and early hominins from chimpanzees and other primates. However, like other areas of Hadza material life covered in this study, the use of plant fibers as a subject in its own right that is yet to be systematically documented by Hadza field research. Nevertheless, the excellent studies that have been done and that are the basis of this work leave no doubt that the use of baobab and other plant fibers is a significant aspect of Hadza life. Because the remains of plants do not preserve well as stone, their significance in human evolution based on the archaeological record is more difficult to theorize. Nevertheless, it is a reasonable assumption that plant use in general and the use of plant fiber in particular would have been fundamental features of the development of hominin material culture.

Chapter 6
The Baobab and Containers

A distinguishing feature of hominin evolution – and an essential difference between human beings and other primates – is a universal dependence on containers. But essential difference is not absolute difference. For example, gorillas and chimpanzees use their body parts as containers, including cupped hands or a combination of the hands or arms held against the chest; they weave a rough container structure in the form of their nest; and they use crushed leaves to make a drinking sponge. However, when we consider the importance of handling containers in the traditional life of the Hadza and San, especially compared with the absence of such containers among other primates, it becomes clear that handiness and container use would have been critical to the development of forager central-place living based on the gendered division of labor (Lee 1979; Marlowe 2010).

The indispensable contribution of containers to the efficient handling of water, food, and other vital resources has made them a defining feature of hominin evolution. Containers are an essential component of the hunter-gatherer way of life including foraging technology and central-place resource transfer. The importance of containers in contemporary life is evident in the extraordinary variety of natural and fabricated forms found worldwide, the diverse materials and techniques for making them, and the fact that they are made to suit innumerable practical and inspirational ends (Beals and Hoijer 1965: 350–365). It is unfortunate that so little attention has been paid to forager container use, in theories of hominin evolution that emphasize central-place provisioning and the carrying of tools, weapons, food, foraging equipment, and infants.

Marlowe (2010:92–93) subsumed his discussion of containers under the more general heading of carrying devices, which is appropriate. With the correlated evolution of handiness and bipedality, the human body would have been from early on an important carrying device for transporting things on the head, around the neck, over the shoulder, on the hip, over the arm, under the arm, and in the hands. But the idea of carrying devices must be clearly distinguished from containers; carrying devices, in addition to containers proper, also include such things

J. Rashford, *Baobab*, https://doi.org/10.1007/978-3-031-26470-2_6

as the human body, cordage, carrying sticks, and leather belts. Moreover, not all containers are carrying devices in the strict sense, since some containers are used for such things as storage, food preparation, and eating. Containers of various forms are of obvious importance in the minimalist material culture of highly mobile hunter-gatherers like the Hadza, given their use in managing essential resources like water, fire, honey, and fat as key aspects of the forager way of life. This chapter discusses the uses of the baobab as a container structure and as a source of handling containers.

The Development of Container Use in Hominin Evolution

The common view in the literature is that container use was a late development in hominin evolution, but there are several reasons why this is untenable. Pre-hominin use of containers would have included the hands, often in combination with other body parts; container associated with foods and other resources such as the indehiscent fruit pod of the baobab; roughly fashioned containers in the form of nest and shelter; and natural container structures in the form of caves and hollow trees. Consequently, container awareness and use cannot be said to have developed late in hominin evolution. Rather, container awareness and use continued and dramatically expanded with hominin evolution, especially in association with foraging containers, carrying containers, and domestic containers such as eating and drinking utensils. With settled life came the need for containers for the storage, preservation, protection, processing, preparation, and use of various things.

The Baobab as a Container Structure

Given the widespread use of hollow baobabs, it is worthwhile to distinguish the value of the tree as a container structure in contrast to its value as a source of handling containers. The baobab as a container structure simply recognizes that old baobabs are full of holes, as earlier noted; these holes range in size from small punctures to huge cavities in the tree's enormous trunk. The ecological significance of the hollow trunk has already been discussed, and the use of the hollow trunk as shelter, refuge, crypt, water container, workspace, storage, resource focal point, rest spot, recreation, and symbolism is explored throughout this work. Other significant structural containers of the Hadza include their branch-framed homes and natural rock shelters and caves.

The Baobab as a Source of Handling Containers

By contrast, handling containers which are also identified as portable containers are those that augment the use of the hands and are an indispensable feature of the practical and inspirational activities of daily life. A largely overlooked aspect of the baobab's importance for the Hadza and other savanna foragers is the use of the fruit pod to make handling containers. Handling containers are an efficient means of gathering, transporting, storing, processing, consuming, and sharing the great variety of products essential to the central-place way of life of nomadic foragers, especially water and food. This no doubt accounts for the variety of containers we find among contemporary foragers. Gathering containers, for example, keep the hands free for other activities, permit the collection and transport of larger quantities of resources than could be used at the time they were gathered, and enhance sharing with kin and other residents of the camp. Water carriers expand the range of foraging beyond what would otherwise be possible based on permanent and seasonal runoff and pools. Storage containers, even if they are only for short-term use, make it possible to stash excess resources for later; they make it possible to even out seasonal fluctuations in the abundance or scarcity of food supplies. Ethnographic studies of contemporary foragers like the Hadza and San suggest that the use of multipurpose containers like fruit pods and baskets would have been a major development in the life of hominins.

The most important plant containers that may have been available to early hominins (and that continue to be used by the Hadza today) would have been the fruit pods of the baobab and those of wild *Lagenaria* species. The latter especially include the ancestor of the bottle gourd, one of the most widely used plant containers in the world (Lathrap 1977). The fact that the baobab is called the calabash tree in parts of Africa is noteworthy in this regard, since the name "calabash" has also been applied to two other important sources of natural containers: the bottle gourd (*Lagenaria siceraria*) of the Old World and the calabash tree (*Crescentia cujete*) of the Americas. Their shared names point to the similar uses of their fruit pods to make bowls, cups, dippers, spoons, ladles, canteens, and basins. According to Austin (2004:252), the name calabash comes from French, but was originally an Arabic name for the bottle gourd. The Latin name, *Lagenaria siceraria*, given by Nicolas Charles Seringe (1776–1858), comes from "the Greek *lagenos*, flask, bottle, or Latin *Lagena*, a flask; alluding to the most common use of the fruits" (Austin 2004:392). This is also one of the most common uses of the pod by the Hadza. As previously noted, natural containers requiring little or no processing are relatively rare in the environment, and no plant sums up better the importance of the container in human life than the bottle gourd. This exceptional vine, long regarded as a native of Africa where all other *Lagenaria* species occur, is the ultimate container plant. It produces pods that are extraordinarily variable in size and shape and in the many uses to which they can be put both as containers and in the making of other products. Although the durable lightweight *Lagenaria* pod is used to make such diverse things as fishing floats and resonators for various musical instruments, its most significant value is as a container, especially for liquids, chief among them being vessels for collecting, transporting, storing, and using water, and the unrivaled source of the household utensils mentioned earlier. What is truly remarkable about the bottle gourd is not only its use throughout the world in pre-Columbian times, but the fact that it shows up in the archaeological record at the very beginning of agriculture in both the Old and New Worlds. The bottle gourd was already in use in Asia as early as 11,000 years ago and in the Americas by 10,000 years ago (Richardson 1972; Erickson et al. 2005; Clarke et al. 2006; Kistler et al. 2014).

Matthiessen (1972:234) noted in his brief account of the Hadza baobab use that "a half shell may be kept to make a drinking cup." Marlowe (2010:78) reported that the "Hadza use gourds for carrying water, honey and coals, and for storing fat" and that baobab pods are used as cups and bowls and also as containers for carrying and storing honey and drinking water. The reports of Matthiessen (1972) and Marlowe (2010) along with the absence of other references, lead us to wonder about the extent to which the baobab drinking cup, bowl, and general container are indeed routine uses of the pod or merely spontaneous occurrences that are the product of convenience.

Marlowe (2010:92) indicated that today the Hadza "often get the gourds from farming neighbors like the Iraqw who grow large ones." However, it is noteworthy, as Marlowe pointed out, that "smaller wild variants grow in the area, and sometimes the Hadza uses these, as they must have long ago." Marlowe did not identify the wild gourds, but they were probably those of the genus *Lagenaria*, of which three species are likely to occur in the environment of the Hadza. The genus *Lagenaria* comprises six species, of which the bottle gourd, *Lagenaria siceraria*, is the only cultivated one (Jeffrey 1978; Ellert 2006). A wild form of the bottle gourd has been described only for southern Africa. There is also *Lagenaria abyssinica*, the perennial dioecious climber and trailer of eastern Africa, occurring from Ethiopia to southern Tanzania; *Lagenaria guineensis* and *Lagenaria rufa* from west and central Africa; and *Lagenaria sphaerica*, which is found from Somalia to South Africa. Ellert (2006:12) has described *Lagenaria sphaerica* as "common throughout most of the country [of Zimbabwe] in riverine and thicket habitat" in association with a vegetation type described as Mopane woodland (*Colophospermum mopane*), also comprising the "baobab (*Adansonia digitata*), *Terminalia*, *Albizia*, *Acacia*, *Pterocarpus* and other tree and shrub species." Writing of what were likely either *Lagenaria abyssinica* or *Lagenaria sphaerica*, Marlowe (2010:92–93) noted: "Occasionally, a man gets some honey but has no container. In this case, he may find a small gourd but more often will find a large baobab pod and cut off about one-third of it so that he can remove the pulp (which he eats) and fill the pod with honey to carry it back to camp." Perhaps the use of the baobab pod as a honey container is not merely "occasional," given the overlap between the harvesting of the baobab fruit and the gathering of honey.

Hadza Container Use in General

There are no specific studies of the history of Hadza container use, and, with the notable exception of Marlowe (2010), the subject has not been systematically treated. Hadza container diversity is small compared with that of traditional food producers, and especially with the simplest household of any industrial society, but it is large compared with that of the gorilla, chimpanzee, and other primates. From an evolutionary perspective, the difference in container diversity between humans and other African apes is particularly significant and provides the basis for understanding the development of the container dependence of our present world system. Containers play a major role in the practical and inspirational activities of the Hadza. Not surprisingly, however, the extraordinary range of Hadza container use was largely invisible in early accounts of their way of life. Bagshawe (1925:117), a British colonial district officer in Tanzania from 1917 to 1923, was clearly focused on the Hadza as "nomad hunters" who spent "most" of their "time and energy" in search of game – a view of hominin prehistory that would become distilled in the 1960s as "Man the Hunter."

Although Bagshawe was not seriously concerned with Hadza container use, we learn, nevertheless, that in the 1920s the Hadza had "cooking pots," but we are not told if they were ceramic or metal. Traditionally, the cultivated *Lagenaria* gourd and ceramic cooking pots, along with scrap metal, tobacco, beer, personal ornaments, and agricultural produce, were among the items the Hadza received from bartering with neighboring food producers, especially the Isanzu, their neighbors to the south. Perhaps given the importance of containers in human life, especially for cooking, it should not be surprising that today for highly mobile foragers like the Hadza, containers are among the trade goods they receive from the neighbors. Natural containers, which are those requiring little if any processing, are relatively uncommon in nature – especially containers that can withstand use with fire. Without the cultivated *Lagenaria* gourd and ceramic and metal cooking pots, early hominins would have been far more dependent on roasting their food, and on the fruit pods of the baobab and wild bottle gourds, than is evident among the Hadza today. According to Bagshawe (1925:117), Hadza cooking pots were "never used except for making beer," this being "the only drink they manufacture themselves … a mild beverage made from the seeds of the baobab tree." Toward the mid-twentieth century, Cooper (1949:10), like Bagshawe (1925:117), noted only that the Hadza had a few earthenware cooking pots, but we learn nothing about their nature or use. Marlowe reported in 2002 that the Hadza no longer got clay pots from trade, but rather metal pots (2002:266). Blurton Jones (2016:49) noted that at the time of his research, the "Hadza used aluminum cooking pots (soufrire), often very old and battered, even patched." There are no current reports of the Hadza use of metal pots to make baobab beer. It is likely that beer-making (assuming that it was at one time common) ended as the Hadza came to have increasing access to the more potent beer produced by their neighbors. Even in the 1920s, Bagshawe wrote of the Hadza, "Though great honey hunters they do not make honey beer." But he noted, "They thoroughly appreciate all intoxicants given to them" (1925:121).

Based on her 6 weeks of observations in the 1930s, the linguist Dorothea Bleek (1931) reported that the Hadza bartered with their southern neighbors the Isanzu for calabashes and clay pots. The former were used as vessels for water and the latter, principally for boiling meat and hard seeds. The kinds of calabashes and hard seeds were not specified, but the former was, no doubt, the *Lagenaria* gourd, and among the hard seeds of importance to the Hadza would definitely have included those of the baobab.

Fosbrooke (1956), as earlier noted, was a government sociologist in the Mbulu District in 1950. He reported that Hadza material culture included several kinds of containers, of which he identified gourds, kaross, leather bags, and cooking pots. Most of Fosbrooke's article was taken up with Hadza bow and arrow technology. He observed that for the Hadza, "the bow and arrow is the point round which the men's life revolves [and] the corresponding tool in the hands of the women is the digging stick, a small implement used for grubbing up edible roots." The many uses of Hadza-produced and trade-acquired containers were clearly not in the same league as the bow and arrow and digging stick, which have become emblematic of the male/female hunter-gatherer way of life. Fosbrooke made no mention of the use of baskets, drinking vessels, mollusk shells, or baobab pods for "mixing poison" (Marlowe 2002:263). It was not until Vincent (1985a) published her important article on plant foods of the savanna, which was a preliminary report of the roots eaten by the Hadza, that we began to learn about the production and use of the digging stick, a tool used by both men and women, though principally a tool of women.

Baobab bark fiber is used to make baskets in various parts of Africa, and although the Hadza do use baobab fiber in several ways, there are no reports of their use of it to make baskets. Marlowe (2010:93) noted that today Hadza women in 3 to 4 days "weave straw into elaborate and well-made baskets … used mainly to carry berries or processed baobab" and they use metal pots to boil meat, ground baobab seeds and fruit pulp, and grain they acquire from food-producing neighbors. They also use mostly impala skins "as shawls or slings for carrying infants and for carrying food" (Marlowe 2010:78).

Hadza men – in addition to their pouches, the meat they hunt or scavenge, and the baobab fruit they gather – also carry their bow, arrows, ax, knife, pipe, tobacco, and fire drill. Their carrying devices are their leather belts for holding their knives, which usually have leather sheathes, and leather pouches for personal items. They also use fruit pods, particularly those of the baobab, for collecting, transporting, and sharing honey. We get some sense of the value of a man's pouch from the following account of Woodburn (1982a:442–443):

> In any large camp men spend most of their time gambling with one another, far more time than is spent obtaining food. They gamble mainly for metal-headed hunting arrows, both poisoned and non-poisoned, but are also able to stake knives, axes, beads, smoking pipes, cloth and even occasionally a container of honey which can be used in trade. A few personally owned objects cannot be staked, because, Hadza say, they are not sufficiently valuable. These are a man's hunting bow, his non-poisoned arrows without metal heads used for

> hunting birds and small animals, and his leather bag used for carrying his pipes and tobacco, arrowheads and other odds and ends. These objects excluded from gambling share two characteristics: first, they maintain a man's capacity to feed and protect himself and secondly, they are made from materials available in every part of the country. In contrast many of the objects used in gambling are made, at least in part, of materials not available in every part of the country.

Women carry their digging sticks and carrying devices, the latter including baobab and bottle gourd fruit pods for carrying coal, water, honey, and fat; they also carry their grass baskets for fruit, especially berries and baobab; leather karosses for infants and roots; and neck pouches for small personal items.

Baobab Resources That Generate the Need for Containers

Carrying devices, especially containers, were essential to the development of central-place provisioning as a crucial aspect of the gender-specialized foraging. When it comes then to considering the relationship between the baobab and Hadza container use, it is not enough to simply recognize that the baobab pod is a useful container. Equally important is the recognition of the ways in which the Hadza's many uses of the baobab are associated with the use of their different containers. Throughout its range, the baobab tree is not only a source of fruit-pod and woven containers and a culturally significant natural and hollowed-out container structure, but it is also a contributing stimulus to the use of different containers and other carrying and storing devices associated with harvest and use of its many resources. Marlowe (2010:93) discussed the importance of baskets; they are used for gathering field-processed baobab pulp in which the seeds are embedded and for gathering berries as well as other fruit, like those of the fig, palm, and tamarind. These baskets are also used to transport small birds, the chicks of which are taken from nests in baobab trees. The rocky whirlpool containers of dry riverbeds are used like flatstones to grind baobab pulp and seeds to make flours. These riverbed stone containers are associated with the kind of environment in which the Hadza like to camp, especially in the dry time of the year when the baobab is in fruit. Clay and metal cooking pots are used for making baobab porridge, an important weaning food. And freshwater shells are used as spoons for drinking baobab porridge. Considering the use of the baobab as a container structure and source of portable containers, and the ways in which the use of baobab resources requires different kinds of containers suggests the baobab may have played an important role in the development of the carrying behavior that so distinguishes human beings from earlier hominins and from contemporary apes and other primates.

Handling Containers as a Hominin-Specific Trait

It is noteworthy that we share with chimps the use of leaf, branch, stick, and stone, although it is clear from ethnographic studies of traditional hunter-gatherers that our use of these materials is far more intensive than that of chimps. We also share container experiences with chimpanzees and gorillas involving the use of the body, the accessing container foods, and the importance of container structures. What truly stands out, however, is that apart from these container experiences, especially body containers and accessing container foods, there are no reports of wild gorillas and chimpanzees using handling containers. Without the container, there could have been no efficient way of managing the great variety of resources associated with central-place living. In the evolution of hominins, with selection favoring handiness, the baobab likely would have been an important catalyst not only for the use of containers but also for the adoption and making of containers. As such, it would have played a role in the development of the gendered division of labor and its associated central-place living and in the reproductive success and intraspecific competition that the advantages of the foraging way of life made possible.

Chapter 7
The Baobab and Fire in Hominin Evolution

Reflecting the general understanding at the time, Christian Kull and Paul Laris (2009:187) wrote, "Humans evolved in Africa in the presence of fire," noting that "Little is known, however, about the interconnections between human evolution and the flammable savanna biome (Keeley and Rundel 2005; Ségalen et al. 2007; Bowman et al. 2008)." Consider the baobab, for example. It has received increasing attention in recent years for its dietary, nutritional, medicinal, and commercial value; the same cannot be said for the intriguing ways in which it enters into the fire-related activities of Africans of the savanna (Attwell et al. 2015). But the value of the baobab as one of the mediating factors in the Hadza interaction with fire is an important aspect of its multipurpose use, and this points to the possible significance of the baobab/fire association in the foraging adaptation of early hominins. As previously noted, hominin evolution is theorized to have involved a transition from an arboriterrestrial ancestor to a committed terrestrial adaptation to the mosaic savanna. The natural fire regime of the African savanna would have been, without doubt, a key factor in the foraging way of life that was the foundation of hominin evolution, and the baobab was likely a part of this history.

Darwin (1871), as indicated earlier, distinguished between the use of tools and the use of fire. Since Darwin's time, many studies have discussed fire as a "tool" (e.g., Fowler and Lepofsky 2011:297). There are those like Holmes (1971:230) for whom "fire is … perhaps man's most valuable tool." By virtue of its contribution to key areas of a practical and inspirational life, it can be said without equivocation that fire is indeed humanity's most valuable tool. Fire has been used worldwide in association with land management, hunting, gathering, agriculture, herding, fishing, cooking, warmth, light, cheer, protection, processing materials, signaling, socializing, smoking, and inspiration. In the making of our now global way of life, the historical importance of the steam engine and internal combustion engine must also be considered. The intensification of resource production has been the dynamic economic basis of hominin reproductive success; fire, more than any other "tool," dramatically intensified the productivity of human beings. The emphasis here,

J. Rashford, *Baobab*, https://doi.org/10.1007/978-3-031-26470-2_7

however, is clearly on "tools" broadly defined as a means to an end. In this general sense, language could be considered a tool of social life, just as weapons could be considered tools for hunting and warfare and boats as tools for fishing.

Human beings have uniquely evolved to be the only species capable of generating fire at will and the complexity of human fire dependence far exceeds that of any other species. In addition to tool, fire as a chemical process is also identified as a force, technology, and vital resource, the dependence on which is a defining characteristic of the hominin lineage. Fire as a natural feature of the environment of hominin evolution and the development of hominin fire dependence have given rise to the view of modern humans as the pyrogenic primate (Russell 1978; Cahoon et al. 1992; Pausas and Kelley 2009; Wrangham 2009; Wrangham and Carmody 2010; David et al. 2011:2223; Archibald et al. 2012).

It is noteworthy that in the presentation of their "Pyrophilic Primate Hypothesis," Parker et al. (2016:61) had to turn to field studies done in Australia to make the argument "for the pyrogenic foraging benefits available to our evolutionary ancestors." Reflecting on Marlowe's work with the Hadza, they wrote:

> [T]he role of fire in subsistence activities among sub-Saharan African foragers living in environments believed to be similar to those in which our genus evolved remains understudied. For example, Marlowe briefly mentions Hadza fire use in the context of making and using fire drills for ignition. He explored new methodologies to examine how hunter-gatherer fires might be detected in the archeological record, but descriptions of fire use away from camps, except for cooking, are absent.

Marlowe might not have gone as far as some would like, particularly with respect to Hadza landscape burning. Nevertheless, his contribution to understanding Hadza fire use is far more substantial than is suggested by Parker and colleagues. Marlowe clearly highlighted the fact that we need more studies and a greater variety of studies to understand both Hadza fire use and, more generally, the role of fire in human evolution.

This chapter looks at the baobab as one of the mediating factors in human evolution toward fire dependence. The baobab's many contributions to Hadza fire-related activities suggest this is yet another important way in which the preeminent multipurpose tree of the African savanna might have been significant for early hominins. The first section of the chapter presents an overview of the significance of fire in hominin evolution, critically reviews the idea of "the control of fire," and discusses the fact that birds and plants, like human beings, have also been identified as spreading fire. The last three sections relate this discussion to the Hadza use of fire and its relation to the baobab as their central-place tree of life. Only a preliminary assessment is offered here, however, since there are no studies that have focused on the baobab's influence on the Hadza relationship to fire or the role of the baobab in hominin evolution as it relates to fire.

The Three Key Moments of the Hominin/Fire Relation

Broadly speaking, natural and anthropogenic landscape fires have three important benefits for human beings. The first is land clearance, which improves visibility, mobility, and resource access. The second is fire-enhanced foraging, cultivation, and herding, especially with respect to the management of fields and pastures. The third is fire-based protection from harmful animals, such as pests, predators, and poisonous snakes; from catastrophic seasonal wildfires at the height of the dry time by judicious burning; and from competing pastoral neighbors by torching land that would otherwise be used as pastures (Laris 2002; Kull and Laris 2009; Paris 2011). The view presented here is that early hominins, especially the lineage from among the Australopithecines that gave rise to the genus *Homo*, were likely beneficiaries of fire as a natural feature of their savanna environment.

The main problem with many conceptual approaches to the human/fire relation is that they do not capture the fundamental fact that the history of hominin "control of fire" is the history of hominization as an evolved dependence on fire, a process that would have started with natural landscape fires. That is, these approaches do not clearly distinguish between early hominins benefitting from wildfires on the one hand and later hominins using fire by spreading, fueling, and generating it on the other. Distinguishing these four components in understanding the relationship between hominin evolution and fire is already implicit in the work of a number of scholars (e.g., Dennell 1989; Gilbert 1989:13; Pyne 1995, 2001; Bowman et al. 2011; Burton 2009; Archibald et al. 2012; Atwell et al. 2015; Gowlett 2014, 2016, 2018; MacDonald 2017). But they have not been systematically treated.

A clear recognition of these three key moments of the hominin/fire relation requires a critical appraisal of the concept of "control of fire" which is widely used in the literature to frame discussions of human fire dependence. But this expression (and its many related forms such as "control over fire," "the use and control of fire," and "the controlled use of fire") is inadequate for theorizing human evolution as the outcome of an intimate association with fire. Used as a subject heading or general concept to mean only the generation of fire, it leaves out of consideration taking advantage of natural landscape fires and the development of methods to spread and fuel fire. For example, the familiar construct of "the control and use of fire" is limited and redundant. Both "control" and "use" can mean the spreading, fueling, or generation of fire. When used to mean only the generation fire – as most often occurs – it is restrictive. It leaves out of focus both the taking advantage of natural fires as the true beginning of the significance of fire in hominin evolution and the initial use spreading and fueling of fire.

The expression "to create and control fire" has proven confusing for many. For example, the discussion of Parker et al. (2016) suffers from equating the two terms. However, the expression is illogical because to "create fire" is to "control fire," but not all control of fire (such as spreading or fueling fires) means the creation of fire. "Control" is the general term of which fire creation is one specific example. It is also important to note that the idea of the "control of fire" (defined as fire spreading, fueling, and generation) still overlooks the idea of taking advantage of wildfires. It would be more accurate to say the initial "control of fire" by spreading natural landscape fires leads to the ability to fuel fire and then to the ability to create it. The ambiguity in the expression "to create and control fire" exists because it is possible that what Parker et al. (2016) mean is that "to create fire" is to generate fire and "to control fire" is to systematically direct its application. Other equally problematic expressions that are used to characterize the significance of fire in hominin evolution include the idea of "the intentional use of fire" which is also expressed as the "deliberate use of fire." This concept leads us to ask in what sense it can be said that fire was used unintentionally by early hominins. Discussions of fire use framed in such terms as the "control of fire," the "creation and control of fire," and the "intentional use of fire" often present in a bifurcated view involving an early period in hominin evolution when there was no "control of fire" and a specific moment in time when fire control is described as having been "discovered" or "invented" or "domesticated" or "tamed."

The Timeline of the Human/Fire Relation

Fire-Benefiting Hominins and Other Species

Long and short chronologies have been presented for the importance of fire in hominin evolution (Sterner 2012:15). Long chronologies in Sterner's work (2012:15) and in the literature generally mean hominin use of fire from the time of the genus *Homo* at 2.5 to 1.7 million years ago, and it is especially associated with *Homo erectus* at about 1.8 million years ago (e.g., Gowlett 1981, 2010; Gowlett and Wrangham 2013; Roberts and Bird 2012; Wrangham 2009). But long chronology is not a complete chronology. A complete chronology must consider the hominin phylogenetic tree in the context of the hominin/fire relationship, starting with benefitting from wildfires. This would be true of the earliest hominins. It is a reasonable assumption that the direct human ancestors among early hominin foragers took advantage of the cleared, resource-enhanced, protected environments produced by seasonally recurring fires (Cahoon et al. 1992). The idea of benefitting from wildfires is here regarded as the earliest relationship of hominins to fire, and it means to gain an advantage from wildfires without actually using fire in the specific sense of spreading, fueling, or generating it.

The fire-produced benefits of space, resources, and protection are the same for humans, chimpanzees, birds, and plants that have evolved to take advantage of fire as a natural feature of their environment. Recurring landscape fires result in a patchy environment comprising not only unburnt and burnt areas but also burnt areas in various stages of succession, and this creates a biologically diverse environment. It cannot be said, however, that the intention of African landscape burners is to produce a pyrodiverse environment in a grand, overall sense. The intention is clearly space, resources, and protection; the biologically diverse human environment is the cumulative outcome of landscape burning for these reasons.

Chimpanzees, for example, provide an example of animals that are fire beneficiaries. Recent studies found they are comfortable with fire, but there is no evidence that for chimpanzees, the primary aim is to forage at the margins of fires or in the immediate post-fire environment (Pruetz and LaDuke 2010). The most important advantage they seem to derive from fire is protection. Pruetz and Herzog (2017:5348) discussed the importance of fire-produced land clearance for chimpanzee mobility:

> [B]urned areas may be attractive because of enhanced predator detection in fire-cleared grasslands (Herzog 2015). In burned areas, vegetation that may have otherwise served as cover for ambush predators is gone. Without sufficient cover, predators may abandon efforts at hunting in these areas, and research has suggested that predators avoid burned habitats (Ogen-Odoi and Dilworth 1984; Eby et al. 2013; Green et al. 2014). At Fongoli, where approximately 85%–90% of the understory is grass, late rainy season and early dry-season travel is difficult. Chimpanzees have been observed taking advantage of recently burned areas as travel corridors, and it is possible that these areas also provide them greater safety in terms of predator detection and reduced ambush sites for predators.

Similarly, a variety of birds are beneficiaries of the space, resources, and protection provided by natural landscape fires. These birds include aerial and terrestrial fire-hunting species and terrestrial fire-scavenging species. According to Grzimek (1970:178), "[F]ire holds a powerful attraction for all birds which live on insects, lizards, frogs, snakes and other small creatures. Storks, marabous, bustards, eagles and buzzards all flock to the scene from far and wide to hunt the lesser game that have to flee from the flames." In addition to benefiting birds foraging at or in the wake of fires, research has shown that landscape burning also increases the diversity of savanna birds. In the first Africa-wide study of the effects of savanna fires on birds and mammals, Beale et al. (2018) found that there were about a 20% increase in the species of mammals and a 30% increase in the species of birds that were associated with landscape fires; they identified this fire-enhanced species increase as pyrodiversity.

Fire benefits for many plants involve various aspects of their reproductive process which are triggered by fire (Maurin et al. 2014). These include fruit dispersal, seed release, seed germination, seedling establishment, vegetative maturation, flowering, and seed production. The most successful fire-dependent plants presuppose the production of an abundance of light, far-spreading, wind-dispersed seeds that track the same path as wind-dispersed fires. A well-known examples is the willow herb of Europe that has earned the name fireweed (*Chamaenerion angustifolium* L. Scop.). Various species of *Acacia* and *Albizia* that grow in many parts of the dry tropics and some pines time the release of their seeds to occur after wildfires. For these plants, wildfires mean space; a fire-enhanced resource environment that includes more light, air, moisture, and nutrients; and protection from competitors.

Animals and Plants That Spread Fire

According to Pruetz and LaDuke (2010:646), "The use and control of fire are uniquely human traits thought to have come about fairly late in the evolution of our lineage and they are hypothesized to correlate with an increase in intellectual complexity." Similarly, Parker et al. (2016:54) wrote, "Members of genus *Homo* are the only animals known to create and control fire." But these assessments are problematic. Here, we can see the theoretical consequences of an ambiguous perspective with respect to the idea of "to create and control fire." Fire control, which is logically the same as fire use, is not unique to humans as often claimed. What is particularly significant is that some birds and plants, like human beings, are also known to control fire by spreading it, but they do not generate fire which is genuinely unique to human beings. Hence, fire control as fire use is not "unique" to humans, as often claimed. Humans and birds spread fire by collecting and carrying fire sticks. Fire-spreading raptors, including the black kite (*Milvus migrans*), whistling kite (*Haliastur sphenurus*), and brown falcon (*Falco berigora*) of Australia, use their talons and beaks to carry fire sticks, and efforts are under way to capture this behavior on film (Bonta et al. 2017). Unlike these "firehawks," humans use containers to collect, carry, and apply live coal. Obviously, plants do not actively spread wildfires as humans and birds do. Nevertheless, pyrophile plants are those that not only resist and recover from wildfires but also take advantage of fire-produced land clearance, enhanced soil fertility, and protection as an essential aspect of their reproductive success. They have also evolved fire-adapted traits that make them highly flammable. In combination with lightning as the ignition source, they serve as natural tinder, kindling, and fuel that contribute to the ignition, intensity, and spread of wildfires. Well-known fire-spreading pyrophytes include eucalypts of Australia, pines of temperate latitudes, cogon grass (*Imperata cylindrica* L. (P) Beauv.) of Southeast Asia, and chamise (*Adenostoma fasciculatum* Hook. & arn.) of California.

Hominins and the Spreading, Fueling, and Generating of Fire

Over time, hominins taking advantage of naturally occurring fires would have led to the initial control of fire in the specific form of intentionally spreading fires, a development likely associated with the Australopithecines. Dennell (1989:11) identified this early spreading and fueling of fire as "ways of storing fire," which he regarded as a critical development for hominins. Fire management would also have meant a growing dependence on the use of fire technology, which is comprised of all the material means involved in the generation, management, and application of fire. These material means include a fire kit comprised of ignitor, tinder, and kindling; fuel, including not only firewood and charcoal, but also oils, resins, and the woody parts of some flowers, fruits, and seeds; fire-handling containers for collecting, confining, carrying, storing, and applying flames or live coal; fire protectors in the form of wind screen and rain cover; and some sort of fire extinguisher, given the well-known danger associated with the use of fire.

The ability to generate fire at will is the final step in the development of the hominin/fire relation. This can be characterized as the hominin mastery of fire. It is identified with the evolution of the genus *Homo* and especially with the Old-World expansion of *Homo erectus* into the cold environments of temperate and polar latitudes (Cohen 1977, 2002; Fagan 1989:127, 128). But distinguishing the residue of anthropogenic fire from that of natural fires is more difficult to determine archaeologically at around 1.8 million years ago than from 400,000 to 250,000 years ago.

The spreading, fueling, and generation of fires have so far been dealt with as three moments in the development of the relation between hominins and fire. But these three moments also occur simultaneously in the lives of contemporary foragers. For example, the Hadza depend on spreading, fueling, and generating fire in their daily lives. For the most part, men generate and fuel fires, and Hadza women spread and fuel fires (Marlowe 2010:92). Household and communal fires in particular burn continuously from 1 to 3 months because they are sustained by spreading and fueling. Radcliffe Brown (1964:258) reported that the Andaman Islanders, in contrast, did not generate fire, depending instead on routine spreading and fueling. He described fire for the Andamanese:

> … as the one object on which the society depends most of all for its well-being. It provides warmth on cold nights; it is the means whereby they prepare their food, for they eat nothing raw save a few fruits; it is the possession that has to be carefully guarded, for they have no means of producing it and must therefore take care that it is always alight; it is the first thing they think of carrying with them when they go on a journey by land or sea; it is the centre round which the social life moves, the family hearth being the centre of the family life while the communal cooking place is the centre round which the men often gather after a day's hunting is over. To the mind of the Andaman Islander, therefore, the social life of which his own life is a fragment, the social well-being which is the cause of his own happiness, depends on the possession of fire without which the society could not exist. In this way it comes about that his dependence on the society appears in his consciousness as a sense of dependence upon fire and a belief that it possesses power to protect him from dangers of all kinds. Indeed, the Andamanese believe it is the possession of fire that makes human beings what they are and distinguishes them from animals.

The same routine spreading and fueling of fire has been reported for the Mbuti. Turnbull (1962:49, 58) observed that for them, "No matter how hot it is, there always has to be a fire." We learn:

> Every woman, when moving camp, carries with her a burning ember wrapped heavily in fire-resistant leaves. None of these Pygmies knows how to make fire. The first thing they do when they stop on the trail for a rest is to unwrap the ember and, putting some dry twigs around it, blow softly once or twice and transform it into a blazing fire… It is damp and cool in the forest, and at any time of the day it is pleasant to sit around the fire; but at night, when it is cold, the fire gives not only warmth but protection, keeping the animals away.

Fire is a fundamental part of the life of the Mbuti. As Turnbull (1962:97–98) notes, "The Pygmies regard fire as the most precious gift of the forest, and by offering it back to the forest they are acknowledging their debt and their dependence." He continued:

> The sacred hunting fire is all that they consider right and necessary, and this is found throughout the forest. It is thought to secure the blessing of the forest which provides the game, and to bring good luck to the entire group… For the net hunters it is impossible to hunt alone. Men, women and children all have to co-operate if the hunt is to be successful… The "Fire of the Hunt" simply involves the lighting of a fire at the base of a tree a short distance from the camp.

Even if the Australopithecines lacked the technology to generate fire at will, these ancestors of modern humans likely benefitted from natural fires, and it is possible that, like the Andaman Islanders, Mbuti, and Hadza women, they too depended on the spreading and fueling of fire.

Wrangham (2009) and colleagues (Wrangham et al. 1999; Carmody and Wrangham 2009) are right to use anatomical/biological evidence from around 1.8 million years in their account of human evolution in association with cooked food, but the archaeologists who have been skeptical of Wrangham's thesis are also right in terms of the archaeological evidence of fire use by *Homo erectus* and modern humans in the time frame of 400,000–100,000 years ago (James 1989; Wood and Collard 1999; Sterner 2012). One key problem is the difficulty in distinguishing in the archaeological record the benefits of natural fires for early hominins and the initial use of fire in the form of the spreading of landscape fires. Yet, these earlier periods of the hominin/fire relation have to be theoretically considered to achieve an adequate understanding of the hominin/fire timeline. As MacDonald (2017:156) noted, "Resolving the chronology for fire use is more difficult than it might initially seem. For most of human prehistory, hunter-gatherers were relatively mobile and put little effort into building structured hearths." The view of Parker et al. (2016:60) is also relevant here. "Modern hunter-gatherers," they have written, "often cook foods over fires that require no prepared hearths and leave no archeological remnant." They go on to "predict that highly mobile, actively pyrophilic ancestors would have done the same." According to these co-authors, "This would explain why the initial harnessing of fire is not marked by hearths. As long as hearths are the requisite indicator of fire control, archeologists will continue to miss a much deeper history of obligate pyrophilia" and "place the acquisition of cooking at a time when nothing dramatic was happening in human evolution." Yet, these earlier periods of the hominin/fire relation involving the spreading and fueling of fire have to be theoretically considered to achieve a more complete understanding of that relation.

The Hadza/Fire Relation

Hadza Oral Traditions and the Kinds of Fires They Make

The obvious inspirational importance of fire in the lives of human beings worldwide is understandable given not only its singular appearance and vital uses but also its real danger. Despite its inspirational importance, however, studies of forager fire use have paid little attention to such things as forager religion, aesthetics, and oral traditions (in contrast, see Wiessner 2014). The inspirational uses of fire include warmth and light for such things as storytelling and nighttime rites; altar fires as places of offerings that are the focal point of shrines and other kinds of religious spaces; the upward movement of flames and smoke that link the earth and the sky in religious cosmologies; the identification of fire with spirits that are central to ceremonial life, especially the Supreme Being, ancestors, and ritually important contact spirits; the ritual disposal of the dead and purification rites; and the use of fire as temporal marker symbolizing renewal, for example, midsummer observances in temperate latitudes and observances associated with the return of the rains marking the renewal of the growing seasons in the dry tropics.

The Hadza account of the origin of their association with fire is evident in their oral traditions. In a passage excerpted from their film *Hadzabe: By the Light of a Million Fires*, Peterson et al. (2013:xvii) reported the Hadza as saying:

> This is OUR story as told to us by our fathers and to them by their fathers by the light of a million fires. Watching the sparks fly into the African night, we listen mesmerized and watch as the elders enact our history – the history of humanity – before our very eyes. They hold our wisdom and their stories – our stories – become alive, attaching themselves to our minds so that they will never be forgotten. And, when the time comes, we will pass them on to our own children as the sparks of another fire pierce the night. These stories are an essential part of our early schooling but when dawn breaks, it brings new lessons and the learning process continues as we tag along behind our parents while they hunt and forager.

Given the practical and inspirational value of fire in their central-place forager adaptation, it is not surprising to learn that a "favorite fireside story" (Peterson et al. 2013:12–14) of the Hadza gives an account of how "The First People" acquired fire and the knowledge of how to use it for cooking and warmth. In the process, they also acquired an understanding of how fire use related to the proper way to have sex. According to this account, the First People of the Hadza were stick people because they could not bend. "They were long, giraffe-like figures who hunted only at night and had no fire because, unable to stoop, they could not light it." One day, God, the invisible (who they call Haines), asked a woman to light a fire:

> The woman tried but the earth was too wet and she failed. Haine turned the world upside down so that the earth would dry and the rain would fall out of the sky. When the earth was dry, Haine made a fire and roasted an eland he had hunted. Then he asked the woman to distribute the meat among all the *Ikanawangube'e kenebe'e* people so that they could taste cooked meat for the first time.
>
> When they had all eaten their fill, Haine first softened and then broke the people's bones so that they could sit by the fire and draw warmth from the flames. That is how the giraffe people became the *Gelanebe'e*, "the people of old".

As the story continues, in those days, there were no children because men had sex with women between their toes. One day, God told a hunter he was doing it the wrong way and instructed him to have sex between the woman's thighs. "Next day when they met again, Haine asked how it had been between the woman's thighs. 'Very sweet', the man answered. Haine told him to keep at it and soon the woman became pregnant."

> Haine provided another eland for the man while he was out hunting and instructed him to tell the other men the secret that lay between a woman's thighs. The hunter called all the men to help carry the eland home, looking for an opportunity to tell them his secret.
>
> As they carried the eland back to camp, !oo!oko, the devil, appeared. He was jealous and didn't want the secret that lay in between a woman's thigh to be known, so he tried to interfere. Fortunately, the *Gelanebe'e* men could now bend and make fire, so they set !oo!oko's head aflame and he ran off.
>
> !oo!oko ran off, but he didn't give up. Wanting the men to be vulnerable, he made the sun set early so that the men would have to sleep in the bush that night at his mercy. But the *Gelanebe'e* built a circle of thorns around themselves and the eland and lit a fire. When all of the men were safe and gathered around him, the hunter told them his secret: "Stop having sex between the toes of the women and go for in between the thighs."
>
> "From that time on," the old man finished, pointing with his pipe towards the laughing faces by the fire, "the *Gelanebe'e* had many children."

And so, God created the conditions for fire by drying the earth. He also made the first game-roasting fire that introduced the "stick-like" First People to the taste of cooked meat and made it possible for them to enjoy the warmth of the fireside where they could come together. It was God's fire that transformed the First People into the *Gelanebe'e*; and it was the ability of "the people of old" to use fire that made it possible for them to protect themselves against the efforts of the devil to thwart the communication of God's knowledge to all regarding the proper way for men and women to have sex.

Hadza fires are associated both with the household, communal, and special activity areas of their residential camps and with their savanna environment (O'Connell et al. 1991; Mallol et al. 2007; Marlowe 2010). In addition to fires for cooking, warmth, light, and camp protection mentioned in the story of the First People, Hadza men and women also produce foraging fires. Women do so for roasting roots and occasionally for roasting game. Game-roasting fires are most often identified with men, but Peterson et al. (2013:24, 98) presented a photograph of a Hadza woman roasting a leopard tortoise, a species primarily eaten by women and children. Hadza women also use fire in association with different kinds of containers, especially cooking utensils, and to make and repair their digging sticks. In addition to game-attracting grass fires and ambush-hunting fires, Hadza men also use fire for other purposes. They make game-roasting fires, especially for monkeys and baboons. They use fire in finishing off the preparation of their fruit pod containers, in shaping wood to make their bows and arrows, in deterring bees when gathering honey, in attaching ax heads to their wood handles, in their smoking pipes filled with tobacco or *Cannabis,* and in managing the landscape to improve foraging both with respect to gathering and hunting.

Natural and anthropogenic landscape fires would have been the most important kind of fire in early hominin history, starting with fire benefitting and later with fire spreading based on transporting fire sticks and by container-collected and carried live coal. Fire-benefitting and fire-spreading suggest that early hominins had knowledge of fire ecology (Caillault et al. 2015). Like our early hominin ancestors, the Hadza set landscape fires for a variety of reasons that include land clearance; attracting game by promoting the growth of grasses; improving the success of tracking;

driving animals in the direction of strategically positioned hunters; protecting themselves from pests, snakes, and predators; and defending themselves against the increasing incursions of their Datooga pastoralist neighbors (Mallol et al. 2007:2037), by destroying what would otherwise be pastures suitable for their animals.

The Consequences for the Hadza of the Fire Needs of Others

In addition to wildfires and to their own contained fires in and out of camp, the Hadza are also exposed to the fire needs of other groups within and at the margin of their traditional territory. This is particularly true of the pastoral neighbors whose overall impact on Hadza foraging is destructive. For example, Madsen (2000:41) wrote:

> Tatoga pastoralists, particularly the Barabaig, represent the front line of Hadzabe land alienation… With large herds of cattle, sheep and goats these pastoralists change the bush land habitat through a variety of means. They use fire to encourage the growth of grasses for their cattle, clear large numbers of trees for house and cattle corral construction (bomas), and dig deep wells in the dry season for watering their cattle.

Other small-scale producers also cut down trees for space and for the commercial production of construction material, firewood, and charcoal.

The Impact of Fire on Baobab Trees

Natural landscape fires are known to seriously damage and even kill baobab seedlings, saplings, and small trees. Nevertheless, an expression of the exceptional vigor of the baobab is its ability to recover from fire as a routine feature of its savanna environment; the tree is often identified as fire-resistant or fire-adapted. Baobab saplings often resprout after fire damage and can overcome both predation and trampling by wildlife and livestock and damage caused by arthropods and microbes. What gives them this resilience is their large water-storing taproot. A remarkable feature of the species, whose tenacious grip on life is part of its status as a manifest tree of life, is its ability to regenerate its bark after fires and even after its bark has been completely stripped from around the tree by intense human harvesting, which is widespread in Africa (Romero et al. 2014; Luckert et al. 2014).

Baobab-Initiated Landscape Fires

In his account of the relationship between the baobab and fire, Watson (2007:70) identified the baobab as a pyrophile. That is, a plant that has evolved to take advantage of fire in its reproductive success. "Ironically," wrote Watson, "the same fire that can devastate the young baobabs may also help break the dormancy of the next crop of seeds. And, still more ironically, mature baobabs may even conduct the lightning that starts the fire that burns their offspring." If it is indeed the case that the baobab can survive wildfires and that wildfires enhance the germination of baobab seeds and eliminate competitors (leaving the baobab free to take advantage of cleared, nutrient-rich, protected environments produced by wildfires), then the baobab could appropriately be described as a true pyrophile tree of the African savanna.

Lightning

The Hadza are clearly concerned with lightning. Petersen et al. (2013:136) reported them as saying that during the rainy season, parents of twins smear charcoal on their foreheads and the bridges of their noses to prevent lightning strikes. The importance of the baobab as a mediating factor in the hominin/fire relation includes the recognition that lightning strikes, spontaneous combustion, and human fires associated with the tree have all been identified as potential ignition sources of landscape fires (Patrut et al. 2010; Watson 2007:70; Wickens and Lowe 2008:158). Because lightning is seasonally recurring and widespread, it is one of the most important causes of wildfires. But according to Romero et al. (2014:98), "The primary cause of death of baobab trees is reportedly related to drought (viz., lack of soil moisture) and less frequently to lightning (de Villiers 1951; Piearce et al. 1994)," and they note too that "Elephants are also known to kill baobab (Swanepoel and Swanepoel 1986)." Mature baobab trees are no doubt susceptible to lightning strikes, but the frequency of baobab-initiated wildfires has not been documented. What is known, however, is that the soft wood of the baobab burns quickly when dry, which makes it a poor source of firewood compared with hardwood trees.

Spontaneous Combustion

Despite the baobab's "remarkable vitality," evident in its ability to resprout after damage and to continue growing after it has been toppled, Hankey (2004:4) wrote "when a tree dies it collapses into a heap of soggy, fibrous pulp. Stories exist of how such quickly decomposing trees spontaneously combust and get completely burnt up." Wickens and Lowe (2008:158) noted:

> It is alleged that the fibrous heap may sometimes spontaneously combust, causing bush fires and much damage to the environment. It has been similarly reported from South Africa that under certain circumstances moisture finds its way into small cavities in the trunk; decomposition sets in, followed by spontaneous combustion. The spectacle of spirally smoke rising from the tree as it slowly but surely burns itself to death appears to bear witness to this tragedy (Cashel 1995). The heating of the wood reported by Diana Mayne [2006, personal

communication to the author] supports the possibility of spontaneous combustion. However, other commentators are skeptical and reject spontaneous combustion in favour of lightning igniting the tinder-like residues.

Although it is worthwhile to consider the possibility of baobab lightning strikes and spontaneous combustion as ignition sources of natural savanna fires, these phenomena are yet to be systematically documented.

The Escape of Human Fires Associated with the Baobab

The clear association between the baobab and human settlement in Africa – including the camps of foragers, the villages of farmers and herders, and the urban landscape of feudal and contemporary states – has long been noted in popular and scholarly accounts. An association between human settlement and the baobab is also a relation between the baobab and anthropogenic fires. This relation suggests the possibility that baobab-linked settlement fires might give rise to landscape fires.

The Baobab as a Place for Human Fires

The Baobab and Hadza Fires

One of the most important ways in which the baobab is associated with the everyday use of fire is the fact that it is frequently the destination of Hadza resource forays and that the Hadza preferentially camp under or near baobab trees for a good part of the year. The focus here is on central-place fires identified as fires of Hadza residential camps in contrast to landscape fires and other kinds of out-of-camp fires. And in this context, it is particularly during the annual fruit harvest of the dry time of the year that the baobab is the place in the Hadza mosaic savanna landscape where fires are most often set and are kept burning for months (Mallol et al. 2007:2037). The baobab camp is a human-created pyrogenic environment because of the continuous presence of household, communal, and special activity area fires and the dispersal of ash from these fires into the surrounding "secondary rubbish areas" peripheral to the swept area of the camp (O'Connell et al. 1991:67).

In their study of the distribution of refuse-producing activities at six "Hadza base camps," which was intended to inform analyses of archaeological site structure, O'Connell et al. (1991:70) reported that for one of these camps, the great majority of camp activities (86% of the total observed) took place in four communal areas, two of which were peripheral and two centrally located. We also learn that "Communal areas were always marked by the presence of one or more hearths and were periodically swept clear of refuse" and that "Communal areas were used for essentially the same range of activities as household areas but were not associated with any particular household or sleeping group (the latter including nuclear families, older women and adolescents)." A broader range of activities was associated with centrally located communal areas, with women accounting for 52% of the activities and men accounting for 49% (which are percentages comparable to household areas). By comparison, only 14% of the activities took place in the 12 household areas. Of household areas, O'Connell and colleagues wrote:

> They were the setting for a wide range of domestic activities, including the preparation and consumption of food, the manufacture and maintenance of tools, clothing, and other equipment. Each was marked by a circle of cleared ground 2 to 6 m in diameter. Each contained a simple shelter and one or more hearths and was flanked by one or more secondary refuse areas.
>
> Household shelters usually contained hearths. These were small features, about 30 cm in diameter, located at the center of the hut floor or just inside the entrance. Outside hearths were also common and were generally positioned within 1 to 3 m of the hut entrance. Like interior hearths, they were relatively small but were often flanked by two to three large stones that served as stands for cooking vessels.

Unlike the household areas of residential camps, the peripheral areas of camps were "defecation" sites and "special activity areas." The latter included an area for men's activities, such as working on their hunting gear, and an area of women's activities, especially associated with "bedrock grinding facilities." These grinding facilities "were large, relatively flat-topped outcrops of granite or schist, whose surfaces were marked by small pecked and polished patches, 30–50 cm diameter, where baobab and other hard seeds were ground. Small handstones were frequently found in association with such features. All residential base camps had these facilities, which suggest their presence was a determinant of camp location" (O'Connell et al. 1991:68). Defecation sites and seed-grinding rocks seem to be the least associated with the routine use of fire in the peripheral areas of the Hadza residential camp.

To avoid ambiguity, it would seem useful to recognize that homes can, and often do, contain more than one kind of fire. That is, not all home fires should be called hearths as is often done. Theoretically, "hearth" means the central home fire used for cooking, warmth, light, processing, and protection. A hearth can be located either inside and outside the Hadza home. But the Hadza also have sleeping fires and other kinds of Hadza home fires, including torch-light fires, protection fires, and disposal fires.

It is worth noting that landscape fires can develop from these camp and foraging fires, whether by accident or carelessness. For example, Mallol et al. (2007:2037) gave a general account of the way in which Hadza men collect honey, including the use of fire to "stun the bees." They wrote that at the end of the honey-collecting process, men "carry some honey off if they have some leftover, leaving the fire smoldering beneath the tree. This fire may burn with flames for about 10 – 15 minutes altogether."

The Platland Baobab and Human Fires

One of the most important documentations of the baobab as a place of fire comes from a study of the fire history of a famous ancient baobab of southern Africa. The article by Patrut and colleagues (2010) reported on the radiocarbon dating of wood samples from two inner cavities of the giant multistem Platland baobab of South Africa. The results indicated growth layers that were several centimeters thick covering "the original old wood" of the inner trunk:

> Four new growth layers were dated before the reference year AD 1950 and 2 layers were dated post-AD 1950, in the post-bomb period. Formation of these layers was triggered by major damage inside the cavities. Fire episodes are the only possible explanation for such successive major wounds over large areas or over the entire area of the inner cavities of the Platland tree, able to trigger regrowth.

What is particularly interesting is the account of the origin of the fires in the Platland baobab Patrut and colleagues presented. They wrote (2010:725):

> The most recent fire episode of AD 1990 is documented and was inflicted by human activity. Workers at the farm set a fire to smoke out a pair of black mamba snakes from the cavities. The fire got out of control and burned badly the inner cavities of the Platland tree. The other fire episodes evinced by the dating results could have been caused by a number of factors, such as natural veld fires ignited by lightning or by drought and excessive heat; anthropogenic veld fires caused by hunting practices, namely to stimulate early grazing; or fires started through honey gathering and as protection against wild animals at the entrance of the cavities.

At the time their study was done, the Platland baobab was estimated to be over 2000 years old. The authors reported (p. 725) that "Artifacts found in the cavities indicate that the Platland tree was possibly used for centuries as a shelter. The tree was along one of the old, possibly ancient ivory and gold trading routes through the region." Against this backdrop, they continued (2010:725):

> The other 5 fire episodes within the cavities of the Platland tree could have been generated by abiotic factors (buildup of biomass, possibly aided by lightning to fuel intense fires to burn also the internal cavities, natural veld fires) or biotic factors (human activity). On the basis of current field observations in Southern Africa for both abiotic and biotic causes and the archaeological and historic record, fires caused by human activity appear to be more likely.

The baobab is indeed a place of human fires. It has likely been so for a very long time in the history of hominin evolution.

The Baobab and Hadza Fire-Making Materials

Fire Drill

The ease with which the Hadza are able to start or spread fire is frequently noted in the literature, but the materials used are rarely specified or systematically discussed. Hadza men and women require fire on a daily basis in various aspects of their practical and inspirational life, and the baobab, as the Hadza's preeminent multipurpose tree, contributes to fire-related activities in a variety of ways, including as a source of the fire drill for starting fires. This Hadza fire drill consists of a small, straight, hardwood stick that is slender and round, a flat base upon which the drill operates made of softer wood, and a hard foundation, such as a stone or knife blade, upon which the drill is placed. According to Marlowe (2010:91), the Hadza fire drill is sometimes "decorated" with "feathers or tuft of hair," but the kinds of feathers and hairs and their aesthetic, symbolic, and religious significance are not specified. In the early twentieth century, Bleek (1931:274) wrote: "Each [Hadza] man carries a bow, arrows, and fire stick in his hand, usually in the left and has a knife in a sheath stuck in the belt which seems to be his most important article of clothing." Bagshawe (1925:122), who, as earlier noted, was a British colonial district officer in northern Tanzania in the early twentieth century, made several trips to Hadza territory from 1917 to 1923. He wrote: "No Kangeju is ever without the necessary [fire-making] apparatus, for he depends upon fire, not only to light his pipe but to protect him at night from carnivorous animals." We learn from Bagshawe (1925:122) that "Fire is made in a few moments by twirling between the palms of the hand a stick of hard wood, the point of which rests upon tinder placed on a flat piece of baobab wood."

Hoebel (1958:220) also described the fire drill as requiring tinder, but the use of tinder is not always mentioned in accounts of the Hadza fire drill. Bagshawe offered one of the few reports of the Hadza use of fire drill tinder, but he does not identify the material used. In a travel blog of a family "safari in Africa" published in the *Christian Science Monitor*, Turner (2002) identified a photograph as "Numbile, a Hadza man, helps 8-year old Josh set an arrow shaft (minus the arrowhead) on a flat piece of wood surrounded by bits of dry grass." It is not clear in scholarly accounts, however, whether the fire drill of the Hadza is such that it creates its own tinder, which it ignites to produce embers that are then placed in hand-held grass kindling and blown into a flame. Nevertheless, once the initial kindling flames, it is either used to ignite additional kindling and fuel that has already been laid out, or it is placed on the ground and a fire built up by the addition of substantial kindling (such as small branches and the baobab fruit pod) before larger fuel is added.

The frequency with which the Hadza use the baobab fruit shell in making their fire drill or as secondary kindling is unknown. As Bleek noted (1931:274) and as Marlowe (2010:91) reported, Hadza men carry their fire "stick with them at all times." Marlowe (2010:91) observed, however, that "Sometimes when a man does not have a fire drill, he uses an arrow as a drill, as Turner (2002) described. The drill is placed into a small piece of wood (*pa'ameko*) the man can find on the spot." Although Marlowe (2010:91) suggested that any kind of wood that is conveniently available is used as the fire-drill base, Peterson et al. (2013:120) specifically identified the "fire stick" and "base" as made from the wood of the *Sacleuxii newii*, *Commiphora schimperi*, and *Markhamia obtusifolia* trees. However, it is reasonable to assume that the baobab fruit pod would sometimes be used as the base of their fire drill and as kindling, since the Hadza preferentially camp at baobab trees for a good part of the year and Hadza women process baobab fruit pulp and seeds in camp. Hadza men make a fire at the base of baobab trees when collecting honey. It should be noted here that the use of the baobab fruit pod in fire-making goes beyond the Hadza. Wickens and Lowe (2008:121), who cited Sébire's (1899) account of Senegal, reported that "A piece of the [baobab fruit] shell is sometimes used as the base in fire making, and drilled with a piece of guinea corn (*Sorghum bicolor*) stalk."

Multisource Fuel Tree

People's choice of fuel reflects the purpose for which the fire is intended and the procurement cost of the fuel in relation to the kinds of fuels that are available. In general, however, the value of fuel is also determined by the effort required to sustain a fire. Grass burns quickly, for example; so sustaining a grass fire in a hearth requires a significant quantity of grass and constant tending. In the literature, the baobab is justifiably identified as a poor source of fuel, because its soft wood also burns quickly (Dalziel 1937:115; Sidibe and Williams 2002:50; Wickens and Lowe 2008:121); a few accounts even described it as "useless" for firewood (von Maydell 1986:93). Although hardwood trees provide universally preferred fuel, researchers today are documenting the baobab's widespread use as a multisource fuel tree whose combustible parts include not only its wood but also its root, bark, fruit pod, seed, and seed oil (Smitu 2005:17; Kaboré et al. 2011:842). Mabberley (1997:11) reported the use of the "fruit and seeds" for fuel, and according to Sidibe and Williams (2002:50), who cited Nkana and Iddi (1991), the fruit pod is used as fuel in Tanzania. More broadly, Wickens and Lowe (2008:121) reported that the "seed pods are burnt as a fuel in both Nigeria and East Africa (Dalziel 1937; Burkill 1985; Pipe-Wolferstan 1988)."

General-purpose and specialty fuels can be described as preferred fuels, while scarcity fuels and convenience fuels can be described as substitute fuels. For example, in their assessment of the multipurpose use of the baobab for sustainable development in the semi-urban fringes of the Dutsin-Ma Katsina State of Nigeria, which was based on 200 interviews involving 40 respondents from 5 settlements, Rabi'u and Rabi'u (2013:490) reported that all regarded the baobab as a source of food, 90% regarded it as a source of medicines, 50% regarded it as a source of fuelwood, and 10% regarded it as a source of fodder. Lisao et al. (2017:452) noted that for the Ovambo people of the Omusati region of northwestern Namibia, the "Exposed roots [of the baobab] are chopped and used as fuel for fire." In various parts of Africa, the use of the fruit pod can be appropriately described as a scarcity, convenience, or specialty fuel (Sanchez 2011:408). Stone and colleagues (2011:4) identified baobab bark as a special-purpose fuel used in stoves, ovens, and pottery kilns. In Venda, located in the Mutale Municipal Area in the northern part of Limpopo Province, South Africa, Venter (2012:13) reported that the fruit pod as a specialty fuel "was used by 7 of respondents [N=60] as kindling." While dry grass seems to provide the primary kindling for the Hadza, it is possible that the fruit pod provides secondary kindling as it does in Venda.

There are no studies documenting the Hadza use of firewood or other kinds of fuel, such as fruit pods, seeds, seed oils, or resins. This is unfortunate since the kinds of fuel available is an important aspect of their mosaic savanna adaptation. It is likely, for example, that the Hadza use the woody fragments of the baobab fruit shell as a convenience fuel, given the importance of baobab fruit pulp and seeds in their diet, especially as weaning food. Although they sometimes process baobab fruit out of camp in order to return with more fruit pulp and seeds, they also collect large quantities of fruit and process them in camp.

Fruit Pod Fire Container

Lagenaria gourds acquired through trade or gathered from the wild for use in collecting, carrying, and storying water, honey, coals, and fat are traditionally the most important containers for the Hadza. Marlowe (2010:92–93) reported:

> Today, they often get the gourds from farming neighbors like the Iraqw who grow large ones. However, smaller wild variants grow in the area, and sometimes the Hadza use these, as they must have long ago. Occasionally, a man gets some honey but has no container. In this case, he may find a small gourd but more often will find a large baobab pod and cut off about one-third of it so that he can remove the pulp (which he eats) and fill the pod with honey to carry it back to camp.

The readily available baobab fruit pod is the Hadza general-purpose container; as with the *Lagenaria* gourd, it is likely that it too has been used for handling live coal. A fire container would have been an important component of the fire equipment of early hominins, who most likely collected fire sticks and live coal to spread landscape fires long before they were able to generate fire. Marlowe (2010:92) said Hadza women told him that they could not make fire, which they regarded as men's work. As earlier noted, the Hadza present a case where men traditionally produced fire and women traditionally spread fire. Marlowe wrote:

> I have not asked them to try [to make fire using a fire drill], but I suspect they could make a fire if they really wanted to. Once I did see a woman trying, but after several minutes she gave up and let her husband do it. Because women often roast some tubers while they are out foraging and these days use matches that I give them, I have asked them what they did before matches. They say they used to carry coals in a gourd. In camp, hearths are used 2 or 3 times a day, so the fire never completely dies out and can easily be rekindled any time. I have occasionally seen a woman pick a few coals out of a hearth before she goes foraging.

The baobab's association with the fire drill, its value as a multisource fuel tree, and the probability of the fruit pod use as a fire container suggest the tree is more significant in the Hadza material culture for fire use than is generally recognized.

The Baobab as a Reason for Hadza Fire Use

Not only does the baobab contribute to the material requirements of Hadza fire use, but it also contributes in various ways to the reasons why the Hadza use fire. These reasons include suppressing bees when collecting honey from hives high up in baobab trees. The baobab is well known throughout its range as a honey tree, and it is the Hadza's most important honey source. The theorized importance of honey in hominin evolution makes baobab honey collecting one of the most important uses of fire. For what is described as a bee-repelling or bee-sedating fire, the Hadza must first build a ground fire which requires their fire-making equipment, primary and secondary tinder, and minimal fuel. As earlier noted, their fire-making equipment includes a baobab fruit pod base to the fire drill and a fruit pod used as secondary tinder and limited fuel. From this ground fire, they make a smoke-producing torch for use at the entrance to the hive. After the smoke has sedated the bees, honey is often collected in a baobab container whose round opening has been fashioned using fire.

The use of baobab smoke derived from burning different parts of the tree is particularly prominent in the literature. For example, in light of the baobab's importance as one of Africa's preeminent honey trees, it is noteworthy that De Caluwé et al. (2009:438) reported "Dried [flower] corollas are burned, and a heavy smoke, which repels insects, is developed and can be used to chase away bees while harvesting bush honey." According to Wickens and Lowe (2008:122), "The dry pulp and pods burn well, giving off an offensive and irritating smoke that has been used to drive away mosquitoes and other insects that torment humans and livestock, and as a hut fumigant." They cited Dalziel (1937), Caius (1941), Adam (1962), Irvine (1962), Weiss (1979), Hardy and LaFon (1982), and the United Nation's Food and Agricultural Organization (1988). Similarly, Lisao et al. (2017:452) reported that the Ovambo people of the Omusati region of northwestern Namibia use the smoke from burnt leaves as an insect repellent.

The baobab's contributions to Hadza fire use also include the fire used by men to secure ax heads to their handles and cooked baobab foods. The latter includes not only of flour from the ground fruit pulp and seeds but also of animals, especially birds and small mammals, taken in or near baobab trees and eaten out of camp. Other fire-associated uses of the baobab that have not been reported for the Hadza include cooked baobab leaves, which are widespread throughout its range, especially in West Africa and the Sudan; cooking oil obtained from baobab seeds; roasted baobab seeds eaten as a nut or ground for use as a beverage identified as a coffee substitute; and baobab ash produced from burning the leaves, wood, bark, whole fruit, fruit pulp, empty fruit pod, or seeds for use in cooking and in soap-making and other health-related activities.

Summary

This chapter shows that the baobab is indeed an important contributor to the various ways in which the Hadza relate to fire in their practical and inspirational life. This suggests the baobab might have been similarly important to early hominins. The theoretical perspective presented is set in the fire regime of the mosaic African savanna as the landscape of human evolution. It is the development of the gendered division of labor that exerted the selective pressure for the evolution of intelligent handiness, bipedality, and uprightness. This development is the context for understanding the impact of fire on hominin evolution starting with early hominins taking advantage of natural fires and then their use of fire in the form of spreading, fueling, and generating it. As the outcome of this development, modern humans as an obligate fire-using species are the sole surviving representatives of the hominin family tree and justifiably described as the pyrophilic primate. Every significant moment in hominin evolution has been associated with an intensifying relationship with fire, including the forager way of life, hominin global dispersal, and the Broad-Spectrum Revolution – a process that culminated in the worldwide origin of agriculture and the making of our present world system.

Part IV
Environmental Considerations

Chapter 8
The Baobab and Hadza Acquisition, Management, and Use of Water

Water, as the source, sustainer, and restorer of life, has been an indispensable resource in hominin evolution, and the water traditions of all people are based on the availability of water in their environment and the ways in which they access, manage, and make use of it in all aspects of their way of life. Because most of hominin history has been based on hunting and gathering, it is important to understand the water traditions of this particular mode of adaptation, especially in relation to the ethnography of historical and contemporary foragers in arid tropical environments. Some have argued that anthropologists have taken for granted the importance of water in human life, but water is too important a feature of the planetary surface and of terrestrial environments – and of any environment considered habitable by human beings – to have been passed over by anthropologists.

Although water was not the featured topic of their discussion, a number of important ethnographers and theorists have dealt with the human relation to water in a variety of ways. We have only to think of the works of Steward and Adams (1955), Wittfogel (1957), and many others who have explored the relationship between hydraulic engineering associated with complex irrigation and the origin of the state in Asia and elsewhere. There is Childe's (1941) Oasis Theory of desiccating global climatic change; this change is said to have produced water-source crowding involving human beings, plants, and animals that triggered the worldwide development of agriculture, domestication, urbanization, and civilization. River-adaptation theories for the single-place origin and global diffusion of agriculture have been put forward by geographers (Sauer 1947, 1952, 1956) and anthropologists (Lathrap 1977). Ecological, maritime, and evolutionary anthropologists have also made important contributions to understanding the centrality of water in human life (Clark 1944; Acheson 1981; Mithen 2010; Rasmussen and Orlove 2017).

Still, anthropologists have not paid sufficient attention to the significance of water in hominin evolution, despite the growing body of literature on the subject (e.g., Braun et al. 2010; Magill et al. 2013). It is noteworthy that Rasmussen and Orlove (2017: 1) concluded "water has become a commonplace theme in

J. Rashford, *Baobab*, https://doi.org/10.1007/978-3-031-26470-2_8

anthropological writings," not because of its importance in theorizing hominin evolution, but because "Shifting consumption patterns, local politics, environmental degradation, and climate change have brought water into a prominent position in many research agendas." These authors identified five themes in their review of water-related articles published in *American Anthropologist* since its inception in 1899. The themes were people, water, and place; production and domestication of water; environmental knowledge; disaster and climate; and an anthropology of water. The themes did not include water in hominin evolution.

In the context of hominin evolution, however, forager water traditions, especially forager water technologies that involve a dependence on water-mediating plants like the baobab, deserve special attention. Access to widely dispersed resources requires not only access to water from diverse sources but also the various devices foragers use to manage water, including the means by which they collect, carry, transfer, store, and use it. This chapter discusses the baobab as an important mediating factor in the water traditions of the Hadza. The first section of the chapter dealing with the baobab's water adaptations is presented as an introduction to this discussion. The two remaining sections explore the possible links between the origin of agriculture and the use of baobab reservoirs and the relationship between handiness, material culture, and water use in human evolution.

Baobab Water Adaptations and Dispersal Strategies

The baobab is often discussed in the literature as if it were exclusively a savanna tree (e.g., Owen 1974: 2; Harlan 1992: 71; Sidibe and Williams 2002: 25). This view overshadows the baobab's remarkable water adaptations not only in terms of its physiology and ecology (Newton 1974; Fenner 1980; Chapotin et al. 2006) but also in terms of its links to Africa's rivers, lakes, coasts, and continental islands, where it has traditionally had a variety of uses for fishing communities. The baobab's water adaptations have made it a versatile tree capable of living in a wide variety of environments. These adaptations include its ability to rapidly absorb water through its wide-spreading shallow roots to take advantage of even the briefest of showers in times of low or irregular rainfall; its capacity to store water in its soft wood and trap rainfall and runoff in the crotches, folds, and cavities of its massive trunk and branches; its water-conserving leaves and its shedding of leaves in the dry time of the year; and its photosynthesizing bark, which makes it productive year-round (Wickens 1982; Armstrong 1983: 151; Watson 2007; Wickens and Lowe 2008; Riedel et al. 2012).

Most of all, however, the baobab's water adaptations involve the spread of its seeds. In his introduction to plant geography, Polunin (1960: 98) noted that "many of the plants that are most successful in colonizing vast areas, resort to more than one means of dispersal." This is true of the baobab, which spreads its fruit by gravity, food bribe, and water. Wickens (1982) expressed the novel view in his influential 1982 study that "It is possible that [baobab] fruits from riverine trees are water-dispersed," and Wickens and Lowe (2008) later described both the fruits of *Adansonia digitata* and *Adansonia gregorii* as locally dispersed by the floodwater of rivers. It is surprising that these authors were tentative in characterizing the baobab as a water-dispersed species considering the amount of information they presented that lends support to this thesis. The influential baobab distribution map of Africa produced by Wickens (1982) based on Lucas (1971) shows a concentration of baobab trees in river basins. Wickens (1982: 179) noted that "Baobabs are present in the Sudan zone streamside vegetation of Nigeria, presumably a natural habitat," and he described the species as also present "around the margins of seasonal pools." We learn that "The northern limit is in the semi-desert scrub and grassland communities to the 100 mm isohyet, with stunted trees fringing the seasonal drainage lines; better grown and often large groves of baobabs occur in the fossil drainage systems where there is probably a fairly high water table" (Wickens 1982: 180). We also learn that "The most westerly record in the Sudan is of a small streamside grove to the north of Zalingei in Anogeissus savanna woodland." Watson (2007: 108) reported streamside baobabs in the Ruaha River valley in central Tanzania.

The Old-World distribution of the baobab and of other *Adansonia* species suggests both successful short-range water dispersal to Africa's continental islands and long-range water dispersal further into the Indian Ocean, even to Australia. The hydrochorous dispersal of *Adansonia* species has received increasing attention over the last 20 years (Wickens 1982; Baum 1995b; Baum et al. 1998; Riedel et al. 2012; De Queiroz 2005; Wickens and Lowe 2008; Cornu et al. 2014; Done 2010). As earlier noted, the baobab is a highly productive tree. A healthy mature tree produces every year for hundreds of years a large number of fruits, each containing a large number of seeds. Air trapped in the baobab's large, woody, indehiscent capsule with its light, dry, chalky pulp in which the seeds are embedded ensures a buoyant fruit evolved for effective long-distance water dispersal (Ridley 1930; Pijl, L. van der 1969). This means that during the annual rains, streams and rivers (as well as ocean currents) can effectively transport baobab fruits locally and over great distances. The buoyancy of the fruit is attested to by the fact that it is used for fishnet floats – most likely a contributing reason for the human dispersal of the baobab on Africa's continental islands and around the rim of the Indian Ocean.

The ability of baobab seeds to germinate following oceanic fruit dispersal was experimentally evaluated by Tsy et al. (2009: 1712). Their evaluation involved four fruits immersed in seawater and four maintained in dry conditions (with a total of 505 and 467 seeds, respectively). "[W]hen sown, 98% and 97% germination rates were obtained," which showed that germination "capacity was not significantly different between the two seed treatments." Kempe et al. (2018: 12) conducted a similar experiment. "Our own experiments with *A. digitata*," they wrote, "in which five fruits were stored in a tub of salty water (3.5% salt content) and maintained a shaking plate to simulate wave action revealed that three fruits decayed after approx. 4–7 weeks (unpublished data). Once the fruit shell cracks, seeds were released and sank. Also, seed shells and seeds of the two fruits that remained floating decayed inside the fruits." They concluded that "long distance marine dispersal in *A. digitata* remains to be verified."

Despite the skepticism of Kempe and colleagues (2018: 22), Baum et al. (1998, 2004) and Tsy et al. (2009) theorized that long-distance marine dispersal was an important consideration based on phylogeography; this was the case "at least for the common ancestor of present baobab species." They did not consider baobab hydrochory and the tree's links to rivers, lakes, and the ocean. However, Cornu et al. (2014) have provided evidence for the dispersal of Madagascar baobabs by marine hydrochory. It is clear that seeds from aquatically dispersed baobab fruits do end up in favorable near-water environments where they germinate and grow; this makes water an important contributor to the multiple dispersal strategies of the baobab. For the baobab as a littoral species, effective dispersal over short and long distances is possible because it readily establishes itself in the saline coastal environments. Watson (2007: 34–35, 60–61, 63, 108) has published excellent photographic illustrations of such water-dispersed baobabs of rivers and of coastal shores and low-lying cliffs.

In the view of Sidibe and Williams (2002: 22–23), the fact that baobab fruits "can also be dispersed by water systems [is] important when considering the disjunct distribution of the genus and the distribution patterns of *A. digitata.*" In support of the idea of transoceanic dispersal of *Adansonia* species, they wrote (2002: 28):

> Until Wickens (1982) discussed the disjunct distribution it had been explained as a very ancient relic of the flora of Gondwanaland, the Mesozoic southern continent which later split to form Africa, Australia, Antarctica and South America (Armstrong 1977). Wickens, however, stressed the possibility of water dispersal of fruits and transoceanic dispersal became a possibility. In a study of the phylogeny of the genus using morphology, molecular analyses of *Adansonia* species and comparisons between *Adansonia* and closely related genera, the evidence produced ruled out Gondwana disjunction but points to transoceanic dispersal. (Baum et al. 1998)

The baobab's oceanic dispersal from Africa to its near-shore Atlantic Islands, to Madagascar and other islands of the Indian Ocean, and to Australia has received increasing attention in recent years (Baum 1995b: 444; De Queiroz 2005: 69; Done 2010: 19; Cornu et al. 2014; Crowther et al. 2016: 228; Abdillahi et al. 2019). Baum (1995b) noted that "In Madagascar, animal dispersal of *Adansonia* has not been documented" and that "it appears that the baobabs of Madagascar are not currently endozoochorous," due most likely to the extinction events that have occurred in Madagascar since human arrival some 1500 years ago. Baum (1995b: 444) concluded that "In the absence of animals as seed-dispersers, it seems that water offers the only means of dispersal available to the extant Malagasy baobabs."

> The buoyant fruit and tough waterproof pericarp of *Adansonia digitata* means that dispersal along rivers, on floodwaters, and along strandlines is probably frequent (Wickens 1978). In contrast, water dispersal may be ruled out for the Australian species on account of its fragile, usually dehiscent pericarp. In Madagascar, water dispersal occurs in *A. perrieri* growing alongside rivers. For example, in Montagne d'Ambre numerous fruit may be found in the Rivière de Makis floating as much as ½ km below the nearest tree (pers. Obs.). Similarly, *A. za* and *A. madagascariensis* fruit, which have tough, indehiscent pericarps and are frequently found along watercourses, are likely to be water dispersed. However, hydrochory is unlikely in *A. grandidieri* and *A. suarezensis* because of their fragile pericarps and in *A. rubrostipa* because it rarely occurs near rivers or streams.

While there is no doubt that baobab dispersal by gravity, non-human food bribe, and water are important, there is growing evidence that human beings (Burtt and Salisbury 1929) have been one of the most important agents for the spread of *Adansonia* species in Africa and to various parts of the world. According to Bell et al. (2015: 2) in their study of the introduction of the African baobab to India:

> Although it has not been previously hypothesized, long-distance hydrochory is a possibility. Baum et al. [1998] provided evidence that *Adansonia* in northwest Australia arrived from Africa via long-distance oceanic dispersal in the Miocene. The fruit of the African baobab has been demonstrated to remain viable after immersion in seawater for six months [Tsy et al. 2009]. This could explain the coastal presence of the African baobab on the Indian subcontinent and other locations around the Indian Ocean. However, the lack of species-level divergence, according to morphological taxonomy, between baobabs in Africa and those found in the Indian subcontinent [Baum et al. 1998; Wickens and Lowe 2008] would suggest a recent dispersal. This makes human-assisted dispersal a more plausible alternative than transoceanic drift.

Rangan et al. (2015) have presented new genetic and linguistic analyses to show an "ancient human influence" on the evolution and distribution of the Australian baobab.

Dispersal by gravity, water, and humans and other animals must all be considered in accounting for the spread of the baobab in Africa and the Indian Ocean. Given that the baobab thrive on Africa's tropical coasts and continental islands and disperses its seeds by ocean currents, one of the most important considerations is the use of the baobab by the coastal foragers and settled fishing-farming communities of Africa and the Indian Ocean. This aspect of the baobab is not well documented, especially for the Indian Ocean. For East Africa, however, it has been reported that the Swahili people have valued the baobab for food, well buckets, hair wash, inspiration, and grave planting. There are reports that Swahili fishermen have used baobab fiber to make boat riggings and rope for mooring and anchoring, fishing lines, and fishnets; and they have used the fruit pods for fishnet floats, bailers, and containers, the latter for handling water, food, bait, and other things. In his account of the useful plants of West Tropical Africa, Dalziel (1937: 115) noted that "wide and light canoes are said to be made from" the baobab. Similar coastal uses have been reported for India (Luxford 1842: 444; Balfour 1885: 22, 1123; Pickering 1879; Mittre 1981; Reddy et al. 2002; Bell et al. 2015; Rangan et al. 2015). Just as important as the baobab's water adaptations is the role these adaptations play in the water traditions of the Hadza.

Hadza Sources of Water

The natural environment of the Hadza, as earlier noted, is the escarpments and rocky wooded hills that form the watersheds for Lake Eyasi and Yaeda Swamp. In this Rift Valley environment, seasonal changes in rainfall and in the availability of surface water from runoff and pooling greatly affect the annual rhythm of Hadza life. Water is abundant during the annual rains from the mid-spring of November to the early autumn of April. Water becomes increasingly scarce in the dry time of year from the mid-autumn of May to the mid-spring of October. September and October, the hottest months of the year, are the height of the dry time, identified in many publications as the late dry season. This yearly wet/dry alternation determines how the Hadza source, manage, and use water, and it exerts a powerful influence on the character of the Hadza terrestrial environment and on the growth and reproduction of the plants and animals on which they depend.

According to Bagshawe's (1925: 120) early account of the Hadza, "Water seems to be a minor consideration with them. They often go for long periods without it and seldom carry it with them…. They know, and when possible carefully conceal, the exact position of every spot where water can be found, and have often shown me where to obtain it in the most unexpected places." Similarly, Cooper (1949: 8) reported that the Hadza "know of many hidden places where water collects, out of the reach of animals, in crevices of the rocks." Bennet et al. (1973: 246) reported that "At Yaida Settlement water was obtained from an irrigation channel fed by a river." However, it is not simply that water is a minor consideration for the Hadza or that they can go long periods without it. Rather, it is that they are not preoccupied with water because it has many natural sources in their environment, sources they know well. Traditionally, these sources have included rainfall; rock seeps and other kinds of springs; runoff in the form of streams and rivers; pools that form on top of large boulders and in riverbeds; groundwater accessed by shallow wells; and plants, most notably the baobab (Tanaka 1969). Baobab water is important for the Hadza; there is one of the best-documented cases of the use of water baobabs in East Africa, especially baobab reservoirs. As the world's preeminent water-storing tree, the baobab is part of the Hadza relationship to the aquatic features of their savanna environment. Presumably, water-storing baobabs would have been important for early hominins as well.

Water-Source Indicator

As earlier noted, one reason for identifying the baobab as a tree of life is its links to water, which is especially valued by human beings and other animals adapted to the dry environment in which it grows (e.g., Hopler 2017). The baobab has a role in sourcing water in four ways: it is a water-source indicator; its moist wood can be chewed; the water-storing roots of baobab saplings can be eaten; and water can be obtained from baobab reservoirs. Like other imagined and botanical trees, the baobab is linked to water in a great variety of ways. Botanical trees of life often grow by springs, rivers, lakes, wells, or the ocean. The sycamore fig (*Ficus sycomorus* L.) of the Egyptians, for example, thrives along the riversides of Africa; the *ahuehuete* cypress or Montezuma bald cypress (*Taxodium mucronatum* Ten.) of the Southwestern United States, Mexico, and Guatemala thrives near springs, streams, and swamps; and the coconut is at home in coastal environments and spreads its seeds by ocean currents. It is not just the trees of life, however. All trees, not just trees of life, can be said to mark the presence of water, however dry the environment in which they grow, and it is not surprising read that "The people of the Kalahari know that they can usually find a source of underground water anywhere a baobab is growing" (O'Meara 2003:6 1). According to Holt (1959:70), however, "The Baobab . . . having a large storage capacity for water seems independent of underground resources, apparently its distribution is controlled mainly by climatic and soil factors." There are no reports of the Hadza use of the baobab as an indicator of groundwater.

Water from Chewing Baobab Wood and Eating the Roots of Baobab Saplings

In his dictionary of plant names and uses, Smith (1882) reported that "The wood [of the baobab] is exceedingly soft, and full of moisture, which it readily yields on pressure, affording a grateful beverage to travelers in the arid places where the trees grow." Although there is no documentation of the Hadza using the taproot of baobab seedlings for hydration, Paul Kirtley (2012) posted a photograph on the Internet taken by Amanda Quaine, with the caption "Hadzabe man chewing baobab for moisture." This use of baobab wood has been reported elsewhere. According to Simitu and Oginosako (2002: 14), baobab wood is chewed for its moisture "by humans and animals in case of extreme water scarcity." In her study of the ecology of the baobab and its sustainable use in Northern Venda, South Africa, Venter (2012: 7.14) also reported that "two respondents (out of thirty) said they ate seedlings when they were thirsty, but very rarely." Elephants are known to chew on baobabs for the moisture contained in the wood, often leading to the complete destruction of the tree and, sometimes, the destruction of both the tree and the elephant on which the baobab has fallen (Wickens and Lowe 2008: 204–206).

Baobab Reservoirs

The idea of the baobab as a water-source tree has long fascinated scholars (Blunt 1923; Sweeney 1969, 1973; Newbold, Owen 1974). Human beings obtain water from plants in a variety of ways. Well-known examples include the water-filled seeds of the coconut, juicy fruits like citruses, succulent fruits like watermelon and other *Cucurbita* species, the hollow culms of bamboo, the water-trapping leaves of bromeliads and the traveler's palm (*Ravenala madagascariensis* Sonn.), the flowers of the African tulip tree (*Spathodea campanulata* P. Beauv.), the stems of some cacti, and water-storing roots, including those of baobab saplings. The aquatic environments of the baobab consist of the water stored in its soft wood and the rainwater and runoff that pools in the depressions, crotches, folds, and holes of the tree's enormous branches and trunk. As earlier noted, the trunk of the baobab often becomes hollow as the tree grows old, and large quantities of water become trapped in its spacious interior. This makes the baobab not only the preeminent water-storing tree in the landscape of hominin evolution but also the preeminent water-storing tree in the world.

The use of water-storing baobabs has been widely reported in Africa, and it seems this use of the tree has developed independently throughout its range. There are those like Welman (1966: 150) who have argued that water storage is the most important of all the baobab uses:

> In the western Sudan these trees are often hollow below the broad fork whence the stubby branches spring, and rain-water, collecting in this natural cistern, refreshes travelers in those thirsty lands. But baobabs offer little else that is useful to man, even when, after the pale flowers have fallen, they dangle oblong green fruits as big as those outsize sausages one sees in Strasbourg shops.

Wickens (1982: 190) noted, "By far the commonest use [of the baobab's hollow trunk] is water storage." Similarly, the International Centre for Underutilized Crops (2002) described the baobab as "a multipurpose tree … probably best known for its swollen, hollow trunk which is most commonly used for water storage." It is doubtful, however, that water storage is "the commonest use" of hollow baobabs. Nevertheless, the tradition of using baobab reservoirs seems to be most developed in the driest parts of its range, where there is no surface water in the dry time and groundwater is only seasonally available or inaccessible via shallow wells.

The importance of the baobab's water storage is evident in the fact that the tree has been variously identified in the literature as a baobab well (Battuta 1969; Watson 2007: 142), water tree (Shepstone 1919: 272), cistern (Porteous 1928: 235), reservoir (Dalziel 1937: 115; Guy 1972: 161; Vaid 1978a: 40), and water tank (Palmer and Pitman 1961: 231). It is not surprising that the baobab should be identified with these descriptors, since they represent the most familiar kinds of structural water containers that have been widely used by human beings. Along similar lines, the bottle gourd (*Lagenaria siceraria*) is valued in Africa and elsewhere as a water-handling container; it is not surprising that the baobab is also identified with the bottle gourd. In her beautiful children's book taken from real life, Kessler and Krudop (2002) told the story of a young girl named Fatima. She learned from experience that it was wise to maintain the baobab reservoir she referred to as her great-grandmother's gourd, even though a new well had been dug in her community of which everyone was proud and for which everyone, save her grandmother, was willing to abandon the water baobab as a tradition of the past.

There are similar reports of natural water storage for other *Adansonia* species in Madagascar and Australia. In this regard, it is noteworthy that in the dryness of the tropical woodland and savanna environments to which *Adansonia* species are especially adapted, water-storing baobabs have traditionally been used by people. According to Sidibe and Williams (2002: 23), "Every species [of *Adansonia*] shows use of the trunks for emergency water supplies and fruits being eaten by humans." Water-storing baobabs offer more than simply an emergency supply as the case of the Hadza and the San makes clear. There is every reason to theorize that the use of baobab reservoirs in the mosaic savanna of the Hadza has had a long history in human evolution.

Reports of Water-Storing Baobabs in the Literature

There is substantial literature documenting the association between human settlement and the distribution of the baobab in Africa, but the specific links between human waysides and the baobab have been little explored. However, the subject is receiving increased attention in recent publications (e.g., Bell et al. 2015), and there are a few reports of baobabs along trade routes in Africa. Paths and roads inevitably would develop in association with the baobab because of its value as a resource environment that provides water and food as well as inspiration, landmark, shade, and settlement site. According to Kew Science (2020), "Common trade routes were often based on the baobab trees growing along the way, and each tree even had its own name." Today we think of a road as a means to a destination, connecting one place with another. With foragers, however, paths are simply the most convenient routes between resources. The baobab grows along the paths of foragers because it is a resource destination situated in a network of other resource destinations. In this context, the pathway follows the naturally occurring baobabs. It should also be noted that there are occasions when the baobab only appears to be a roadside tree because a road has been cut through a landscape of baobab trees, as is the case with the famous Baobab Avenue of Madagascar. In addition to pathways following naturally occurring baobabs or passing through stands of baobab trees, it must also have been the case that some wayside baobabs are the consequence of human incidental dispersal resulting from people eating the fruit pulp and discarding the seeds, especially when they are on the move (Watson 2007: 37). Another possibility that should not be overlooked is that baobabs along transit routes were intentionally planted. And sometimes the baobab only appears to be a roadside tree because it is cultivated in residential environments (Blier 1987: 69–71).

Mali

The Muslim scholar Ibn Battuta (1969: 322) provided one of the earliest accounts of the use of baobab reservoirs in the Western Sahel. This account was based on a visit to West Africa that was part of his extensive travels through the known Muslim world of Asia and Africa between 1325 and 1354, and he wrote the following of his experience in Mali:

> On the way there are many trees, and these trees are of great age and girth; a whole caravan may shelter in the shade of one of them. There are trees which have neither branches nor leaves, yet the shade cast by their trunks is sufficient to shelter a man. Some of these trees

> are rotted in the interior and the rain-water collects in them, so that they serve as wells and the people drink of the water inside them. In others there are bees and honey, which is collected by the people. I was surprised to find inside one tree, by which I passed, a man, a weaver, who had set up his loom in it and was actually weaving.

Despite Battuta's early account, Owen's (1968: 55) assessment was that the "water storage property of the baobab seems to be little known in West Africa." Owen, however, was curious to learn more about the use of water-storing baobabs in West Africa, and he sent out a request for information. "It is only comparatively recently that deep wells and bore holes have eased the lot of the nomad in the drier parts of West Africa and I should be interested to learn from readers of the Newsletter whether or not the baobabs of Senegal, Ghana, Niger and Northern Nigeria have, in living memory, been put to a similar use as those in Western Sudan." We do not know if Owen received any responses to his inquiry, but in Mali, Gustad et al. (2004: 582) identified "water trapped in tree" as one of the uses of the baobab in their discussion of the cultural and economic value of trees in the parklands of the municipality of Cinzana.

The Sudan

Water-storing baobabs have also been reported for the eastern Sahel, and one of the best-documented traditions is in the Sudan. Sweeney (1974: 54) noted that in the deserts of Darfur and Kordofan as well as in the Nuba Mountains, he "discovered many of these trees to be hollow, some holding as much as three hundred gallons of water. In the more arid regions, some trees were well-known by the nomadic Arabs as being the only sources of water for many miles and regular halting places for camels." Arab pastoralists were not the only ones to use water-storing baobabs, however; these natural and human-excavated tree reservoirs were often the chief source of water for farming communities and travelers who passed through during the dry time of the year (Blunt 1923: 116; Newbold 1924a, b; Owen 1970: 26; Simoons 1998: 272; Watson 2007: 142).

Natural and artificial water baobabs have made it possible for people to live in places that would otherwise have been uninhabitable. In addition to providing water for domestic use, irrigation, and livestock, these baobabs also served as boundary markers and fruit trees. In the case of artificial baobab reservoirs, some branches are cut to prevent the tree from splitting and to channel runoff into the trunk. A water-trapping ditch is also dug around the base of the tree, and a container is used to collect and transfer water from the catchment to the trunk of the tree. Blunt (1923: 116) reported that "the siting of 90% of the Hamar villages in Western Kordofan must have originally been determined by the location of the Tebeldi [as the baobab is called in the Sudan]" and he estimated that "there were 30,000 water-holding baobabs in the western Kordofan Province of the Sudan with a storage capacity of seven to half a million gallons." These baobabs are named and their names, not surprisingly, are "invariably of the feminine gender" (Blunt 1923: 116) and associate the tree with mother, flowing water, tears, honey, and birds. Wickens (1982: 170) noted that "In the Sudan the trees are regarded as personal property that may be inherited or sold and the ownership of the various trees is kept in local government registers" (see also Wickens 1979; Jahn 1979). As Owen (1970: 26) reported, "This [water-storing] property of the tebeldi gives it a very high status in commercial village life. Not infrequently, individual trees became involved in boundary disputes, the settlement of which called for considerable perception and experience on the part of the District Officer."

Southern Africa

There have also been reports of water-storing baobabs in southern Africa. Palmer and Pitman (1961: 232) noted that "many [European] hunters and early settlers in the lowveld have had cause to bless these natural water reservoirs." According to Wickens and Lowe (2008: 105), in northeast Zimbabwe, they were "used by the Portuguese travelling from their trading post in the Mutoko area to their Masapa post at Mount Darwin (Cashel 1995)." In his discussion of the plant lore of the San, Story (1964: 88) noted that "There are records from the early European travelers that the Native (Bantu) tribes also used the rainwater collected in certain Baobab trees (*Adansonia digitata*), and this is still true today." In his account "Trees in South Africa," Mogg (1950: 14) wrote:

> Then there are those series of old historical [baobabs] ... stretching chain-like east to west across the Kalahari to South West Africa, spaced at about 60 mile intervals. These were invaluable to travelers for their water storage. They are to all intents and purposes a set of living tanks buried in a mass of shady foliage. It has long been a tradition that it is a crime, punishable by death, if a traveler leaves the bung out of a trunk allowing the stored precious water to escape wastefully. Such a grossly inconsiderate act might prove fatal to subsequent weary, footsore, parched travelers.

In theorizing hominin evolution, the most relevant accounts of water-storing baobabs come from the Hadza in eastern and the San southern Africa (Lee 1965, 1972a, 1972b). Lee (1979: 120) identified three components to the natural water supply of the San, and he ranked them as follows: First is standing water consisting of ten permanent water holes; the semi-permanent water holes that retain water through most years but may dry up in a drought year; and seasonal large and small summer water holes or pans. "Next in importance," Lee wrote, "is tree water, which collects in the hollow holes of mongongo and other trees during the rainy season." Third is succulent water-bribing fruits and water-storing fruits: "At the bottom of the hierarchy of water sources are the water-bearing roots of such plants as *Fockea monroi*, *Raphionacme burkei*, *Raphionacme hirsuta*, and *Ipomoea verbascoidea*. These roots are dug during the dry season and enable groups to camp for short periods in areas that completely lack standing water." Water-storing roots, Lee noted, "is the least desirable of the water sources because of the added work involved in searching daily for the roots and in digging them out of the hard molapos soils."

It is noteworthy that Lee listed tree reservoirs as the second most important source of water for the Kung San. In addition to water-storing baobabs, he identified *Boscia albitrunca* and the mongongo tree, *Ricinodendron rantanenu* (now a synonym for *Schinziophyton rautanenii* (Schinz) Radcl.-Sm.). The water is obtained by using the hollow stems of *Kalanchoe rotundifolia, Leonotis microphylla*, and *Panicum kalaharense* as "sipping sticks" (Story 1958: 35, 1964: 87; Van Der Post and Taylor 1984: 43–44). *Boscia*, mongongo, and baobab are the same trees that had earlier been mentioned by Story (1964: 87), who was particularly concerned with the loss of San traditional plant knowledge. Story wrote of the "extermination" of the "Bushman of the Cape Colony" and noted, "It seems inevitable that [those that continue to exist in what is now Botswana and Namibia] … will have to give up their ancient way of life as their country becomes encroached upon by the Europeans and Bantu and consequently that most of their knowledge of the native plants will soon be lost." This situation led Story (1964: 87) to write: "I was therefore glad to accept an invitation from the Smithsonian-Harvard-Peabody Expedition in 1955 and again in 1958 to investigate the plants which supply them with much of their food and water and which make it possible for them to live where anyone without their specialized knowledge would die within a few days."

Story (1964: 88) began his discussion of San plant lore with a section titled "The Quest for Water." He identified the same water-storing trees mentioned by Lee and noted that "the most important among them in the/gikwe country [of the central Kalahari] being *Boscia albitrunca* which is often hollow and whose lower branches often die back to leave knot-holes through which the rainwater can enter." Not all the San groups have access to the baobab because in the Southern Hemisphere, the tree reaches the limit of its natural distribution in the northern Kalahari, especially in the area that includes southern Angola, northeastern Namibia, and Botswana. In presenting an account of his botanical survey that covered northeastern Namibia and central Botswana, Story (1958: 35) reported that "the southern most [baobab] specimens were seen in the neighborhood of Ghanzi, and from there it occurred sporadically to Gutscha Pan, the farthest point north reached by the expedition. There were big gaps in its distribution and it was rare throughout, but in some parts of the Transvaal it occurs abundantly and may have an unbroken distribution over many square miles of country."

Although the baobab is more important for the Kung compared to the Central Kalahari San, there is no doubt that for the Kung, the most important water-storing tree was the mongongo, which, in addition to its water, is especially valued for its seeds. Story (1964: 88) wrote:

> It generally has several stems rising separately from the level of the ground and where they rise the wood often decays away until an underground chamber is left. This fills after good rains and the *Ricinodendron* groves, well supplied with drinking water and with an abundance of good food, are ideal homes for the Bushmen as long as the supplies last.

Along similar lines, Lee indicated that "The amounts provided [by these mongongo reservoirs] are minuscule: A single tree may yield from 1 to 20 liters per season, compared with hundreds of liters for even a small summer pan such as !Kwidum. Yet these trees are strategically placed in food-rich sandy soils where standing water is unavailable, and therefore their small quantities are eagerly sought." Tree water is obtained by using the hollow stems of *Kalanchoe rotundifolia*, *Leonotis microphylla*, and *Panicum kalaharense* as "sipping sticks" (Story 1958: 35, 1964: 87; Van der Post and Taylor 1984: 43–44).

The G/wi San of the Central Kalahari have the most challenging water situation among African foragers. Silberbauer (1972) indicated that in the absence of year-round surface water and tree catchments, the G/wi San are especially dependent on succulent melons and water roots at the height of the dry time. Plants represent the major source of water for the G/wi San during the dry time, with individuals obtaining approximately 4.5 liters of water per day. According to Tanaka's (1976) estimate, over 90% of the water requirement for Central Kalahari San comes from plant sources, including melons, berries and beans, and roots and tubers. The most important water-storing fruit for the G/wi San is the tsamma melon, which contains about 90% water. Unlike most other fruits, the tsamma melon is available during the dry time and can be stored for short periods. There is yet another way in which the baobab is useful in water management of the Kung. Lee (1979: 123) reported that in addition to ostrich eggshells and empty whisky bottles obtained in trade, the Kung also made use of the "large pods of the baobab tree" to make water containers.

East Africa

Water-storing baobabs have also been reported for East Africa. According to Wickens (1982: 170), in the Northern Frontier Provinces of Kenya, baobab reservoirs made it possible for "slave and ivory raiders" from Ethiopia to "cross otherwise waterless country."

Unlike the central Kalahari, the baobab is patchily distributed in the wetter environment of the northern Kalahari, especially in association with water holes. The baobab is relatively common, however, in the mosaic savanna of the Hadza. This means the water-storing baobab is more than an incidental occurrence in the Hadza landscape and the use of baobab water goes well beyond an emergency supply as suggested by Matthiessen (1972: 234). Baobab water is important for the Hadza. As the world's preeminent water-storing tree, it mediates the Hadza relationship to the aquatic features of their savanna environment. Presumably, water-storing baobabs would have been important for early hominins as well. Woodburn listed water storage among the Hadza's important economic uses of the tree and noted that these baobab reservoirs are "sufficiently important to the Hadza as to have a special name, *kesati* (1970: 40)."

Hadza Management of Water

In addition to water sources, Hadza water traditions also involve containers for collecting, carrying, storing, and using water. As earlier indicated, Woodburn (1970: 40) documented the use of the baobab fruit pod as "a rough and ready drinking cup" and so did Marlowe (2010: 78), who also identified the use of the pod as a bowl. Other reported Hadza uses of the pod include a tree-water dipper and a convenient container for honey.

The most complex Hadza water-collecting activity is associated with the baobab reservoir. It involves a peg-climbing technology and a cordage-and-container technology to make a dipper in order to extract water from the tree. The Hadza use of the fruit pod as a tree-water dipper seems to be specific to accessing baobab reservoirs in a convenient way. It is worth noting, however, that in coastal East Africa, the fruit pod is also used to access water from wells (Hobley 1922: 76; Irvine 1952: 186; Weiss 1979: 40). Writing from personal experience, Pickering (1879: 819) reported that in "Zanzibar, the tree appeared to me introduced, the natives using the shell of the fruit for water-buckets." Similar uses of the baobab fruit pod have also been reported for other parts of Africa where, in addition to water containers, the pod is also made into kitchen and eating utensils described in the literature as bottles, bowls, saucers, cups, spoons, and ladles (Lee 1979: 123; Hobley 1922: 76; Irvine 1952: 186; Weiss 1979:40; Arum 1989: 22; Sidibé et al. 2002).

It is likely that the fruit pod as a water container was more frequently used by the Hadza in the past as a dipper and as a canteen for carrying and storing water. Today they also have access to industrially produced containers of various kinds. However, the prehistoric ancestors of the Hadza would not have had access to cultivated *Lagenaria* water gourds or to earthenware cooking pots, which the Hadza have traditionally obtained from their agricultural and pastoral neighbors; it is likely they would have made much greater use of the baobab container than has so far been reported.

The Hadza Use of Water

In addition to sourcing and managing water, Hadza traditions include the ways in which they make use of the aquatic features of their savanna environment. All discussions of the human uses of water involve a categorization of these uses, and such systems vary widely in accordance with practical or theoretical interests with which they are associated. The Hadza are engaged in the five basic categories of water use common to human beings everywhere: travel efficiency, preferred settlement site, domestic needs, resource acquisition, and inspiration.

Travel Efficiency

In terms of travel efficiency, much attention has been given to the development of water crafts and water transportation that greatly expanded the movement of people and the range of available resources. The proximity of aquatic features such as gorges, river valleys, floodplains, and coasts also contributed to the ease of moving through the environment.

Settlement Site

Irrespective of the environment and mode of adaptation, temporary or permanent human settlements always require an available supply of water. Its links to travel, domestic needs, resource acquisition, and inspiration make water a preferred place to settle, especially with regard to the domestic use of water in association with camps and, later, villages and urban areas. Obtaining water is one of the most significant resource goals of Hadza foraging, which is why they normally establish their camps close to water. Although temporary or permanent human settlements must always be close to an available supply of water, these settlements need not be a settlement at watersides. Foragers like the Hadza have good reasons for not establishing their camps at water holes. There is "fear of driving away the game, and also to lessen the danger of unexpected visits by strangers" (Bagshawe 1925). Not driving away game is no doubt linked to the Hadza's use of water holes for nighttime ambush hunting in the dry time. Other reasons for not camping at water holes include avoiding conflicts with neighboring groups and needing safety from predators. The Hadza prefer, instead, to carry water to their camps which are ideally established close to water, but in rocky elevated locations with good views that are secure, dry, sheltered, windy, and mosquito-free. Woodburn (1968a: 50) pointed out that the Hadza "consider that about three or four miles is the maximum distance over which water can reasonably be carried and camps are normally sited within a mile of a water source." According to Blurton Jones and colleagues (2005: 216), women and children go to fetch water between 5:00 p.m.] and 7:00 p.m. when "the temperature has fallen to pleasant levels and most people are at home." Unlike Kung children, Blurton Jones (1993: 316–317) noted "Hadza children appear to be given many errands and to perform useful tasks, bidden and unbidden." One is that they "commonly are sent to fetch water and sometimes firewood." That the Hadza routine is central-place water provisioning is well documented. What remains unexplored is the extent to which this water is not only for household use but also for distribution to kin or to other members of the camp.

It is a reasonable assumption that hunter-gatherers of the African savanna would always establish their camps based on the convenient availability of water as well as the maximum opportunity for acquiring other resources. Even when camps were moved specifically to improve access to water, such moves still had to be made in consideration of maximizing other resource opportunities. And when a move was for resource maximization, it still had to take account of water. Even in humid environments with high rainfall and abundant permanent runoff and pooling, foragers still require access to widely distributed water sources. According to Marlowe (2010: 41), water is the "main constraint on camp locations" and in "the rainy season, drinking water often comes from pools of water on top of big flat rocks and from running streams." Given the dramatic seasonal changes in the number, variety, and distribution of water sources in the Hadza's environment, Woodburn (1968a: 52) reported that their camps "are commonly small and widely dispersed in the wet season, large and concentrated near the few available sources of water in the dry season." It is clear, however, that Woodburn did not view the Hadza as water-tethered in the strict sense that Lee described for the Kung. He did not identify water scarcity as the exclusive reason for the larger camps of the dry time, noting instead: "Perhaps the most important single factor creating the larger camps in the dry season is that at this time of the year more large game is being killed… In the wet season, when few large animals are killed, there is less incentive for people to congregate in large camps and the divisive effects of quarrels tend to keep the size of the camps small" (Woodburn 1968b: 106).

It was Woodburn's view (1972: 202) that in the hunter-gatherer literature "great stress is commonly laid on the shortage of food and water as if these were the sole cause of nomadic movement." He wrote, "For the Hadza at least, and I would imagine more generally among hunters and gatherers, too much emphasis on this type of interpretation is misleading." According to Woodburn (1972: 203), "Camps are certainly sometimes moved because food supplies have run short in a particular place and quite often because water sources are exhausted. Members of a Hadza camp may live on a very meager supply of water in a hollow in a baobab tree or in a puddle under a rock. When the water is finished and none remains elsewhere within reasonable distance of the camp, a move is regarded as essential."

For the Hadza, temporary water sources are important in both the wet and dry times of the year because they make it possible to expand the range of resource acquisition away from permanent runoff and pools. Water-storing baobabs affords them greater flexibility and efficiency in undertaking resource forays and choosing camping locations. Lee (1993: 26–27) described a similar scenario for the Dobe Ju/'hoansi. Their water sources, as earlier noted, included permanent waterholes, seasonal pools, tree reservoirs, and water-bearing roots. "With these sources," wrote Lee (1993: 26–27), "the Ju/'hoansi plan their annual round, spending the winter season close to the permanent waters and the summer months ranging widely at the secondary and lesser water sources." As with the Hadza and probably early hominins as well, the baobab reservoir and tree reservoirs generally are among the diverse and widely distributed water sources on which they have traditionally depended.

Water's Domestic Uses

Unlike Bagshawe (1925: 120), Bleek (1931: 278) had a limited appreciation of the water features of the Hadza environment and their significance in Hadza water traditions. She noted only that water was "scarce," was "their only beverage," and was "only carried for drinking and cooking." Eighty-five years later, Blurton Jones (2016: 14) expressed a similar view when he wrote "Water is an issue for foragers, farmers, and herders alike; but foragers merely need enough to drink, and to cook with." But this states only the minimal needs. Although Hadza water traditions have not been systematically documented, we get glimpses of it in the literature. In addition to onsite drinking and the in-camp use of water for drinking and cooking, Hadza also use water domestically for washing and bathing. Water's ability to dissolve and disperse other substances makes it a powerful solvent, and this means washing is one of the most important categories of water use. Marlowe (2004: 692) noted, for example, that "Women often sit and groom children, removing lice, washing them, and blowing their noses." Marlowe (2004: 692) also noted that "When anyone is injured and cannot forage for a while, their close kin usually attend to them and bring food back for them until they recover." Presumably, they also bring back water. However, water exchange as an aspect of Hadza water traditions has not been studied.

Water as a Resource Environment

Within each of the five basic water-use categories, there is a great deal of variation, and this is particularly true of the link between water and resource acquisition. Anthropologists have long regarded rivers, lakes, and oceans as resource-rich environments that provide such food sources as fish, mollusks, crustaceans, and amphibians; opportunities for hunting aquatic, near-water, and water-visiting birds and mammals; and fruits and other resources from aquatic and near-water trees, shrubs, and herbaceous plants. It is worth recalling here that the foundation of Alister Hardy's (1960) original formulation of the aquatic ape theory was the resource abundance of the sea accessed by wading, swimming, and diving. "Obviously," wrote Washburn and Lancaster (1968: 294), "water was important for drinking, for richer vegetation along rivers and lakeshores, and for concentrating animal life." Nevertheless, these authors seem to have questioned the general idea of water as a resource-rich environment for early hominins. They (1968: 294) wrote in their classic paper on hunting and human evolution:

> [W]ater must have been a major physical and psychological barrier and the inability to cope with water is shown in the archaeological record by the absence of remains of fish, shellfish, or any object that required going deeply into water or using boats. There is no evidence that the resources of river and sea were utilized until this late preagricultural period, and since the consumption of shellfish in particular leaves huge middens, the negative evidence is impressive. It is likely that the basic problem in utilization of resources from sea or river was that man cannot swim naturally but to do so must learn a difficult skill ... For early man, water was a barrier and a danger, not a resource.

However, water as a barrier and a danger versus water as a resource environment presents a false set of alternatives; and the idea of the Broad-Spectrum Revolution as simply the results of preagricultural humans learning how to swim is not persuasive. The global demographic model (Cohen 1977) centered on "the food crisis in prehistory" suggests it was the need for intensification following human global expansion correlated with large-scale megafaunal extinction that led to the greater aquatic adaptation associated with the Broad-Spectrum Revolution, which involved a more intensive use of water generally; and the Broad-Spectrum Revolution was the immediate precursor to the origin of agriculture worldwide. "Water use" is not just the value of water itself as a substance apart from other considerations. The importance of water in hominin evolution has to be more broadly conceived than Washburn and Lancaster presented it. This is so when we consider the efficiency of traveling through gorges, along river valleys, and across floodplains. From the broader perspective, water's full value as a resource environment for the Hadza also includes the water-eroded features of dry river beds, such as water-holding rock basins; smooth rounded hammer stones and anvils; and grindstones and flat-rock surfaces used to process baobab fruit pulp and seeds into flour, which is eaten raw or cooked with water to make porridge.

The water features of the Hadza homeland are resource environments associated with gathering, fishing, birding, and hunting. Watercourses are places where one can gather the much loved *tafako* (*Salvadora persica*) and other berries, as well as figs, which are not as highly prized. The Hadza occasionally fish. According to Marlowe (2010: 63):

> The Hadza do not normally eat fish … In 1998, when weather conditions created a giant lake complete with giant catfish in the Yaeda valley and Bantu fishermen caught, ate, and sold the fish, some Hadza men took to walking out in the lake and whacking the fish on the head with their bows. They ate the catfish but complained that they did not taste good. They say that fish is bad, "like snake," which they also do not eat. In fact, they eat no reptiles except for Tortoise (*k'oloko*; *Geocholone pardalis*).

In addition to gathering and fishing, the Hadza also target large migratory aquatic birds. As a resource environment, one of water's greatest value for the Hadza is nighttime ambush hunting at waterholes in the winter dry season; this is a kind of ambush hunting which also occurs in association with animals that are attracted to feed on fallen baobab flowers or acacia seedpods (Patterson et al. 2013: 35; Rashford 2015: 214). In addition to water as a resource environment, it is worth noting here that the link between Hadza water use and hunting also includes the making of arrow poisons (Peterson et al. 2013: xvii, 43).

Water and Resource Production

In addition to water as a resource-rich environment, there is also the idea of water as a requirement in resource production. A critical development in water management occurs with the origin of agriculture when it expanded from drinking water to include the irrigation of crops and the watering of livestock. For food producers, the baobab serves as a seasonal marker. The deciduous baobab puts out new leaves and begins flowering just before the annual rains in early spring, and some cultures use these changes as indicators of the approaching spring rains, as a planting cue, and as a harvest forerunner. Describing the baobab's spring flowering as a seasonal marker for the Mbeere of Kenya, Riley and Brokensha (1988: 197) reported that "The nocturnal blossoming of these trees, taken in association with (*mw*)-ARANGE (*Delonix elata*), is used as a weather indicator for planting at the start of the long rains in March–April." Not surprisingly, the baobab tree itself is also associated with rainmaking ceremonies (Ferry et al. 1974; Watson 2007: 49; Wickens and Lowe 2008: 52–53). Baobab leafing out and flowering are important seasonal timers for farmers, but this use of the tree would not be expected for the Hadza as foragers.

With agriculture as population-driven global intensification that gave rise to feudal states, hydraulic engineering is evident not only in complex irrigation systems but also in works associated with flood control and domestic water supply. Understandably, the context for current water use is our global industrial way of life in which water has become a highly contested resource. Industrial-scale irrigation and other water-management technologies coupled with the increasing demands of commercialization and rapid urbanization now involve major interventions into the natural hydrologic cycles. These interventions include enormous water-storing reservoirs, power-generating dams, river-divergence projects, and watercourse canalization.

The global intensification in the human use of water and the competition it has brought about are clearly impacting the Hadza, whose access to the natural water supplies, traditionally been the basis of their foraging way of life, is changing in destructive ways (Madsen 2000). Hadza water rights have been largely ignored by their farming and herding neighbors, newcomers to Hadzabeland, and government officials. Natural water holes have been taken over by commercial interests or degraded by cattle and other livestock. The Hadza told Peterson et al. (2013: xxvi):

> We don't dig out the water holes so that water can be collected by the bucketful. Instead, we drink directly from springs and return for more when we are thirsty. Digging out the springs for crops and livestock lowers the water table, making it impossible for the wild animals to drink. If the wild animals find no water, they will be forced to move on and where would that leave us?

As Peterson et al. (2013: 201) noted, "Populations of water dependent wildlife grazing species such as wildebeest, zebra and buffalo [and we might also add eland] have plummeted due to competition for dry season grazing and water."

The environmental impact of farmers, herders, and other newcomers is also altering watersheds. One of the dramatic ways in which this is occurring is the cutting of trees to make space for farms, livestock, and artificial water holes; to produce wood for livestock enclosures and homes; and to provide firewood and charcoal. Irrigation schemes, especially in association with the increase in commercial farms, have drained water away from Yaeda Swamp, transforming what was a rich resource environment into a dry lake bed. The Hadza of the Mangola area now have limited access to some of their traditional water holes. These developments – digging wells, felling trees, and restricting access – also raise questions about the future of Hadza water-storing baobabs and about their continued access to these baobabs.

Baobab and Water in Hadza Inspirational Life

The significance of water as a feature of Hadza adaptation to the aquatic features of their mosaic savanna environment is evident in the ways in which water is associated with their oral traditions, religion, and recreational activities. Hadza told Peterson et al. (2013: 136):

> The parents of twins must always cross a flooding stream last when in a group because we believe that once they cross, the water level will rise. During the rainy season the parents must smear charcoal on their forehead and nose bridge to prevent lightning strikes. This they must do for the rest of their lives.

In narratives of the ancestral Hadza couple, we learn that rock pools and baobab water were the sources of the watercourses that today form the Lake Eyasi basin. Tradition has it that the elephant-eating giant monster Duduk'we captured Ts'ikayo, the founding ancestress of the Hadza, when she was unintentionally left behind in a gathering expedition. Duduk'we promised to spare her life if she became his daughter and she agreed, but she was afraid of him. At his cave, Ts'ikayo climbed a rock that the spirits had made grow tall, well beyond the reach of Duduk'we. After Duduk'we returned from a hunting expedition with elephants attached to his arms, waist, and legs, we learn from the account presented to Peterson et al. (2013: 17–18) that:

> He was still hungry, so he skinned and butchered the elephants himself, sorting the meat from the fat and cooking it all in huge earthen pots. He ate all the meat and all the fat, and when the earthen pots were empty, he drank dry all the cisterns in the rocks and baobab trees. Satisfied, *Duduk'we* fell into a very deep sleep.
>
> "In his sleep," the old man [telling the storying] continued, his brow wrinkled in thought, "In his sleep *Duduk'we* peed."
>
> He peed and peed and peed. He peed out all the water he had drank from the rock cisterns and the baobab hollows. He peed so much that torrents ran from in between his legs creating all the ravines and watercourses that today form the Lake Eyasi basin.

The river is featured in another Hadza story presented by Peterson et al. (2013: 9). It is creation narrative that accounts for their origin as a people in association with the origin of the baboon and of their traditional agricultural and pastoral neighbors. The Hadza recognize baboons as their "greatest competitors" for various foods, including the baobab. The told Peterson et al. (2013: 3) "We sing an old song that describes a year [in the spring] when very little rain fell, and when the tubers were thin, dry and shrunken like the people were and the baboons finished the baobab fruits." The Hadza say they were baboons in the beginning until God, whom they called Haine, sent some baboons to gather food and some to collect water. The group that gathered food became the Hadza whose bestowed way of life was hunting and meat-eating. The other baboons ended up "frolicking in the water" and forgot to collect it. Because of this, they remained baboons. This story is set in the context of the routine activities of the Hadza. In his study of hunter-gather infancy and childhood among the Kung, Hadza, and others, Konner cited Blurton Jones (1993: 316) who wrote:

> Between the ages of 3 and 8, Hadza children seldom accompany their mothers, but do not always do so.... The children, usually in sizable mixed-age groups, may spend some hours out of camp. Sometimes they are at a favorite play site or at the water hole.

In addition to religion and oral traditions, there are also reports of Hadza water use associated with death. We learn from Peterson et al. (2013: 171):

> In the past when a person became very old, there would come a time when he or she would call everyone together and say "My children, I will not be making the move to the next camp with you. You have my blessing." This was the old people's way of saying that it was time for them to die. Before we made the next move, we would ensure there was a gourd of water, meat and plant food left for the elder. We then would say goodbye and walk away. These days, we bury our dead and, if possible in the place they were born.

"There is some fear of pollution from handling a corpse," Woodburn (1982b: 192) wrote, "and one of the rare occasions on which people wash their hands is after participating in a burial." Woodburn (1982b: 189) also described how water entered into the internment process itself.

> The surface of the grave may sometimes be trodden down after water has been poured on it, in order to make a hard surface so that disturbance by hyaenas is less likely. The treading of the grave, when it is carried out, seems usually to be done by the men and to be treated as an entirely straightforward, practical matter. The women bring containers of water which are poured out over the surface of the grave. The men mix the water with the clay soil and tread it down to form a flat surface which will soon dry hard in the sun.

Skaanes (2015) noted that the Hadza use water and riverside mud or "clay taken from the center of termite mounts" to make special dolls described as having "power" associated with fertility. Skaanes (2015: 252) also reported that the Hadza make different kinds of dolls including "toy dolls, made by little girls out of edible roots, pieces of wood or other ready at hand materials, and made into dolls by wrapping pieces of cloth around them and carrying the around."

Hadza Protection from Water's Influence on the Environment

Bagshawe (1925: 117) offered the following description of the Hadza and their environment prior to the arrival of large numbers of migrants:

> Their country is an inhospitable wilderness, full of game but heavily infested with tsetse fly and very short on drinkable water, and, excepting for some nomad "Dorobo" to the north of [Lake Eyasi]... they have it to themselves. Very little is known of this area. No roads pass through it and, though it affords good shooting, no food is obtainable in it excepting the meat of game. During the rains much of it is almost impassable black mud, and during the dry months, excepting for a few places, such as Jaida Swamp, water is not only scarce but

> dangerous, for many of the springs and drinking places appear to be impregnated with something which causes a severe and persistent diarrhea.

The Hadza, like all human beings, relate to water not only for its many uses but also because they have to protect themselves from its potentially harmful effects. For foragers, this includes such things as not camping at water holes to avoid mosquitos and water-hole predators, choosing a camping site not likely to flood or become excessively muddy during the annual rains, digging shallow wells to filter polluted water, and building water-repelling dwellings. In the dry time, the Hadza do not build substantial grass dwellings, but in the rains, they do so using construction materials including baobab poles, branches, and wood. These dwellings must be sufficiently watertight to withstand heavy downpours. Another means of protection from rain is to move to rock shelters where these are available. We can conclude, then, that water use and water protection determine Hadza camp movements and both require the management of water.

The Origin of Agriculture and the Use of the Baobab Water

Given its extraordinary size, it is natural that the earliest accounts of European explorers, botanists, colonial officials, military personnel, and missionaries were particularly concerned with the baobab's age and the use of its hollow trunk to store water. It is now more than a hundred years since these accounts first appeared, and the information they provided was made available to a wide audience through publications like *Kew Bulletin*, the *Geographical Journal*, and *Scientific American*. An influential article on the water-storing baobabs of the Sudan was published by Blunt (1923: 114 and 116), who expressed the view that the baobab "is one of the oldest, if not the oldest living trees in the world … of great interest in the botanical world," and he noted that "the oldest trees are naturally hollow." Douglas Newbold (1924a) responded critically to Bunt's argument for the longevity of the baobab, and he wrote "I cannot believe that a tree which soon becomes a hollow shell and is admittedly shallow-rooted could stand 2,000 Kordofan *kharifs* with their gales and storms."

In his contribution to the debate between Bunt and Newbold, Parr (1924: 118) wrote:

> I cannot … admit that the tebeldis "soon becomes a hollow shell": while in Western Kordofan the inhabitants were unanimous that tebeldis had to be hollowed by themselves, and that it took about three weeks satisfactorily to hollow a normal tree. This proves that the core is soft, but I think disproves the idea that it becomes hollow naturally.

In a response to Parr's article, Newbold (1924b: 122) wrote:

> As regards [water-storing baobabs] … it is true that the Hamar have had themselves to do the hollowing of the water-holding tebeldis but I have always understood that this hollowing was more in the nature of finishing-off what nature has begun, viz., the extraction of the soft core and paring of the inside of the walls, rather [than] the actual excision of a whole

solid cube of wood from the tree. However, it is only a question of degree and I am open to correction if I have exaggerated. But I must still confess that the lateral spreading roots and rotting core of the tebeldi make me skeptical of its ability to stand intact in a sandy soil for 2,000 years.

Today we know that the baobab can live for millennia and that natural baobab reservoirs are widespread in Africa. We can conclude then that the baobab reservoir must have been just as important to early hominins as they are today to the Hadza and San and the people of western Sudan.

It is important to keep in mind, however, that although all three groups make use of water baobabs, foragers like the Hadza and San take advantage of baobabs that naturally trap water in their trunk, while farmers of western Sudan hollow out their water-storing baobab trees. This distinction between natural and artificial baobab reservoirs is theoretically important. For foragers, baobab water is essentially for drinking and domestic use, especially when away from natural springs, streams, and pools. For the farmers of western Sudan, baobab water is not only for drinking and domestic use but also for irrigation, watering livestock, and sale. A self-filling baobab is useful as a natural water source, but an artificially filled baobab is useful in the management of water. The latter expresses a greater use of the baobab's water-storing potential associated with agriculture as a more intensive mode of adaptation.

The use of baobab cisterns in the Sudan represents this process of intensification, which accompanied the rise of agriculture some 12,000 years ago. According to Wickens and Lowe (2008), it takes 3–4 days to excavate a baobab reservoir, and there is the added recurrent cost of making and maintaining the water catchment at the base of the tree and transferring water from the catchment into the trunk of the tree. It is noteworthy that baobab water is used to grow crops, especially watermelons, which El-Bushra and El Sammani (1977: 40) reported are used "for supplying water to their animals" and which J. H. H. (1926: 486) noted are "invaluable when men are out tapping [gum arabic] in distant lands." According to Doxiadis Associates (1965: 117), watermelons were "cultivated in fields alongside the food crops to provide a 'water crop' to drink during the harvest work, and later."

In addition to water, the intensification in the use of the baobab that accompanied the development of agriculture probably also included pastoralists in many parts of Africa. They pollard baobab trees (and in some cases cut them down) to feed their livestock, which also prey on baobab seedlings, saplings, and small trees. Watson (2007: 30–31, 146, and 179) presented striking illustrations of this use of the baobab as a fodder tree.

Handiness, Material Culture, and Water in Hominin Evolution

Instead of the vague idea of "the freeing of the hands," the fundamental discussion this chapter presents is that hominin hands as the preeminent tool of intelligence are an evolved specialization for the intensifying development of material culture in the context of the gender-specialized division of labor, the foundation of the reproductive success of the human species. It was earlier thought that tool use, culture as learned adaptation, and language as the means for complex social organization were exclusive differences between humans and other primates; but these ideas have long been abandoned in light of newer findings (Boesch 2012: 5). We should not, therefore, think of water traditions and the material culture for managing water as exclusively human. Like the San and Hadza, the chimpanzees of Senegal (Galat-Luong and Galat 2000) and Uganda (Hunt et al. 1999) dig wells to access and filter water from runoff or pools; chimps have also been widely reported to use leaf sponges to draw drinking water from water-trapping trees and the wells they dig (Boesch 2012: 90). Also noteworthy is Boesch's (2012: 54) report that "The Fongoli chimpanzees [of Senegal] were also seen to immerse themselves in cool water during the very hot hours of the day, a rare sight indeed in a species famous for avoiding water." Even with these overlapping traditions, however, water management by chimpanzees is minimal compared with what is theorized for early hominins, who were exceptionally handy and container-dependent and who operated with a gendered division of labor.

The selective pressure favoring the correlated evolution of handiness and bipedality linked to material culture was essentially the development of the forager way of life as the true distinguishing feature of hominin evolution. Material culture, as earlier indicated, includes the fabricating tools, gathering equipment, and hunting gear necessary for forager adaptation, a way of life that especially involved carrying things and devices for doing so. Water would certainly have been among the important things carried from early on, and water management meant the use of containers and conduits to control the pooling and flow of water. More broadly, containers and other kinds of carrying devices were indispensable for handling the variety of resources that have increasingly become an important part of the practical and inspirational life of human beings worldwide. Consequently, water containers, especially water-carrying containers (along with containers to carry food, infants, crafting tools, foraging equipment, hunting gear, house-building resources, and fire-producing materials), would have been among the most important mutually reinforcing considerations that made handiness and carrying significant contributing factors in the selective pressures that favored bipedalism as the fossil marker of hominization. Water management, particularly with respect to sourcing, collecting, carrying, transferring, storing, and using water, reinforces the idea that carrying was important to the evolution of bipedality – provided that carrying is understood to be a modality of handiness in support of the forager adaptation.

Around 12,000 years ago, the water use associated with hunter-gatherer adaptation was intensified with the development of farming and herding; by 7000 years ago, this intensification continued with the worldwide rise of feudal states often involving a complex hydraulic technology which, in addition to agriculture, also aided craft production and trade. With the development of long-distance mercantile trade associated with feudal states, merchants were transformed into capitalists whose worldwide industrial production of commodities would become the economic foundation of our present world system. This system had its early history in oceanic trade routes, with a significant part of its energy coming from wind and hydropower involving sailing ships and waterwheels; it has developed to include hydroelectric, hydrothermal, nuclear, and solar power. The major current concerns with water today involve scarcity, competition, and hydropolitics. It is clearly the outcome of the ongoing intensification of human water use which has produced an urgent need to increase water supplies by improved and expanded reservoirs and groundwater sources and by better voluntary and compulsory conservation, the latter including mandatory adoption of more efficient technology and price increases. In addition to human needs, however, current concerns must include the environmental stewardship of water to support the biodiversity of nature. This imperative has long been a strong argument against the folly of river divergence, artificial canalization, and water pollution.

Summary

This chapter considered the baobab's contribution to Hadza water traditions in the context of similar uses of the tree throughout its range. The baobab is the premier water tree of the Hadza and this suggests the baobab would likely have been an important mediator in the relation between early hominins and water.

Chapter 9
Baobab Seasonality

The planetary seasonal cycle governs human life everywhere. For foragers like the Hadza, seasonal synchronization is necessary to ensure the success of their strategic mobility. However spontaneous it might seem, it is a fundamental requirement of their camping-based hunter-gatherer way of life. Seasonal reckoning is also essential for cultivators and herders, since it is the basis for anticipating optimal conditions for the growth and reproduction of the plants and animals on which they depend. The need to be in sync with the seasons gave rise to the calendar in association with the development of intensive agriculture, an indispensable organizing feature of large populations with their complex divisions of labor and centralized state governance. In all modes of production, the link between seasonal food availability and sociability has long been noted, especially the influence of seasonal food abundance on the annual timing of such things as religious activities, puberty rites, weddings, mortuary practices, and festive occasions. The importance of seasonality in human adaptation is evident in the methodology of ethnographic fieldwork. Researchers are expected to include an account of the annual cycle of practical and inspirational activities as an important part of the framework of their discussion. How well we understand the seasonality of the baobab is an important condition for understanding the tree's contribution to the seasonal adaptation of the Hadza. This chapter presents the seasonal framework necessary for understanding the baobab's contribution to Hadza life. It is divided into three parts. The first and second offer critical assessments of current approaches to the seasonality of Hadza life and of the baobab, and the third provides a brief overview of the Hadza seasonal year.

Current Approaches to the Seasonality of Hadza Life

Although there are no studies specifically about Hadza seasonality, the subject is mentioned by most researchers and is often important to their discussions. However, the lack of a systematic approach creates a number of difficulties. For example,

J. Rashford, *Baobab*, https://doi.org/10.1007/978-3-031-26470-2_9

Woodburn (1968a) reported that the "wet season"/"dry season" polarity dominates Hadza life and this seasonal oscillation considered apart from the annual seasonal progression has been treated as natural and unproblematic. However, in view of the arbitrary nature of the "wet season"/"dry season" perspective, it is not surprising to find significant variation in the seasonal timing and relative duration of the "wet season" and "dry season." In different accounts, the dry time can be 4 months (Hawkes et al. 2001a), 5 months (Mabulla 2003:47), 6 months (Vincent 1985a:134; Woodburn 1968a; Marlowe 2010), or even 7 to 8 months. Sherry and Marlowe (2007:111) clearly recognized the ambiguity in the wet season/dry season perspective when they represented "the long dry season" as June–November/December and the wet time as December/January–May. This view is out of sync with most of the general literature, especially in defining the dry time/wet time as June-Dec/Jan-May.

In their evaluation of the nutritional composition of wild plant foods and honey used by the Hadza, Murray et al. (2001) – citing Woodburn (1968a), Vincent (1985a), Bunn et al. (1988), O'Connell and Hawkes (1988), and Hawkes et al. (1989) – concluded that "there is no general consensus concerning the seasonal availability of Hadza food" because of "inter-and-intra-annual variability." Marlowe (2010), however, attributed the absence of a general consensus to the lack of phenological studies. These explanations are important, but the fundamental reason for the lack of consensus is the absence of a systematic approach to correlating the astronomical seasons with the dry and wet times of the year and with the calendrical months. The seasonal framework presented in this chapter is based on a unified approach to the divergent views on Hadza/baobab seasonality. Hadza researchers have focused on the wet season/dry seasonal polarity, but "wet season" and "dry season" are not true seasons. To avoid ambiguity, it is necessary to situate this annual climatic oscillation within the framework of the true astronomical seasonal sequence.

The planetary seasons – the annually recurring sequence of interrelated astronomical, meteorological, terrestrial, and biological changes by which we distinguish the passage of spring, summer, autumn, and winter – result from the 66½ degree tilt of the earth's axis in relation to its orbit of the sun. The beginning and end of each season is marked astronomically by the solstices and equinoxes, and the months in which they occur are the seasonal transition months of December, March, June, and September. The duration of each season is divided into four periods; in the Southern Hemisphere, for example, the period from the vernal equinox on September 22–23 until the end of September is the beginning of spring, October is early spring, November is mid-spring, and from December 1 to December 20 is the end of spring. From the summer solstice on December 21–22 to the end of December is the beginning of summer, January is the early summer, February is mid-summer, and from March 1 to 20 is the end of summer. The same format applies to autumn and winter.

To avoid ambiguity, the concept of "seasons" should only be used in reference to the planetary seasonal cycle. What are generally referred to as the "wet season" and "dry season" are more accurately identified as the wet and dry times of the year. The wet time, which is to say the time of annual rains, is from the mid-spring of November until the early autumn of April, and the dry time is from the mid-autumn of May to the early spring of October. Just as the equinoxes and solstices mark the beginning and end of each season, November and May mark the transition from the dry to the wet time and from the wet to the dry time. Woodburn (1968a:50), citing data from the African Meteorological Department, says rain "falls almost exclusively during the period December to May with peaks in December and March," and Marlowe (2010) follows Woodburn in this regard. This means the dry time is June–November and the wet time is December–May. However, November or May could be assigned to the dry or wet time of the year. November is here assigned to the dry-to-wet transition – the start of the annual rains – while May is assigned to the wet-to-dry transition and marks the beginning of the dry time, which lasts through the early spring of October.

Assessing the Seasonality of the Baobab

Although the baobab is distinctly seasonal, the duration of its harvest period has been variously identified as a "few months," "several months of the year," "throughout most of the year," and "year-round." Some authors identify the baobab as a fruit of the dry time, while others describe it as available during the autumn rains. Those who regard it as a dry-time fruit give different months of the dry time, and those who identify it as year-round give different reasons for doing so. Still others identify it as available year-round, but only intensively utilized at a particular time of the year. There are a number of factors that make it difficult to determine accurately the reproductive seasonality of the species, which is likely why Hadza researchers like Woodburn (1968a) and Marlowe (2010) were unable to give a clear account of the period over which baobab fruits ripen. Instead of a simplistic assumption that these differing accounts are either completely right or wrong, the aim here is to regard them as aspects of a more coherent theoretical understanding. The key to reconciling these differences involves greater attention to the eight problematic areas discussed below.

Baobab Flowering

The Time of the Year When the Baobab Begins Flowering

The baobab is generally reported to flower during the annual rains, which suggests the timing of flowering initiation is partly determined by the climatic regime in which the tree grows. The baobab puts out its new leaves starting sometime between the early spring of October and the mid-spring of November, and shortly thereafter, it also begins flowering; the beginning of flowering before new leaves appear has also been reported. Peak flowering occurs from around the end of spring in December to mid-summer in February, after which, in favorable locations, flowering continues to decline degrees into the early autumn of April. This conception of the baobab's flowering period seems to be what Wickens had in mind when he wrote, "Flowering is said to occur just before or at the start of the rainy season, although my own observations in the Sudan suggest that flowering may occur throughout the year except at the height of the dry season – January to March [in the northern hemisphere]." Unfortunately, this is not the way Wickens's notion of the baobab's flowering seasonality has been interpreted in the literature, largely because his account is ambiguous in that it draws no clear distinction between the initiation of flowering and the duration of flowering.

For example, Baum (1995a:331), who cited Wickens (1982:185), wrote, "*Adansonia digitata* usually flowers at the start of the wet season (the timing of which varies across Africa), but there are many populations that flower at other times of the year." By conflating flowering initiation with flowering duration, and by contrasting flowering "with the start of the wet season" and flowering "at other times of the year," Baum (1995a:331) gives the misleading impression that individual baobabs and different baobab populations flower at different times throughout the year. When Sidibe and Williams (2002:17) published their widely referenced book, they formulated Baum's interpretation of Wickens (1982) in the following way: "The [baobab's] flowering time varies greatly; in general flowering can occur anytime except during the height of the dry season." I agree that flowering initiation does vary across Africa, but it is mostly reported to vary between the early spring of October and the start of summer in December. However, flowering continues for a period thereafter, depending on the rainfall regime. This means flowering duration must be taken into account.

The Duration of the Flowering Period

The duration of the baobab's flowering period has been variously identified as spring (Onderstall 1979:76) or early summer (Cowen 1952; Randhawa 1965:133); as early spring to the beginning of summer (Sidibe and Williams 2002:17–18; Watson 2007; Kamatou et al. 2011:910) or to the beginning of winter (Wickens 1982:185); as mid-spring to either the beginning of summer (Sidibe and Williams 2002:17–18), early summer (Gebauer et al. 2002:157), or the beginning of autumn (Assogbadjo et al. 2005:51); and as the end of spring to early summer (Davis and Ghosh 1976:471; Duxoux 1983:229; Dhillion and Gustad 2004:90) or mid-summer (Jaeger 1950:369). Climatic variation in the flowering period is of obvious importance, but this variation expresses an underlying commonality. Contrary to the view that baobabs randomly flower through the year, there is growing recognition that under favorable conditions, flowering begins in the spring or early summer and continues up to the winter dry season, as Wickens (1982) observed for the Sudan.

The Seasonal Timing of Fruit Maturation

According to Sidibe and Williams (2002:17), whose view has been widely adopted in the literature (e.g., Sanchez 2010:13, Venter and Witkowski (2011:632), baobab fruits "develop 5-6 months after flowering." This means if flowering initiation is October, then ripening initiation 5 to 6 months later would be the end of summer in March to early autumn in April, which is the period of the autumn rains. If, however, flowering initiation is in November, then fruits would ripen in the early autumn of April or in the mid-autumn of May, which is the onset of the dry time. The 5- to 6-month maturation period is rejected here as a general characterization in favor of a 6-month maturation period, because the baobab's ripening period is widely associated with the onset of the dry time and because 5 to 6 months is inconsistent with the most reliable published accounts of baobab phenology. For example, based on their field research in Benin, Assogbadjo et al. (2005:55) have reported that fruit maturation begins "approximately 6 to 7 months after the beginning of flowering." November flowering produces ripe fruits in May. This means that in the environment of the Hadza, flowering during the rainy time from November to April would result in fruits ripening from May to October, and this is a much better fit with published accounts.

The Correlation of Flowering Duration with Ripening Duration

The duration of the baobab's flowering period correlates with the period of time over which the fruits mature, ripen, and fall from the tree. The 2006 baobab field manual produced by the International Centre for Underutilized Crops advises that the color of the fruit "does not change rapidly with maturity and individual fruits mature at different times, so harvesting should be carried out selectively." The simplest explanation for this is durational ripening, which is one important factor in the baobab's long harvest period. Fruits ripen and fall from the tree in the order in which the flowers appear through spring, summer, and autumn. May–June is the time of year when ripe baobab fruits first become available, with peak quantities from the early winter of July to the mid-winter of August, which is the middle of the dry time.

It is possible that gathering baobab fruits at the start of the ripening period in the early dry time is limited for the Hadza because only a small number of fruits would fall from the tree at this time. At the start of the ripening period, the Hadza would be able to harvest fruit by hand from low branches that can be reached from the ground, as well as by using a reaping stick, by throwing sticks to dislodge the fruit, and by obtaining fruit from trees that are easily climbed without technological aids. The peak baobab harvest is the middle of the dry time from the early winter of July to the mid-winter of August. By providing winter fruit, the baobab extends the major harvest period of the wet part of the year until the mid-winter in August. This is the most likely time when the gathering of baobab fruit from the ground determines the camp movements of the Hadza.

The Seasonal Timing of Durational Fruit Fall

Giant baobabs with their smooth trunks are difficult and dangerous to climb in the absence of accessible low branches or vines or neighboring shrubs or trees (Thomas 1959, Rashford 1987a, Blurton Jones 2016:163). The Hadza do so, however, to harvest honey, fruit, water, and fiber and because they use the tree as a viewing platform. Honey, their best-loved and most nutritious food, is commonly identified as the reason Hadza men risk the climb. The Hadza have been climbing the baobab for a long time. According to their oral traditions, their original ancestors traveled from the sky down to the earth via the trunk of the cosmic baobab. It is noteworthy then that the "greatest number of severe accidental injuries is to Hadza men who fall from trees they have climbed in search of wild honey" (Bennet et al. 1973). Blurton Jones and Marlowe (2002:207) have reported "two recent deaths and two serious injuries resulting in permanent disability from falls from baobab trees." According to Marlowe (2010:160), "Men do not live as long as women. Old men are the most likely to fall out of tall baobab trees to their deaths, since they continue to try to collect honey into old age. They usually continue to forage until they die, but get foods that are easier to acquire, such as berries and baobab." The difficulty and danger of climbing baobab trees means that most fruits are probably harvested from the ground, and while durational ripening is the equivalent of a potential harvesting time, it is not exactly the same as the actual harvest time, which is a function of the period during which the fruits fall from the tree.

Differentially Delayed Fruit Fall as the Source of the Baobab's Extended Harvest Period

Ripe baobab fruit in the May–October dry time is of great seasonal importance for the Hadza. The main harvest, however, is from the mid-autumn of May to the mid-winter of August, with peak abundance in the early to mid-winter dry season of July and August. During the September–October height of the dry time, yields of ripe fruit diminish. And fruit is least available during the November–April autumn rains. If fruit fall was strictly the result of durational flowering, then the end of durational fruit fall 6 months after the last flowers bloomed in April would be in October, at the height of the dry time. The reason for not accepting the simple idea that durational fruit fall is solely the outcome of durational flowering is that some fruits remain on the tree well past October, which is when one would expect fruit fall to end based solely on November–April durational flowering. Fruit fall during the November–April rains is identified here as wet-time differentially delayed fruit fall.

Technically, differentially delayed fruit fall would also occur during the baobab's May-to-October durational harvest time, and this makes it necessary to distinguish between the differentially delayed fruit fall of the dry time and of the rains. If plants that flowered in November produce the earliest ripe fruit, and yet not all of these fruits fall at the expected time, which would be 6 months later in May, then the period from June to October is what is referred to here as dry-time differentially delayed fruit fall. The concept of durational fruit fall, correlated with differentially delayed fruit fall in the May–October dry time and in the November–April annual rains, is necessary to fully understand the baobab's peak winter harvest along with its extended harvest period. It is durational fruit fall augmented by the May–October differentially delayed fruit fall that results in the peak fruit fall in the early winter of July and the mid-winter of August. But it is the November–April period of differentially delayed fruit fall that accounts for the availability of baobab fruit for most of the year.

For example, there are reports of the Hadza harvesting baobab fruit in the summer. In a table that presents the percentage of different foods the Hadza brought into camp in early summer, Marlowe (2010:127) indicated that 67% were berries, 14% meat, 10% baobab, 8% tubers, and 1% honey. The same table presents data on 24 camps surveyed over 5 years. Berries still accounted for the largest share at 28.5%, with meat at 27.1%, tubers 24.7%, baobab 11.5%, and honey 8.1%. It is noteworthy that the percentage of baobab brought into one summer camp almost equals the percentages that "represent 5 years of data in 24 camps spanning 11 years" (Marlow 2010:127). It is the wet-time differentially delayed fruit fall that significantly extends the baobab's harvest time following the end of the May-to-October period of durational fruit fall. And it is the dry-time and wet-time differentially delayed fruit fall that obscures the precise nature of the baobab's near year-round availability.

The Persistence of Ripe Baobab Fruits on the Ground

How long does the baobab's dry indehiscent fruit remain useful after falling from the tree? Ground persistence is clearly variable. During the rains, the pod rots quickly and is invaded by insects that feed on the pulp and seeds. Long-term ground persistence is also limited in areas where large animals feed on the fruit.

The Natural Storability of Baobab Fruits

With its hard shell, dry pulp, and durable seeds, the baobab's indehiscent fruit can be kept for a long time without special treatment. The Hadza do not engage in long-term storage, however, and we can assume that for them baobab storage has a limited impact on extending the availability of fruit throughout the year.

Explanations for the Near Year-Round Availability of Baobab Fruits

One of the many things that make the baobab an exceptionally useful tree for foragers like the Hadza is the availability of its ripe fruits for what some regard as "most of the year" and others as "throughout the year." Murray et al. (2001:12) identified the long-term availability of ripe fruit as the baobab's most important quality compared with the other preferred foods of the Hadza. In comparing the nutritional composition of some of the Hadza's wild plants and honey, these scholars reported that although baobab seeds are inadequate in three essential amino acids, they are, with their high fat content, equivalent to honey in the amount of energy they provide. What is particularly noteworthy is that while baobab seeds and honey have equivalent energy content, baobab seeds are available "over the majority of the months of the year," while honey "is available in quantity only for a few months [in the 'wet season'] each year, as is true for non-baobab fruits" (Murray et al. 2001:12).

The reasons the baobab should be considered a long-term source of ripe fruit are not always made clear, however. For example, in his emphasis on the importance of a seasonal polarity in Hadza life, Woodburn (1968a:50) identified the extent to which the gathering of berries and roots and the hunting of large and small game occurred in the wet or dry time of the year, but he provided no specific information on the timing of wet or dry periods for the baobab harvest. Nevertheless, Woodburn gave the impression that ripe baobab fruits could be had throughout the year when he claimed "Vegetable food – roots, berries, the fruit of the baobab tree, etc. – though not often obvious to the casual observer, is always abundant even at the height of the dry season in a year of drought" (1968a, b:50).

Some scholars explicitly support the idea of year-round ripe fruit availability, but they differ in the factors they identify that are responsible for this pattern. For example, Tomita (1966:161) attributed year-round harvesting to ground persistence. This factor does play a role in the duration of the harvest period, but, as already noted, it is insufficient to account for the baobab's near year-round availability. Mabulla (2003:39), who presented a table of the spatiotemporal distribution of plant food in the Eyasi Basin adopted from McDowell (1981a), offered a unique perspective on the year-round availability thesis. His table indicated that the baobab was harvestable year-round, though it was a "rarely utilized food, but available" from the start of spring to early summer; from the mid-summer to early autumn, it was a "minimally utilized food." But from mid-autumn in May to the middle of mid-winter in August, it was an "intensively utilized food." What Mabulla's table presented was not peak fruit fall, but the main period during which the Hadza use the fruit, which he identifies as the period of peak Hadza "dependence" on the baobab. Elsewhere, Mabulla (2003:48) identified "May to late June" as the period when the Hadza "depend more on baobab." There is general agreement in the literature that the peak fruit-fall harvest of the baobab is from the early winter dry season in July to the mid-winter in August, but the idea that the baobab is available year-round and is "rarely" or "minimally" utilized at certain times of the year does not square with ethnographic reports of the Hadza. It is hard to imagine a time of the year when baobab fruits are available to the Hadza but are not harvested.

Marlowe (2010:115) presented data indicating that the Hadza gather baobab through the year, even though in his recent ethnography he described the fruit as available "throughout most of the year." He attributed this long harvest period solely to the individual differences between trees in their seasonal fruiting. This, he said, suggested significant genetic variability; individual trees flower and fruit randomly throughout the year, ensuring that some fruit is available at most times. With respect to Marlowe's perspective, an important factor that complicates our understanding of the baobab's harvest seasonality is the existence of both fruiting and non-producing trees, the latter identified in ethnographic accounts as "male trees." This is a well-recognized phenomenon. Non-producing trees are those that do not flower; those that flower profusely but do not set fruit or abort all their fruits prematurely; and those that produce only small quantities of flowers and mature fruit each year. Because the baobab produces perfect flowers, it is difficult to explain the existence of so-called male and female trees (Rashford 2015:220–223; Swanepoel 1993; Venter et al. 2017:263). Assogbadjo et al. (2008:77) found that tree fertility was 1 of the 21 traits West Africans used to differentiate baobab morphotypes. The female/male variable was noted by 59.2% of the 98 interviewees from Benin, 19% of the 92 from Burkina Faso, 26.8% of the 92 from Ghana, and 16.2% of the 128 from Senegal.

Non-producing baobabs give the false impression that throughout the year there are always some trees fruiting while others are not, a perceived condition mistakenly attributed to genetic differences in the actual fruit-fall harvest times of individual trees. Because ripe fruit in seasonally varying amounts is available almost year-round, it is perhaps not surprising that the Hadza single out a baobab in song for not having fruit on or under the tree. This could describe a "male" baobab. As noted earlier, however, neither ground persistence nor variability in seasonal fruiting adequately accounts for the extended harvest seasonality of the baobab when these two factors are considered apart from other aspects of baobab phenology.

From what has already been said, we can conclude that the baobab does indeed have a long ripening period. However, the baobab cannot be said to fruit year-round because the tree flowers and fruits throughout the year. The fruit's long-term ground persistence must also be considered. A more accurate theoretical perspective than year-round production, one that is in keeping with the literature, is that the baobab is decidedly seasonal throughout its range in Africa, but with a potentially long harvest period. This reflects a reproduction strategy based on the combination of durational and differentially delayed fruit fall. In sum, May–June is the start of the ripening period, July–August is peak harvest, and September–October is the beginning of the decline of the harvest. November–April is the slow ending of the harvest as the culmination of differentially delayed fruit fall. This means the autumn rains mark the time when baobab fruit would be scarce or absent.

Overview of the Hadza Seasonal Cycle

With the exception of the dry-to-wet and wet-to-dry transitions, the seasonal changes of the Hadza's tropical environment are subtle compared with the dramatic changes at temperate latitudes. But in contrast to Woodburn's emphasis on (1968a) an annual wet/dry polarity, a true seasonal perspective makes it possible to combine the wet/dry oscillation with the seasonal progression. Dorothea Bleek suggested that the Hadza conceive of the seasonal cycle in three parts. In "the division of the year," she wrote (1931:278), the Hadza "borrow Bantu terms, but have a name of their own for the cold dry season following the [peak autumn] rains. Otherwise," she tells us, "they merely say, 'it is hot' or 'rain falls'." Bleek's statement suggests the Hadza's conception of the seasons is more involved than the oscillation perspective suggests. For example, Edenmyr (2004:15) defined the meaning of the Hadza noun, *hongola-pi?i*, as "rain season Oct. to Dec," which is the early spring to the beginning of summer.

The height of the dry time – the hottest time of the year, from the end of winter in September through the early spring of October – is an important seasonal moment for the Hadza. With the winter/spring transition in September comes the earliest signs of plants flowering and putting out new leaves, coinciding with the return of the annual rains, especially beginning toward the end of early spring in October. Then occasional showers of late October indicate that the rainy half of the year, starting in mid-spring of November, marks the beginning of a new annual growing cycle.

Honey and baobab become scarce from the September–October end of winter to the early spring, and at that time of the year, the Hadza are especially dependent on meat from large game, the start of the *Cordia* berry harvest, and the year-round availability of roots. The Hadza congregate in larger camps then than at any other time of year, because the intense dryness at the end of winter and early spring makes it more likely that men will be successful in ambush hunting, taking advantage of a seasonally restricted number of game trails that lead animals to dry-time water sources. The mid-spring of November to the beginning of summer in December is the time of the spring rains, which peak in December. The rains produce an overall greening of the landscape, the result of the renewed growth of herbaceous plants, the full leafing out of deciduous trees and shrubs, and the flowering of the baobab. This is the time when insects become plentiful, bees and termites swarm, and birds nest.

The mid-spring of November until the mid-summer of February is the height of the general fruit harvest and honey harvest, with a number of fruit trees declining in the autumn rains. Ripe baobab fruit of the previous year become increasingly scarce even as young baobab fruits are developing. The spring rains are followed by the summer dry season, when *Cordia* berries become plentiful. During this period, the Hadza live in large berry camps. The peak autumn rain is from the end of summer in March through early autumn in April. And during this period, the Hadza live in small camps. In addition to small game, food through the autumn rains is mainly gathered honey and roots.

After the annual rains, the sun shines brightly during the May–June dry time, as the days of bluer skies grow cool and windy. Winter is the coolest time of the year, when deciduous trees and shrubs shed their leaves, grasses and other herbaceous plants turn brown as they die or go dormant, rainy-season runoff and pools dry up, and the ground hardens – the myriad greens of the wet November–April growing seasons have now become the myriad browns of the May–October dry time. This is the time of year for the ripening of fruits and dispersal of seeds, timed to take advantage of the return of the spring rains in November. The May–June beginning of the dry time is the harvest period for the Hadza's traditional farming neighbors and commercial farms. May–June is also the start of the baobab harvest, which peaks in July–August. By providing winter fruit and honey, the baobab extends the major harvest period until mid-winter in August. The Hadza particularly like this dry, cool time of year when baobab fruits and winter berries are abundant and game is easier to track.

Summary

This chapter, which is based on a perspective that unifies the concepts of astronomical seasons with the annual wet/dry polarity and the calendrical months, provides a seasonal framework for understanding the baobab's contribution to Hadza life. The Hadza are a roots-and-fruits culture, and the baobab is one of their most important fruit trees. The seasonal nature of its flowering and fruiting is a significant factor in the tree's value to the Hadza. However, the baobab is much more than a fruit tree. It is the Hadza's all-purpose tree of life. The seasonality of the Hadza/baobab relation briefly introduced in this chapter hints at the great variety of ways in which the baobab is practically and inspirationally useful to the Hadza throughout the year. This particularly includes its practical value as a multisource food tree, especially at the time of year when other food resources are scarce.

Part V
Hadza Baobab Resources: Food, Health, and Exchange Benefits

Chapter 10
The Hadza's Preeminent Fruit Tree

Of the many ways in which the baobab is a tree of life for the Hadza, none is more important than its year-round value as a multisource food tree. Published accounts indicate that for many Africans, however, the baobab is preeminently a fruit tree. For others, baobab-specific foods (which are those that come directly from the tree itself) have also included its root, bark, leaf, flower, seed, and seed sprout. In some parts of Africa, the leaves (Rashford 2019) and seeds are dietary staples, but there are few accounts of the consumption of the tree's root, bark, flower, and seed sprout (Rashford 2015). Apart from the baobab and various berries, the fruit trees most often mentioned in the literature on the Hadza are the fig (*Ficus sycomorus* L.), doum palm (*Hyphaene thebaica* (L.) Mart.), desert date (*Balanites aegyptiaca* (L.) Delile), tamarind (*Tamarindus indica* L.), and marula (*Scelerocarya birrea* (A. Rich. Aubrev). None of these comes close in value to the baobab as both the Hadza's most useful fruit tree and most important multipurpose food tree. This chapter explores the significance of the baobab as the Hadza's preeminent fruit tree. In doing so, it stresses the importance of a more systematic understanding of the overwhelming Hadza dependence on plants, including the importance of fruits in the Hadza diet. Following a discussion of Hadza climbing baobabs for fruit and other vital resources, the second part of the chapter looks at the various ways in which the fruit of the baobab is consumed, the use of the fruit for feeding children, and the baobab's importance for children at the September–October height of the dry time.

Characterizing Hadza Dependence on Plant Foods

The Hadza can aptly be described as a fruits-and-roots people. This characterization identifies two of the most important categories of food plants in hominin evolution. As a prerequisite to highlighting this evolutionary trajectory, it is important to recognize that the terminology employed in the literature discussing the nature of fruits and roots is not systematic. It is necessary, therefore, to specify how these terms are

J. Rashford, *Baobab*, https://doi.org/10.1007/978-3-031-26470-2_10

used in this work. Fruit, from a culinary perspective, is the reproductive structure of a plant that has a sweet fleshy pulp that can be eaten in its natural state when ripe and is suitable for making such things as drinks, preserves, sweets, and deserts. Fruit defined in this way involves enlisting the unwitting support of animals in the dispersal of its one or more seeds. Botanically, however, fruit, as a plant's reproductive structure, includes not only sugary succulent fruits but also what is usually identified as seeds (which as embryonic plants are the essential part of a fruit), nuts (which botanically are fruits), and some kinds of vegetables, such as leguminous fruits (Spjut 1994).

The problem with the use of "fruit" and "root" is that, if not clearly defined, the terms confound rather than clarify the discussion. This is evident, for example, in the influential 1981 article of Peters and O'Brien published in *Current Anthropology* titled "The Early Hominid Plant-Food Niche: Insights from an Analysis of Plant Exploitation by *Homo*, *Pan*, and *Papio* in Eastern and Southern Africa." At the time, these authors noted that in the discussion of human evolution, "Less consideration has been given to plant foods, despite their importance to the higher primates and to historic African gatherer-hunters" (1981:127). In organizing a study that would ultimately include 461 genera of plants used for food by humans, chimpanzees, and baboons, Peters and O'Brien (p. 129) offered the following classification:

> The general food-item types that we have taken from the literature and used in our analysis are flowers/inflorescences, fruits (including berries or nuts), seeds/pods, leaves/shoots, stem/stalks, bark/cambium, and underground storage organs. We have reassigned "fruits" identified as belonging to the Leguminosae and items termed "beans" to the category seeds/pods. Likewise, we have included the items called "pith" in the category stems/stalks. The six instances of "gum" or "sap" are not included in the analysis presented here; nor are mushrooms, which are exploited by all three primates but which were not identified generically.

That this system of classification is problematic is evident in Conroy's (1981:134) response to the use of the terms "fruit" and "seed" in the article:

> A description of some of the characteristics of "fruits" found in the savannah might also be helpful. The term "fruit" conjures up images of Florida oranges and Granny Smith apples! I would assume that some of the "fruit" available to Plio/Pleistocene hominids had a consistency that would place them in the "seed" hardness category. Maybe someone should develop a "fruit-hardness" index? This should be a *sine qua non* for future microwear studies.

But how does one go about "reassigning" "fruits" botanically speaking, and are fruits and seeds separate things? We can theorize that early hominins consumed botanically defined fruits in the form of berries and other pulp-producing fruits and in the form of nuts. In some cases, the seeds of pulpy fruits would also have been eaten, as is the case of the baobab, while with nuts, the seed is the only part of this kind of fruit that is eaten. Sometimes, the whole fruit is eaten, as with immature legumes, which, from a culinary point of view, are categorized as vegetables. For example, the Fabaceae ranks as one of the most important families of plants for human beings (Harlan 1992:14–15; Masefield et al. 1969). Some fruits of this family, like the tamarind and the arillate fruits of species of the genus *Pithecellobium*, are "fruits" in that they have hard seeds surrounded by a sweet pulp; some fruits like the peanut are classified as nuts. Some, like green beans, involve the whole fruit or just the pods or seeds eaten as a vegetable.

Similarly, ambiguities arise in discussions of "roots" as plant foods. Ungar et al. (2006), in their evaluation of recent studies based on archaeological and paleoenvironmental evidence, nutrition research, and comparative primate studies, stressed the role of dietary changes in the African origin and early evolution of the genus *Homo*; they used the phrase "underground storage organ" to encompass what are biologically roots, tubers, corms, and bulbs. Instead of this phrase, many researchers prefer "tubers" as a general term, and Coursey chose the technical term "hypogeous plant foods." In the Caribbean, starchy food crops are identified as "ground provisions." For coherence and overall consistency, all these underground plant parts are herein identified as "roots," with the understanding that the term is intended to include true roots, tubers, corms, and bulbs.

To simplify the Peters and O'Brien classification in a way that avoids the kinds of ambiguities in the use of "fruit" and "roots," we can identify the fundamental categories for comparison as roots, stem, bark, leaves, flowers, and fruit. That is, roots include true roots, tubers, bulbs, and corms; stems include stalks and piths; bark includes cambium; leaves include shoots; flowers include inflorescences (which are flower clusters on a fixed floral axis); and botanically defined fruits include berries, nuts, and legumes and the seeds of all three (bearing in mind that some seeds are ingested as part of a succulent fruit and pass through the digestive system intact, while other seeds are intentionally selected for food, since seeds are one of the most nutritious components of the human diet). The baobab was likely a major contributor to the fruit diet of early hominins, but it is possible that it also provided edible root, bark, leaves, and flowers. The importance of a clear distinction in the kinds of plant foods highlighted in this work is relevant to the discussion of the fate of baobab seeds resulting from fruit consumption by humans, chimpanzees, and baboons presented in Chap. 24.

The Significance of Fruit Among the Staples of the Hadza Diet

Although the Hadza prize meat and honey above all other foods, Woodburn (1968a) reported that an estimated 80% of their staples are derived from 10 species of plants, including four species of tubers, five of berries, and the baobab – the only large tree of great importance in the Hadza diet. But Woodburn (1968a:52) gave the false impression that all Hadza plant foods were equally unsatisfactory when compared with meat:

> The low opinion which Hadza have of the vegetable food which makes up the bulk of their diet is not surprising when its unpalatability is taken into account. Roots are, in general, tough, fibrous, and have little taste; many of the berries are hard and dry and contain large stones which are swallowed whole; *undushi* berries leave a dry, sticky residuum in the mouth; *k'alahai* berries split the lips and tongue if eaten in quantity. Very little of the veg-

> etable food is eaten with much enthusiasm. But the advantage of vegetable food over meat (or honey), and the basic reason why it constitutes the bulk of the diet of the Hadza, is that it can be obtained quickly and, above all, predictably. When they go out to look for vegetable food, the women can be sure that they will find some type without undue effort.

It is surprising that in this passage, Woodburn did not go on to indicate the Hadza's opinion of the baobab. For example, Woodburn (1970:42) wrote in his account of Hadza material culture: "Both baobab pulp and kernels may be, and often are … eaten raw, either dry or mixed with water." And it is clear from his many publications and from the work of others that the baobab is important in the Hadza diet and held in high esteem. The baobab is justifiably identified as the foremost food tree of the Hadza, given the variety of foods it provides, especially honey, ranked first among the top five foods preferred by both men and women, and the highly nutritious pulp and seeds, ranked third by both men and women (Marlowe 2003).

Efforts to identify the staples of the Hadza diet are evident in the earliest accounts of their forager way of life. Using the criterion of Hadza preference, Woodburn (1968a) saw meat as representing 20% of the Hadza diet, with fluctuating amounts of honey throughout the year, along with baobab, berries, and roots. Marlowe's study of Hadza in-camp foods (2003:366) clarified and broadly confirmed Woodburn's view; data on Hadza in-camp foods supported the view of the Hadza as a fruits-and-roots people. The six basic categories of Hadza in-camp foods were roots (23.5%), meats (19.3%), baobab (19.2%), berries (17.2%), honey (14.2%), and grain (6.6%). Forty-two percent of in-camp foods were biologically defined fruit, the largest categories of camp foods, comprising baobab (19%), berries (17%), and grain (6%). The baobab represents the largest category of fruits brought back to camp; it is the Hadza's most important fruit, more so than berries and grains.

But there is another way of looking at this data that brings the baobab to the fore. Besides fruits, the other major category of in-camp plant food is roots. Roots (23%) and baobab (19%) vie for constituting the highest percentage of in-camp foods, depending on how certain Hadza foods are categorized. If we consider both baobab fruit (19%) and honey (14.%) as baobab-sourced foods, their combined share is 33.4%, greater than the percentage of in-camp food represented by roots. The breakdown would be baobab foods at 33.4%, followed by roots (23%), meat (19%), berries (17%), and grain (6%). In this work, I argue that honey should be categorized as a plant-sourced food, since it largely comes from beehives in the baobab's hollow trunk and branches. If we add to the 66.5 percent plant-derived foods the 14.2 percent honey (as a plant-sourced food), then plant foods represent 80.7 percent of Hadza in-camp foods with fruit (43%) and roots (23.5%) making up by far the largest portion.

Climbing Baobab Trees for Fruit and Other Resources

The baobab's true value for the Hadza only comes into full view when we consider its importance in their practical and inspirational life. Based on the generalized man-the-hunter and woman-the-gatherer framework, for example, we might superficially conclude that the baobab is irrelevant to meat-eating and that the tree's primary food value is in honey taken by men and ripe fruit gathered by women. But this is not correct. The seeming anomaly of men returning to camp with significant quantities of baobab fruit deserves an explanation, since the men do not, it seems, concern themselves with collecting berries and other fruits or roots (Marlowe 2003:220). Men also cooperate with women in gathering baobab fruit. They harvest fruit that can be reached from the ground by hand or with a reaping stick, and they throw sticks to knock down higher fruit. They also climb the tree to pick the fruit by hand, as illustrated by Peterson et al. (2013:94), or with a stick, and to dislodge fruit by shaking accessible branches. Nothing reveals the importance of a comprehensive view of the baobab more than the paper of Kraft and colleagues assessing the significance of tree-climbing in human evolution. They wrote (2014:105):

> Committed terrestrial bipedalism is considered a defining feature of the human lineage. This prevailing view underlies the arboreal-terrestrial dichotomous framework that has informed form-function inferences in the fields of comparative primate anatomy and paleoanthropology (Kimbel and Delezene 2009). For example, similarities in ankle and foot traits between humans and some early hominins (e.g., *Australopithecus afarensis*) have led some to reject any significant amount of arboreality for these hominins (Latimer et al. 1987; Latimer and Lovejoy 1990a, b; Ward et al. 2011). This conclusion rests on the assumption that humans do not climb trees, and if they do, they are clumsy and incompetent (Latimer 1991). Thus, when a human-like morphology is observed in a fossil hominin, it is assumed to indicate similar incompetence.

For Kraft et al. (2014:105), "This general approach, and thc dichotomy upon which it is based, has been criticized as an oversimplification," and in their conclusion, they noted (2014:115):

> In Africa, rainforest and savanna foragers climb to acquire honey, baobab, meat, and fruit among other things. Because it is unquestionable that these resources also existed during the Plio-Pleistocene, we suggest that hominin climbing behavior, if and when it occurred, could have occurred in pursuit of similar food resources as that of modern hunter-gatherers.

What is significant is that the Hadza climb the baobab for all the reasons Kraft et al. (2014) mentioned, and although it has not been reported that they climb for leaves and insects as some groups do (which Kraft et al. do not mention), they also climb the baobab for water, for fiber from the branches, in pursuit of game, and as a surveillance platform (Watson 2007; Peterson et al. 2013).

The Ways of Consuming the Baobab Fruit

Away from camp, the Hadza eat the fruit pulp and spit out the seeds. When fruits are brought back to camp, however, it is usually for processing. There are different descriptions of the Hadza fruit processing and consumption. It is not clear whether these differences reflect variations in the place or time of the observations, variations in the relative completeness of the descriptions, or variations that are essentially contradictory descriptions.

There are two ways in which fruit pulp with embedded seeds is processed for consumption after it has been removed from the pod. To separate pulp from seeds, the pulp is either soaked in water and stirred or pounded and crushed. The water-soaking method is not mentioned for the Hadza who pound the fruit to produce pulp-and-seed flour, or to separate the pulp and seeds, or to produce both pulp and seed flour. However the fruit is processed, the dissolved pulp, pulp flour, seed flour, and pulp/seed flour are eaten as they are or used to make beverages, weaning paste, porridge, and sweets.

Jelliffe et al. (1962:25) reported that a "thin gruellike mixture [is] made of the uncooked powder or the ground seeds of the baobab fruit … mixed with water." This is contrary to Cooper (1949:9) who, as earlier noted, said the pulp and seeds were pounded together and cooked. We are left to wonder if pulp flour and seed flour were always eaten separately or sometimes eaten together. Necessity sometimes dictates that they are eaten separately. Though the Hadza suck the pulp from the seeds and then spit the seeds out, they will also gather the seeds in baboon dung (Woodburn 1970; Murray et al. 2001), which they wash, dry, and process into flour. Contrary to Jelliffe (1962:25), however, Woodburn (1970:40) reported that the pulp and seeds were processed together to produce a combined pulp-and-seed flour. Woodburn wrote (1970:4041):

> To prepare food from baobab fruit, the women extract the pulp and the seeds from the pods and pound them up with a stone usually in one of the hollows in the flat rocks that are found in many parts of Hadza country. The pounding stones are not usually carried about but are left by the side of the hollows. The shells of the seeds are winnowed out using a soft piece of leather and the mixture of pound kernels and pulp is then cooked off with water to make a tasty type of porridge which is eaten by everybody but is valued especially as an infant food.

Since the time of Jelliffe and Woodburn, the literature has been unclear about the ways in which the Hadza process and consume baobab fruit. For example, Murray et al. (2001:5) presented a view similar to Jelliffe et al. (1962:25), reporting that the "seeds are removed from the dry fruit pulp which is eaten separately." Wickens and Lowe (2008:72), on the other hand, supported Woodburn's (1970:40) observation that a cooked porridge is made from the pulp-and-seed flour; they also cited Murray et al. (2001:5), without dealing with the difference between the views expressed by Woodburn and by Murray and colleagues. Frank Marlowe (2010:120) supported Woodburn's view in his Hadza ethnography.

Baobab fruit pulp is rich in vitamin C (Nicol 1957; Woodburn 1970:40) and calcium (Murray et al. 2001) and baobab seeds are one of the tree's most nutritionally important foods. They provide the Hadza with the most complete source of energy among plant foods available during the dry season.According to Murray et al. (2001:12), "Baobab seeds are a good source of protein and, due largely to their high fat content, their energy content is equivalent to that from honey… In addition," they noted, "baobab seeds are a more dependable energy source than honey." Berbesque and Marlowe (2009:4) presented a table comparing the nutrient content of honey, meat, tuber, berries, and baobab as the five most important Hadza foods; it was compiled from the work of eight researchers including Weiner (1973), McCullough and Ullrey (1983), Vincent (1985a), Schoeninger et al. (2001a), Murray et al. (2001), Cordain et al. (2002), Van Zyl and Ferreira (2004), and Galvin et al. (ND). The table showed baobab fruit was highest in fat, second to meat in protein and kilocalories, second to tubers in carbohydrate, and third to honey and berries in sugar.

Men as Providers of Baobab Fruits

Men supply baobab fruits to their wives as supplemental food to make up for their reduced foraging because of baby care, to provide toddlers with weaning food, and as routine food for small children. In a study of the implications of male provisioning for human pair bonding at the critical period when women had dependent children, Marlowe (2003:220) reported that women brought back to camp almost all the tubers (38.8% to the men's 6.3%) and almost all the berries (24.8% to men's 6.3%), while men brought back almost all the meat (39.6% to women's 1.2%) and almost all the honey (30.2% to women's 0.7%). However, both women and men brought back baobab fruits in large quantities, even though women brought back almost twice as much as men (25.6% to men's 13.8%). It is noteworthy that Hadza men brought meat, honey, and baobab fruit back to camp, since honey and baobab fruit are regarded as ideal infant foods.

Hawkes et al. (2001a:683) observed that "Hadza men emphasize big-game hunting but also take other resources. On most days, they collect plant food snacks and (less often) small game for themselves (Berbesque et al. 2016). They occasionally bring home small game and sometimes baobab fruit." In his comparative view of hunter-gatherer infancy and childhood in which he cites Blurton Jones (1993:316), Konner (2004:43) reported that with the Hadza "even 3 year olds try their hand at digging or picking up baobab pods and processing them." In a table that showed "how different are the foods targeted by women and by men," Marlowe (2010:127) indicated that "the only substantial overlap [in the foods acquired by men and women] is in baobab." Contrary to the view of Hawkes et al. (2001a), however, men collecting baobab fruits appears to be more than "occasional," based on Marlowe's (2010:116) account: "When men are heading back to camp empty-handed after hunting or checking beehives," Marlowe reported that "they often stop and load up with many baobab pods so as to take something back to camp." This begs the question of why, in comparison to other plant foods, the baobab should be the clear exception to the gender distinction.

Given the value of the baobab's fruit pulp as infant food, we would expect, as Marlowe argues in his critique of the provisioning explanation for human pair-bonding, that a significant portion of the baobab and honey brought back to camp is brought by men whose wives have dependent children. Citing Ember and Ember (1979), Marlowe pointed out that "cross-species analysis of birds and mammals revealed that species in which parental care interferes with a mother's foraging were more likely to form pair bonds," and this is the explanation Marlowe (2003) presented for human pair-bonding. In measuring the mean foraging returns of the Hadza as measured by daily Kcal of food brought into camp, Marlowe (2003:221) concluded: "Among those with an offspring under three years old, men accounted for 58% of daily Kcal, and 69% of daily Kcal among those with an offspring under 1-year-old. These data alone suggest that men might be provisioning in a way that benefits their wives when they have young offspring" – they bring a lot of honey and baobab fruits back to camp. A Harvard University website posted a photograph taken from behind of a Hadza man carrying in each hand about 20 baobab fruits held together by their stems. The question is "Why do they bring back so much baobab fruit?"

Explanations for Male Baobab Provisioning

A number of reasons have been given for why men bring back substantial quantities of baobab fruit to camp, and these reasons are more often complementary rather than contradictory (Marlow 1999; Marlowe, 2010; Blurton Jones 2016). Men acquire reputations for their prowess as hunters and honey providers, but equally important – though unsung – is the fact that over the course of a lifetime, they are steady providers of baobab fruits.

Wife's Supplemental Provisioning

The baobab fruit is loved by all and is especially important for children at the height of the dry time and in the spring hunger season. Wives' supplemental baobab provisioning benefits a woman's health and her ability to care for nurslings, toddlers, and older children. Weaning foods, including the contribution of the baobab, benefit a woman in shortening her interbirth period (thus increasing her reproductive rate) because weanlings can be left with alloparents in camp. This benefits the fitness of a woman, her husband, and their children. If it was just a matter of taking something – anything – back to camp (rather than return empty-handed), then men should also return with more roots and with more berries and other fruits than has been reported (Marlowe 2003:220). Berries take time to harvest and roots take time to dig. In addition to nutritional value, men supply baobab fruits to their wives because they are easily gathered from the ground and carried in large quantities by their stalks.

Baobab as a Weaning Food

Reports in the literature make it abundantly clear that the Hadza use baobab fruit as a weaning food of major importance, with implications for the reproductive success of individuals and kin groups and for their overall reproductive success as a people. It is well known that humans wean earlier than other African apes; Kennedy (2005:140) explained the evolutionary basis for this in terms of selection for intelligence. His theoretical perspective is within the framework of "man-the-hunter," and this means that for weaning food, he emphasized only meat:

> This argument [for the significance of early weaning in hominin evolution] supports Barry Bogin's (1997) suggestion that in early societies, the weanling could best be supported on a diet including marrow, brain, and other animal tissues acquired initially through scavenging and later by hunting. It seems logical that the beginning of this pattern appeared with the origin of more active meat-seeking behaviors, by ca. 2.5–2.6 mya. This shift provided the nutritional support for the initial expansion of the brain, and the origin of the genus *Homo*, which the fossil record suggests occurred soon after that time.

Kennedy's account overlooks the importance of the baobab's fruit, whose value as a common weaning food of the Hadza is its desirable taste, high nutrient value, seasonal abundance, and easy procurement.

Babies and toddlers are completely dependent. For Hadza babies, the shift from breast milk to regular food begins when they are about 6 months old; they are weaned when they are 2–3 years old. Weanlings are usually left in camp with caretakers such as older children or elderly Hadza because they are too large to be carried easily and too young to walk long distances where the women forage. But as Crittenden (2016:321) indicated, Hadza children "are active foragers in their own right and are able to collect approximately half of their daily caloric intake by the time that they are 5 years old (Blurton Jones et al. 1989, 1997)." Crittenden continued, "Children tend to focus on resources that are relatively easy to collect (for example berries, fruit, nuts, and/or small game animals and birds) and located close to camp (Crittenden et al. 2013; Crittenden and Zes 2015)."

Marlowe (2010:120) identified baobab-sourced honey and baobab fruit as two of the most important Hadza weaning foods. In their account of Hadza weaning foods, Jelliffe et al. (1962:25) reported that "Rendered soft fat, as from the zebra, and bone marrow, both raw and cooked, are introduced in the early months," and this is "followed by a thin gruellike mixture made of the uncooked powder, or the ground seeds, of the baobab fruit (*Adansonia digitata*) mixed with water." And when "the infant has 2 to 4 teeth, prechewed meat will be fed by the mother, and by the age of 18 months the full adult range of foods will be consumed."

Baobab fruit pulp and seeds are preferred foods for children of all ages, and according to Woodburn (1970:40–41), the cooked pulp-and-seed flour "is valued especially as an infant food." The account of Jelliffe et al. (1962:25) suggested that the "thin gruellike mixture of baobab flour" made from the fruit pulp and seeds is not cooked. Peters (1979:269) cited the work of Jelliffe et al. (1962:25) and referred to "uncooked [baobab] gruels" that Hadza women prepare "for all to eat." But for this important use of the tree, baobab food as "a thin gruellike mixture" or "a gruel" is clearly not an adequate conception. The Hadza are weaned mainly on baobab milk and other beverages, on baobab paste, and on baobab porridge. Tomita (1966:169) reported that the pulp is placed in water and stirred by hand to hasten its dissolution. "This juice the Hadza call … Hadzapi milk and [they] give it to their babies… It is given to babies in place of milk." It is because of the baobab that the mosaic savanna environment can appropriately be identified as the hominin ancestral landscape of "milk and honey."

Provisioning of Small Children

Current research suggests Hadza men bring baobab fruit back to camp to provision wives who have dependent children. Men's baobab provisioning of their households declines when older children, including teenagers, are capable of collecting their own baobab fruit. The teen years, in particular, mark a transition from receiving care from others to providing care to others; this includes the collection of baobab fruit.

Convenience

Baobab fruit is mostly gathered from the ground, and Hadza men are able to pick it up in large quantities. They bring baobab fruit back to camp in part because it is easy to collect and carry. The men do not have to go out of their way, since they already visit the baobab to collect honey and water, to harvest bark fiber, and to target game. Unlike berries and roots, baobab fruit can be carried without containers, using the long, firmly attached stalks as handles. This is consistent with the finding of Codding et al. (2011:2502); their study comparing Aché, Martu, and Meriam showed that "men's and women's foraging interest converge when high-energy resources can be reliably acquired, but diverge when higher-energy resources are associated with higher levels of risks."

Self-Provisioning Old Men

Marlowe (2010:160) noted that old men collect baobab fruit and berries, "often just enough to feed themselves." But women told Blurton Jones (2016:421) that husbands hunt, collect honey, trade, and rest. "While grandmothers were described as digging, or by one woman as baby-sitting, grandfathers were described as sitting and growing old or getting baobab for the toddlers."

The Baobab's Importance at the September–October Height of the Dry Time

Many things can affect the duration and magnitude of each year's baobab harvest including weather, disease at any point in the process from flowering to fruit maturation, and fruit competition from baboons, neighbors, and commercial traders. In addition, interannual variation in supply and changes in harvesting technique through the year impact Hadza access to the fruit. Gathering baobab fruits is likely hardest at the mid-autumn start of the baobab harvest since the fruits are just beginning to ripen and do not readily fall from the tree. At this time of year, a limited number of fruits can be gathered from the ground, where they are picked by hand or with a reaping stick or by throwing sticks to dislodge them. In the early dry time of May–June, it is possible that the size or shape of the tree or the presence of neighboring trees might allow Hadza men to harvest some fruit by climbing the tree. This is probably also the time of year when baobab fruits are most harvested by men using freshly cut sharpened pegs to climb the tree. We know that the Hadza routinely climb baobab trees for honey; it is often assumed that they do so only to obtain honey. But Sherry and Marlowe (2007:111) reported that they also climb the tree using climbing pegs to harvest the fruit. At the height of the dry time, from the end of winter in September through the early spring in October, it is likely that the seasonal decline in the availability of fruit might lead to climbing to harvest sufficient fruit. During the peak winter harvest, however, fruits are readily collected from the ground by women, men, and children.

The period identified in the literature as "the late dry time" is generally the September–October height of the May–October dry time, and the seasonal passage of this period marks the annual transition from winter to spring including dry to wet conditions and old to new growing seasons. Despite the extreme dryness of September and October, this is when the first signs of spring, in the form of flowers and leaves, signal the approach of the November–December spring rains. According to Wickens and Lowe (2008:79), "The fruits [of the baobab] are ripe by the end of the dry season." But as noted earlier, a ripening in September and October is not consistent with current field research. It is more likely, as has been argued, that the period of September and October is the end of durational and dry-time differentially delayed fruit fall. The falling of baobab fruit starting in November reflects the extent of a declining differentially delayed fruit fall through the wet half of the year.

Contrary to Wickens and Lowe's (2008:79) identification of September–October as the peak harvest time of the baobab, this is when the baobab's fruit-fall harvest begins to decline. Reports suggest this period is when the amount of ripe fruit available varies most from year to year. The assumption here is that when durational fruit fall and dry-time differential fruit fall end in the September–October period, wet-time differential fruit fall continues, but in declining amounts, and the procurement costs of gathering the fruits would likely increase. Procurement costs are higher because camps are sited to take advantage of water-hole ambush hunting, and all baobabs within reasonable distance are harvested first. The increasing travel distance associated with intensifying baobab forays means that instead of bringing back whole fruits to camp, the Hadza would extract the pulp with the embedded seeds from the pod in the field, allowing for a greater quantity of fruit to be taken back to camp. The September–October decline in the amount of ripe baobab occurs along with that of winter berries; during this time of year, the Hadza are essentially dependent on the meat from large game acquired from water-hole ambush hunting, on roots as the most dependable but least favored of their five preferred staple foods, and on the start of the spring-summer berry harvest.

Within the framework of the "seasonal polarity" perspective that Woodburn emphasized (1968a), Marlowe (2010:108) seemed to suggest that the time of the year when the Hadza would be expected to be "seasonally stressed" would be the dry time generally, but in doing so, he overlooked the importance of the baobab harvest from May to October, especially the main harvest period from May to August. He wrote: "One might think the dry season would be a lean season, but people do not lose weight during the dry season, mainly because at that time men take more meat through ambush hunting at water holes."

According to Hawkes et al. (2001a:683), however, the September–October height of the dry time is a period of "rigor" for the Hadza. In a later publication, they reported that "The sharpest seasonal contrast [in Hadza life] was between the late dry, when hunting returns were best (largely because of nighttime ambush hunting success), and all other seasons, when hunting was relatively poor but plant foods more widely available." The key point that Hawkes et al. (1997:557) made was this:

> Almost all of the variation in children's weight changes is associated with [the September-October height of the dry time]... and [November as the start of the annual rains] ... During [the September-October period] ... children lost weight. With the onset of the wet season and the related increase in their own foraging efforts (as well as those of their mothers and grandmothers), they made larger weight gains than at any other time of the year.

The onset of the spring rains in November, Hawkes and colleagues reported (1997:557), marks "the beginning of a pattern of long-distance trips to collect Cordia (ondishibe) berries." They wrote:

> Children routinely accompanied adults on these trips; their foraging times jumped sharply relative to those of the preceding late dry as a result. They foraged less in [December as the turning month marking the transition from the end of spring to the beginning of summer] ... than in [the November start of the annual rains] ... but more than in any other season, a pattern that may reflect continued recovery from the rigors of the late dry.

Baobab and Children's Productive Activities in September–October

Documenting the annual variability in the September–October baobab harvest is clearly important in understanding the ecology of Hadza life. This is evident in the discussion of children's productive activities. Evolutionary scholars interested in the foraging activity of hunter-gatherer children (e.g., Blurton Jones et al. 1994, 1997) have compared the Dobe !Kung and Hadza (Blurton Jones et al. 1993). Broadly speaking, both groups are of the African savanna; containers, digging sticks and poisoned arrows are key components of their traditional productive activity; and they share the use of many plants, including the baobab. In putting the foraging of Hadza children into an evolutionary perspective, Hawkes et al. (1995:688) wrote:

> The assumption that children of mobile foragers provide little of their own food is a common and important (if often implicit) element of many arguments about major developments in human prehistory, from the origin of bipedalism through the development of agriculture. Its principal support comes from the Dobe !Kung. Other, less well-known but equally mobile, groups display different patterns, with children foraging actively and contributing significantly to their own support. These "exceptions" challenge the common assumption and any arguments that depend on it. More important, they provide an opportunity to identify the determinants of variation in children's foraging practices. The more successfully that variation can be explained, the better the foundation for hypotheses about its effects in the past.

Hawkes et al. (1995:688) argued for an alternative view based on their fieldwork. They report, "Among the best-known exceptions to the proposition that children of mobile hunters do not forage are the Hadza, whose children have been described as active foragers throughout this century (e.g., Obst 1912; Bleek 1931; Kohl Larson 1958; Jelliffe et al. 1962; Woodburn and Hudson 1966; Blurton Jones et al. 1989)."

According to Alvard (2003:138), "Observations show that !Kung children forage little until they are well into their teen years. In contrast, Hadza children are active foragers from an early age (<5 years)." Crittenden (2016:321) noted that Hadza children "are able to collect approximately half of their daily caloric intake by the time that they are 5 years old (Blurton Jones et al. 1989, 1997)." She also reported that children tended to focus on easily obtainable resources close to camp including hunting small animals and birds and gathering berries and baobab fruit (Crittenden et al. 2013; Crittenden and Zes 2015). In accounting for the foraging difference between Kung and Hadza children based on his evaluation of the literature, Alvard (2003:138) wrote:

> Using a detailed analysis of empirical and experimental foraging data, these researchers have shown that ecological differences, primarily the spatial distribution of food and water resources, and the differences in the processing requirements of the different foodstuffs gathered by these groups, explain the observed behavioral variability. There is a lack of nearby resource opportunities for !Kung children comparable to the easily accessible Baobab patches available to Hadza children, and return rates for the Kung children are correspondingly low (Blurton Jones et al. 1997).

Alvard's (2003) perspective is not an accurate account of Hawkes et al. (1995:694), because they were not emphasizing proximity of baobab fruits and other resources to camp as the determining factor in foraging by children. They presented "the simple hypothesis that children seek to maximize their mean rate of nutrient acquisition while foraging" (Hawkes et al. 1995:694), and this is where the seasonality of the baobab harvest becomes important, especially in the September–October height of the dry time when the harvest is declining.

In their initial 1986 report on Hadza children (5 and older) foraging in the late dry season near camp in a locality called /ilo, Hawkes et al. (1995:694) indicated that they gathered *makalita* [*Eminia atenullifera*] berries, *//ekwa* [*Vigna frutescens*] roots, baobab fruit, and a kind of honey called *kanoa*. In the late dry season of 1988, however, things were very different:

> In the 1988 study period Hadza youngsters had a limited set of options: staying near camp and gathering *makalita* and/or *//ekwa* or traveling with adults to the distant berry patches and gathering *tafabe* [*Salvadora persica*] and/or *ondishibe* [*Cordia* sp.]. The resource that figured so prominently in the 1986 sample, the baobab, was essentially unavailable to children foraging near Mbea and Ilomo during the 1988 observation period. There were many baobab trees near both camps but little fruit in that season … Given only the tuber options near camp, the choice is straightforward: *makalita* yield 436 cal/hr., *//ekwa* 267 Cal/hr. Not surprisingly, children generally chose the *makalita* patch.

Reports of the baobab harvest during the September–October height of the dry time suggest it is highly variable – abundant in some years and scarce in others. Even in leaner years, the baobab harvest in the September–October period and through the spring, in combination with the September–October big-game feasting, is important to the seasonal year of the Hadza.

Summary

The baobab's long fruiting season, coupled with the abundance of dry storable fruits it produces, with their nutritious fruit pulp and seeds, has made it the Hadza's most important fruit tree. This use of the tree would, no doubt, also have made it the most important fruit tree for early hominins of the African savanna. While this chapter focused on the baobab's food value as a fruit tree, the next four chapters explore the food value of the baobab as a source of drinks, honey, birds, and meat.

Chapter 11
Baobab Beverages

Beverage, even if only in the indispensable form of plain water, is a staple in the diet of all people. This is especially true in the tropics, where many kinds of beverages have developed in response to high temperatures and the need for hydration, and the abundance of fruits from which drinks can be obtained naturally, or by pressure, dilution, liquidation, or maceration. The baobab provides a variety of beverages. In addition to water, and to honey consumed as a liquid, any drink with baobab in it, whether as the principal ingredient or a contributing ingredient, is considered a baobab drink. This chapter suggests the baobab can justifiably be identified as one of the most important beverage trees of the African savanna; it introduces the baobab as a source of different kinds of drinks generally and among the Hadza in particular.

Because the terms used to discuss baobab drinks have never been systematized, their use in ambiguous ways is unavoidable. A simple system of beverage categorization is provided here as a framework for understanding the significance of the baobab as the Hadza's beverage tree and to compare Hadza baobab drinks with baobab drinks in other parts of Africa. The categories used are standardizations of what already appears in the literature and popular media, with modifications made where necessary to provide a consistent view. With this in mind, the four most recognized classes of drinks worldwide are thirst quenchers, nutrient drinks, alcohol, and stimulants, and the baobab is a source for all four. There are, broadly speaking, eight kinds of baobab drinks, two of the most important of which are water and honey. With stored water and honey, the baobab is a mediating factor rather than the direct source of these drinks. The other six baobab drinks are obtained from some part of the tree itself. Two of these involve the flower in the form of flower drinks and floral nectar; the remaining four are obtained from the baobab's fruit pulp or seeds, often in combination with other ingredients.

J. Rashford, *Baobab*, https://doi.org/10.1007/978-3-031-26470-2_11

Baobab Beverages Derived from Fruit

The acidic pulp of its fruit is one of the most important food resources of the baobab. Its use to make drinks reflects both the nutritional value of the pulp itself and its augmented nutritional value when combined with other ingredients such as baobab seeds, other fruit, cereals, milk, or honey. In scientific and popular accounts, references to the use of the baobab fruit in drink-making usually focus on the pulp and a single beverage, often identified as "baobab drink" or "baobab juice." The difference, if any, between these two names is never made clear; judging from the discussion associated with these terms, they are in most cases synonyms. For example, in an approach that underestimates the baobab's importance as a beverage tree, Adedayo et al. (2011:83) presented a conception of the baobab as the source of a single drink of highly variable form, which they identify as "the baobab fruit drink":

> One of the Nigerian local refreshing and nutritious drink is the baobab fruit drink. The baobab fruit drink, being a direct replica of the fruit pulp in water is a natural fruit juice with all the necessary vitamins, minerals and other essential nutrients required by the body in the right proportion that meets the daily body requirements of those nutrients. Usually the drink is made by dissolving the fruit pulp in either hot or cold water or mixed with milk hence this study is carried out to ascertain the nutritional values of the baobab fruit drinks made by dissolving the fruit pulp in cold water (at ambient temperature).

Not recognizing that the baobab is the source of different kinds of drinks depending on the amount of water added and the combination of ingredients used obscures the fact that even the water-based pulp drink is variable. Chadare et al. (2009:347) noted that in Benin, "The consistency [of a commercial water-based baobab drink] is adjusted according to consumer preference," and they go on to point out that various beverages are also "prepared with a cereal to which baobab is added." Freedman (2018) was aware of the problem of identifying the baobab as producing a single kind of beverage when he noted that the seeds are "used in [the] preparation of a beverage" and the "pulp [is] used [as] a beverage base."

Instead of a single drink of variable form, the approach here is to recognize the need to clearly distinguish the great variety of baobab drinks reported from all over Africa. From the perspective of the four classes of drinks that provide the framework for this discussion, the kinds of baobab drinks suggested above by Adedayo et al. (2011:83) could be classified as a baobab thirst quencher and baobab nutrient drinks in the form of baobab juice and baobab milk. But the distinction between thirst quencher and juice is not made clear by Adedayo and colleagues, and it should be. It is hinted at, however, when the authors report that their study was focused on "the baobab fruit drinks" made with water, even though they never say what these different water-based baobab drinks are; they describe instead the lab procedures followed to determine the mineral composition of what could either be considered a baobab thirst quencher or baobab juice. Baobab drinks differ based on the use of the fruit pulp, seeds, or a mixture of the two; the ratio of baobab ingredients to water or milk, affecting the thickness of the drink; the variety of other ingredients used, including honey, sugar, or other sweeteners; whether the drinks are consumed cold or hot; and whether they are fermented.

Baobab Thirst Quencher

In the literature and in popular accounts, the plain baobab thirst quencher made from the fruit pulp is one of the most frequently mentioned kinds of baobab drink in Africa. According to Lanzara and Pizzetti (1978:77) and many others, the pulp is said to "produce a drink like lemonade." In fact, the pleasantly tart fruit pulp is the reason the baobab is also called the lemonade tree, sour gourd tree, or cream of tartar tree. It also accounts for the taste-associated names that identify the baobab in the Caribbean and in India as a kind of tamarind, the tamarind being a sweetly acidic fruit that is widely used throughout the tropics to make refreshing drinks (Rashford 1994a; Groote et al. 2007; Leach 2011). Moiselet (1998) described the use of the baobab fruit pulp in Senegal to make "drinks for children," and he noted that the pulp "mixed with water … is similar [in appearance] to coconut milk with a taste of lime." Coconut milk, a white liquid produced by mixing the grated endosperm of the coconut with water, is a common ingredient in cooking. Based on this comparison, what Moiselet (1998:4) described is more like a nutrient drink in the form of a pure baobab juice, than a baobab-flavored thirst quencher (and it is worth noting that "thirst quencher" is a more apt name for coconut water, which is the liquid endosperm of the immature coconut).

The baobab thirst quencher is a flavored and often sugar-sweetened drink that is more watery than other baobab fruit drinks. In this sense, it is similar to the well-known fruit-flavored drinks made from lemon, lime, orange, and other citruses. Tamarind and lime are widely used for thirst-quenching drinks throughout the tropics, but not for making a thick juice or very thick "smoothie." Unlike the fruit of the baobab, which can be eaten at all levels of concentration, the fruits of the tamarind and lime are too acidic to make drinks; they have to be greatly diluted. Theoretically, however, it is useful to make a clear distinction between a baobab-flavored thirst quencher and the thicker baobab juice and baobab milk nutrient drinks.

Baobab Nutrient Drinks

The most important consumer beverages around the world today are thirst quenchers, hot stimulants, and alcohol. The tart fruit pulp of the baobab is especially associated with the global commodification of the fruit, and published accounts suggest thirst quenchers and nutrient drinks are far more familiar than baobab stimulants and alcohol. However, the category of nutrient drinks is a rapidly emerging interest, given current health concerns. The making of nutritious drinks is one of the most important uses of the baobab's fruit pulp and seeds, and they include baobab juice and mixed nutrient drinks, the latter represented by such drinks as baobab milk and baobab smoothies. Baobab juice and mixed nutrient drinks have more fruit pulp than the baobab thirst quencher, but not as much as the baobab smoothie.

In contrast to the pulp combined with the seeds or with cereals, baobab drinks are also made from the seeds alone combined with other ingredients. An example of baobab/cereal mixed nutrient drinks involving a seed/cereal combination is given by Dalziel (1937:114) who reported that "a fine meal – Hausa gavin kuka – of the seed kernels mixed with meal of millet, etc., is made into thin gruel or drink." Dalziel's use of "etc." suggests there are other ingredients used with the seed flour in making porridge and drink.

True Baobab Milk Drinks

Because most accounts note that the baobab fruit pulp is dissolved in water or milk to make beverages, it is important to identify a specific category called baobab milk. There are already baobab drinks identified as "baobab milk" in publications and on the Internet. However, only such combinations involving real milk are categorized in this work as baobab milk, and these baobab/milk mixtures can involve pulp or seeds (or both), as well as other ingredients such as cereals and honey. According to Wickens and Lowe (2008:75), for example, "Pounded seeds mixed with milk, boiled and flavoured with honey make a sweet drink in Zimbabwe (Monteiro 1875; Hedrick 1972; Brenan 1949; Roodt 1995)." Pander (2015) has told the story of Levison Gombachika from Chirimba, a township of Blantyre in Malawi. "He remembers an easy recipe he particularly liked as a child because it tasted like a very good dessert to him. Levison took Baobab power, mixed it with milk and added a little bit of sugar."

Baobab milk is a mixed nutrient drink that includes pulp/milk and seed/milk mixtures, fermented baobab milk (which is to say, baobab sour milk), and baobab milk smoothies. An early influential account of a kind of baobab milk was presented by Nicol (1957:287) who, in addition to noting that the cattle-owning Fulani of northern Nigeria produce a baobab milk drink, wrote:

> Milk and baobab fruit juice mixtures are popular drinks with Hausa farmers. This drink is always available if there is money to purchase it, particularly during the hot time of the year (October-April) when new farms are being cleared or hoeing of old farms is taking place, preparatory to sowing.

Wickens and Lowe (2008:73) cast doubt on the idea that the water-dissolved baobab liquid was used as an adulterant, and citing a number of published sources, they noted that "other pastoral [groups] … in West Africa, Zanzibar and Tanzania are reported to use the pulp to curdle milk." Touching on this point, Bankole (2006:173) described the Nigerian drink called "nono" as "a product derived traditionally from naturally fermented milk," and he noted that it "has considerable economic importance in the northern part of Nigeria as it is the main fermentation product of fresh milk appreciated by indigenes." Far from being an adulterant, Bankole has explained that "In the norther parts of Nigeria, 'kuka' (cream of tartar of baobab fruit) is often added to 'nono' to thicken 'nono' for sale and as a spice to enhance its taste" and that this commercial baobab milk drink was also a product of fermentation.

Look-Alike Baobab Milk Drinks

In addition to genuine baobab milk drinks, the name "baobab milk" also identifies look-alike baobab milk drinks made by dissolving the pulp in water, with other ingredients often added, especially baobab seeds and cereals. According to Bosch et al. (2004), for example, the plain pulp "is made into a milk-like drink called 'gubdi' in the Sudan." Writing of a Malawi drink, Owen (1970:31), who cited Williamson (1955:12), reported that the seeds, "when scraped and pounded, mixed with the fruit pulp and boiled with water … make a good sweet 'milk'." Obizoba and Anyika (1994:157) have noted that in Nigeria a highly nutritious "baobab milk" can be made from combining the baobab fruit pulp with the flours from the baobab seeds and acha (*Digitaria exilis*). In a nutritional analysis of this look-alike milk of pulp, seed, and cereal, they wrote:

> The mixtures were composed of 70% acha and 30% baobab [pulp/seed] flours (70:30 protein basis). Baobab milk contains more protein (1.5%) and mineral (Fe: 17.8 mg; Ca: 134.2 mg) than human milk (protein: 1.3%; Fe: 0.2%; Ca: 30 mg) or cow milk (Fe: 0.1 mg; Ca: 1.20 mg) and most leading national commercial infant milk powder formulas e.g. Cerelac (Fe: 10.0 mg). The composite flours contained more nutrients than the baobab or the acha flour alone.

Chadare and co-workers (2008:347) identified two kinds of pulp/millet drinks in their discussion of baobab food products in Benin. One was described as "mixing cereal dough with water to which baobab pulp is added," and the other as "baobab pulp, locust bean pulp and water" combined.

Baobab Smoothie

As a class of drinks, the smoothie is identified here simply as a thick nutritious drink associated with health and overall wellness made with fruit combined with water, fruit juice or milk, and sometimes yogurt or ice cream. The plain or mixed baobab smoothie, compared with the thin thirst quencher and the moderately thick juice and mixed nutrient drinks like baobab milk, is the thickest baobab drink of all. In Malawi, a bottled baobab drink called Malambe Super Baobab Fruit Juice is produced. Based on the description presented in his taste review, Giglio (2010) identified this drink as a bottled pulp smoothie, characterizing it as more like a liquid meal than a refreshing drink. Unlike other researchers, Giglio's view is that "baobab juice" is always thick and therefore more like a smoothie. If this were true, there would be no baobab thirst quencher or nutrient drinks. The thickness of baobab beverages is an empirical question, however, especially when the making of drinks might be a matter of tradition or of mixing to preferred consistency and taste.

Other Kinds of Baobab Drinks

Baobab Flower Drinks

Baobab flower drinks and baobab floral nectar constitute one of the least explored categories of baobab drinks. Atkin (1951:459) reported that the flower was "sometimes crushed with sugar and water to make a refreshing and unusual beverage." Eating the raw flower and using it to make a beverage have also been reported for the Sudan (Gebauer et al. 2002:158) and for Benin and Burkina Faso (Wickens & Lowe 2008:95, 70). According to Wickens and Lowe (2008:95), "herd boys [in Namibia] seek out the nectar (Toutain 1978; Mshigeni and Hangula 2001; Gebauer et al. 2002)."

Baobab Coffee

Baobab thirst quenchers and nutrient drinks are more commonly reported in the literature than baobab alcohol or baobab stimulants. The category of mood-altering stimulants is one of the most commercially important kinds of drink worldwide and includes the familiar coffee, tea, cocoa, and yerba mate. The baobab is the source of a hot stimulant produced from its seeds, which are roasted, ground, and brewed with hot water to make a coffee-like beverage (Olaitan et al. 2014). There are many references in the literature to seed-roasted baobab coffee. According to Gebauer et al. (2002:158), for example, it is a coffee substitute in "some areas" of Sudan; Wickens and Lowe (2008:75) noted that "Well-ground roasted seeds are … widely used as a coffee substitute." Accounts of baobab coffee are often too general to determine if baobab coffee is drunk in addition to real coffee, as an ingredient mixed with coffee, or as a substitute for coffee. However, the Slow Food Foundation (2018), a group assisting small-scale producers in over a hundred countries "to promote a model of agriculture that is based on local biodiversity and respect for the land and the local culture," reported on their website that a substantial baobab coffee tradition in East Africa:

> Mambuyu is locally referred to as "our wild coffee" by people of the Kamba tribe in the dry area of Machakos, in southern Kenya. It is made from seeds of the baobab tree, locally called the mbuyu tree. The seeds are harvested twice a year, in February and in December, and often children, who are attracted to the color of the seeds, assist in the harvesting. Women then dry and grind the seeds, and are the ones who prepare the mambuyu.
>
> To make this product, the powder of the ground seeds is boiled with water to produce a hot beverage drunk in the mornings like coffee. The seeds can also be fermented to create an alcoholic drink enjoyed by elders. Pregnant Kamba women once commonly drank mambuyu for its nutritional value. Mambuyu is also among the traditional gifts brought to a home after a baby is born. The drink is said to be an excellent source of energy, providing this population with the energy to travel long distances for trading.

> Because coffee used to be expensive the Kamba people believe that mambuyu is a gift from God, and so the locals used to water the baobab trees every morning to take care of them. Today, approximately 300 kg of mambuyu is produced annually, mainly for personal or community use, though the prepared drink can occasionally be found at local markets.

The *Manica Post* (2017), the largest provincial newspaper in Zimbabwe, posted an article on its website titled "Baobab coffee producers think big." It reported the commercial production of baobab coffee had been proceeding over the past ten years.

Baobab Alcohol

Citing Pipe-Wolferson (1988), Booth and Wickens (1988:13) reported that baobab flowers are "used to make a liquor." In Tanzania, baobab pulp has also been identified as an ingredient in the fermentation of sugarcane juice to make beer (Fleuret 1980). Wickens and Lowe (2008:72) noted:

> The Shambas of the Usambara Mountains, Tanzania, use baobab fruit pulp together with the fruits of the sausage tree (*Kigelia aethiopum*) and the wild olive (*Olea europaea* subsp. cuspidate, syn. *O. chrysophylla*) as a fermenting agent for sugar cane juice. This concoction is reputed to produce an alcoholic beverage with 7% alcohol content in 24 h. The brew is regarded as very important for marriage ceremonies and cooperative work groups as well as being a valuable source of income for the cane growers.

According to Jelliffe et al. (1962:25), the Hadza consume honey in its natural state and berries and other fruits are eaten raw. But meat and roots are roasted since "only older women are permitted to have cooking pots in which they boil meat." Jelliffe and colleagues seemed not to have been aware that Bagshawe (1925:121) had earlier reported, "Cooking pots are never used except for making beer." Bagshawe went on to note that "Though great honey-hunters [the Hadza] … do not make honey-beer. They thoroughly appreciate all intoxicants given to them, but the only drink they manufacture themselves is a mild beverage made from the seeds of the baobab tree." There are no reports today of the Hadza making baobab beer. It is likely that the making of a baobab beer has declined or ended because of the ready availability of the locally produced alcoholic drink called "pombe," which is made from corn and sold in nearby villages.

The Bouye Drink of Senegal and Gubdi Drink of Sudan

One of the best-known traditional baobab drinks of Africa is bouye of Senegal. The Senegalese word "bouye" is frequently identified as a kind of "baobab fruit drink," and so it is. But bouye is not a single kind of drink; it is the general name for different kinds of traditional and contemporary baobab drinks made in Senegal today. According to Cox (2008), bouye uses the baobab fruit pulp to make "a refreshing beverage similar to lemonade." By contrast, the author of an article on the local juices of Senegal on the website "Eat Your World" (which presents itself as a "global guide to local food") described bouye as a fruit-pulp drink that was "thick, sweet, and tangy, with the slightly chalky consistency of a banana shake" (Anonymous 2018a). If bouye can also include a pulp/milk combination, as some sources have suggested (Anonymous 2018b), it would not be similar in density to lemonade as a sweetened, water-based thirst quencher, but would be more akin to a baobab milk shake. Other descriptions of bouye identify it as a pulp/seed nutrient drink. Judging from published descriptions of bouye, it appears to be a popular name for different traditional baobab drinks of Senegal, represented by baobab thirst quenchers, and for nutritious drinks such as baobab juice and baobab mixed drinks, the latter including baobab milk, and baobab fruit pulp and seeds combined.

Gubdi is a drink widely associated with the Sudan, and, like bouye, it appears to be a general name for different kinds of baobab drinks. According to Nicol (1957:287) and Wickens and Lowe (2008:71–72), gubdi is a Hausa word. The water-based pulp drink that Wickens and Lowe identified as gubdi could be considered a plain baobab thirst quencher or juice or a mixed nutrient drink. According to Wickens and Lowe (2008:71–72), "*Gubdi*, supplemented with hungry rice (*Digitaria exilis*) flour, is used as a milk substitute for babies."

The Hadza and Their Baobab Beverages

Not only is water not the Hadza's "only beverage," as Bleek (1931:278) claimed, but making drinks is one reason for the use of containers to bring water back to their residential camp. In the heat of the tropical savanna, nothing is more refreshing than shade, a breeze, and a cool drink. Because the baobab is associated with all the traditional drinks of the Hadza, it can justifiably be identified as their most important beverage tree; this is especially so when we consider that in addition to baobab-derived drinks, the tree is also a source of water and of honey. The baobab is likely one of the longest-serving beverage trees for human beings. To live with baobab trees as the Hadza do today and as ancestral hominins did is to live in a land of water, milk, and honey, three of the many important resources the baobab provides.

As earlier indicated, the fruit pulp of the baobab can be used to make a variety of refreshing drinks, depending on the ratio of water to fruit pulp and the addition of other ingredients. In his account of a Hadza baobab drink, Tomita (1966:169) noted that after the pulp is removed from the pod, it is "squeezed and crushed by hand in water to make juice," which he identified as Hadza milk. This, of course, could be the plain baobab thirst quencher, or the more concentrated baobab juice, depending on the ratio of fruit pulp to water. For the Hadza, according to Marlowe (2010:120), "Honey is eaten raw, though it is sometimes mixed with the baobab," and he has suggested the Hadza made a single baobab drink in the form of a variable mixed nutrient drink:

> When [the baobab] … is taken back [to camp], the pulp and seeds are pounded into flour with a hammer stone. When a little water is added, it becomes paste, which is the main weaning food. When honey or certain berries are available, they are sometimes added to the tart baobab pulp and water to make a delicious sweet and tart drink.

It is not clear from the current documentation if Hadza baobab drinks also include the pure fruit-pulp thirst quencher in contrast to juice or if the drink Marlowe described should be considered a baobab smoothie because of its thickness. The Hadza are known to collect baobab seeds from baboon dung. Given the absence of the pulp, it would be interesting to know if these seeds are ground to make a pure seed drink or a drink of the seeds mixed with other ingredients, such as honey and berries.

The New Superfruit and Baobab Drinks Today

Insofar as drinks and drink-making was a feature of forager adaptation, baobab drinks would likely have included baobab juice and mixed nutrient drinks made from the fruit pulp and seeds combined with berries and honey as major ingredients or as flavorings and sweeteners. With the development of agriculture came an elaboration of baobab nutrient drinks. Pastoralists added baobab milk to the mix and farmers added their cultivated cereals. Given the growing concern with healthy living, baobab fruit is now regarded as a new "superfruit" alongside the now familiar açai berry (*Euterpe oleracea* Mart.), blueberry (*Vaccinum* spp.), cranberry (*Oxycoccus* spp.), goji berry (*Lycium barbarum* L.), mangosteen (*Garcinia mangostana* L.), and pomegranate (*Punica granatum* L.). "Superfruits" is a marketing term without official definition in major consumer markets or among regulatory authorities. Nevertheless, in the world of commerce and in the popular imagination, superfruits are fruits judged to be especially nutritious because they are high in antioxidants, fiber, vitamins, and minerals; they are now commercially promoted worldwide as preventing disease, improving overall well-being, and increasing longevity. Jackson and Maldonado (2015:49) wrote:

> When comparing the micronutrients found in baobab to those found in superfruits, one must take into account that values for baobab are for the dried fruit pulp, whereas the data for superfruits are for raw foods. The vitamin C content of baobab dried fruit pulp is up to five times higher than that of raw blueberries and 15 times higher than that of pomegranates. It has much higher levels of niacin (vitamin B3), slightly higher levels of vitamin B1, and about the same amount of vitamin B2 as the selected superfruits. Baobab dried fruit pulp has also been found to contain greater quantities of calcium, magnesium, and iron. It is worth noting that removing water from fresh fruit concentrates the nutrients, so it may be an unfair comparison. However, as baobab is natural dry, the situation is unavoidable.

Today "baobab fruit powder" is sold on the Internet and in stores. It is advertised as ideal for making thirst quenchers; nutrient drinks, including juice and milk; and plain and mixed smoothies (Yu 2016). Commercially products in which baobab fruit pulp is the chief ingredient or a contributing ingredient already include baobab-flavored drinks, sodas, sports drinks, smoothies, and yogurt. In addition to beverages, Gruenwald (2009:212) notes that the pulp can be used in "fruit filling, jams, sauces, puddings, and deserts. Clearly," he wrote, "the possibilities for product development using Baobab fruit pulp are numerous." But if the baobab is now being presented to the world as a newly recognized superfruit, we can conclude it would also have been a superfruit in the beverage tradition of early hominins.

Summary

This chapter has highlighted the use of the baobab as a beverage tree not only because of the value of its fruit pulp and seeds but also as a source of water and honey. While real baobab milk drinks and baobab coffee, and perhaps also baobab alcohol, probably have historic uses, early hominins could have met their need for hydration with a baobab thirst quencher, juice, or smoothie, especially because the tree provided its own ready-made mixing and drinking cup, water, and sweetener.

Chapter 12
Africa's Honey Tree

One of the most powerful means flowering plants evolved to enlist the unwitting support of animals as pollinators and seed dispersers is an energy-rich sugary reward, often in combination with bright colors and attractive scent. Nurtured by our primate ancestry, the love of sugary foods, including floral nectar and ripe fruit, has led human beings everywhere to regard honey as the epitome of sweetness. Honey has likely been important throughout hominin evolution, judging not only from the love of honey evident among human beings worldwide but also among chimpanzees, gorillas, and orangutans and among a variety of monkeys, such as baboons and macaques. But as Crittenden (2011:257) noted, "It has been suggested that honey may have been an important food source for early members of the genus *Homo*, yet the importance of meat and savanna plant foods continue to be stressed as the most relevant foods in dietary reconstructions." With respect to theorizing the significance of honey in hominin evolution, what is remarkable about the baobab is its contribution to hominin dietary requirements as a source of water, meat, important plant foods, and large quantities of honey.

The baobab's status as a honey tree has long been recognized, and it is not surprising to learn that in addition to the tree's size, longevity, shade, stored water, and spacious hollow trunk, honey was one of the things that caught the attention of the Muslim scholar and traveler Ibn Battuta, who encountered the baobab while traveling in Africa in the fourteenth century. Baobab honey is collected from wild bees that establish their hives in the tree; an old baobab may have two or more hives in its hollow trunk and branches. The baobab/honey connection also includes honey produced by beekeeping groups that hang artificial hives in baobab trees. Honey producers in Kenya hang as many as 20 hives in a single baobab tree. Consider, for example, the people of the mountainous Usambara region of northeastern Tanzania, who generally hang artificial hives in *Acacia* trees. They "rarely keep bees," wrote Fleuret (1980:328), because their "beer is brewed from sugarcane, bananas or maize." However, on the plains surrounding the Usambara Mountains, honey is the major ingredient in brewing beer. Fleuret continued:

J. Rashford, *Baobab*, https://doi.org/10.1007/978-3-031-26470-2_12

> Beehives are most commonly made of bark or hollowed logs. *Dombeya leucoderma* (*keyu*) is preferred, as the bark can easily be slipped off a section of log in one piece. The ends of the resulting lightweight hollow cylinder are plugged with rounds cut from the log, and the finished product is hung in a tree – Adansonia *digitata* (*mbuyu*, baobab) in the lowlands … Some baobab trees in the plains may contain 15–20 such hives.

The Hadza are well known for their love of honey, and honey baobabs represent yet another essential aspect of their relationship to the baobab as their tree of life. The focus here is not only honey consumption in human evolution, but the importance of honey baobabs in human evolution. It is important to keep in mind, however, that it is the baobab's overall value as a resource-rich environment, reflecting its status as an ecological tree of life, that is the basis for its fundamental importance in human evolution. In terms of the baobab as a resource destination, this is particularly true of its unique position as the Hadza's preeminent fruit and honey tree. The chapter discusses the Hadza sources of honey, emphasizing the fundamental relation between the baobab and its bee ecology, including the honeyguide and other birds. This is followed by a discussion of Hadza honey-collecting equipment, honey collectors who fall to their death from baobab trees, Hadza use of honey, honey-collecting seasonality, and tool use and the efficiency of honey collecting.

Hadza Sources of Honey and the Baobab/Bee Complex

Matthiessen (1972) was mistaken when he wrote it was only "sometimes" that the baobab provided the Hadza with honey. This suggests baobab honey acquisition is a chance occurrence of limited importance. Current research makes it clear that the baobab is indeed the Hadza honey tree and honey is important in many aspects of their practical and inspirational life. The Hadza collect honey from seven species of bees (Woodburn 1968a; Marlowe 2010; Marlowe et al. 2014) that they identify by name as belonging to two groups (Peterson et al. 2013:80). The first is the common stinging honeybee, *Apis mellifera adansonii*, which produces the most honey. The Hadza call this bee ba-alako. The remaining six species they identify by phrase names as being kinds of ba-ala bees. These six ba-ala bees are represented by five species belonging to the genus *Trigona* (known by the English common names sweat bee and stingless bee) and a species of the genus *Lestrimellita*, which is also identified as a sweat bee.

Although these seven species colonize various features of their environment, *Apis mellifera adansonii* is especially associated with old baobab trees. Woodburn (1970:40) was unambiguous in stating "Honey is frequently obtained from bees' nests in the large natural hollow in their trunks." However, the baobab is more than just a secure nesting site for bees. Although the baobab flower is chiefly pollinated by bats, it is also visited by bush babies, moths, ants, wasps, and bees. Wickens and Lowe (2008:222) reported that bees "take advantage of the minute opening of the sepals in the late afternoon before the flowers have fully opened, to enter the bud and steal pollen (Guy 1971, 1972; Jaeger 1950; Humphries 1982)." They can also be seen visiting the flowers in early morning. In addition to gathering pollen, bees drink water from baobab catchments. So intimate is the baobab/bee association that Guy (1972:161) described the baobab as providing "many of the facets of a natural system centered around the African honeybee." He wrote: "Here in fact was a natural apiary. In attendance were the ubiquitous fork-tailed drongo (*Dicrucrus adsimilis*) and the greater honeyguide (*Indicator indicator*). Flocks of bee-eaters (*Merops* sp.) also paid regular visits."

The Honeyguide Bird

Honeybees typically establish their hives in hard-to-reach places, the most common being tall trees and cliff faces, and around the world, people go to extraordinary efforts to raid them. For foragers like the Hadza, harvesting wild honey in large quantities is a technical process that requires knowledge of honeybees, including the ability to locate hives and determine if they are ready to be harvested. The Hadza, like other East Africans, are aided in their search for honey by the honeyguide. This bird feeds on beeswax, but it depends on humans to gain increased access to its favorite food. It was long thought that the honeyguide/human relation developed from an earlier coevolved relationship between the honeyguide and the honey badger, as reported by Marlowe (2010) and Crittenden (2011), but this view has been questioned in recent years. Spottiswoode et al. (2016) and Dean et al. (1990) have presented a persuasive argument that the honeyguide's behavior most likely developed in association with hominins. The Hadza communicate with the honeyguide by whistling, and the bird responds. It leads them to the hives of the *Apis mellifera* by its agitated behavior, changing vocalization, and flying from tree to tree (Marlowe 2010:117). After raiding the hive, the Hadza leave some honeycomb at or near the baobab to reward the honeyguide. According to Marlowe (2010:117), after a Hadza hunter "opens up the hive and extracts the honey, the bird gets to clean up the leftovers." But the Hadza told Peterson et al. (2013:89): "The honeyguide eats wax and larvae . . . We are careful not to leave much wax at any one harvested hive so that the honeyguide remains hungry enough to lead us to other hives."

Based on mitochondrial DNA evidence, Spottiswoode and colleagues (2016) suggest that the honeyguide dates back some three million years and would therefore have been present in the environment of hominin evolution. It is noteworthy that the origin of the honeyguide dates to around the time when manufactured stone tools first appear in the archaeological record. The development of increasingly sophisticated stone tools would have meant a growing ability to access new resources. Stone tools, particularly the hand ax, would have provided greater access to honey, especially after the mastery of fire associated with *Homo erectus*.

The efficiency of Hadza honey harvesting made possible by the honeyguide and honey-collecting equipment allows them to access significant quantities of honey on a regular basis. It is this copious consumption of honey that has likely had a significant influence on hominin evolution as evolutionary ecologists have argued (Crittenden 2011; Wrangham and Machanda 2013; Kraft et al. 2014; Marlowe et al. 2014). In fact, honey-collecting equipment associated with the rise of the genus *Homo* some 2.5 million years ago may have been responsible for the coevolved relationship between the honeyguide and hominins. Chimps and baboons take honey when they can. However, the honeyguide does not lead chimpanzees or baboons to hives in trees. Humans eat honey, but not beeswax, while the honeyguide eats beeswax, but not honey. The fact that hominins were able to access large quantities of honey on a regular basis is likely the foundation for the development of this mutually beneficial relationship between the honeyguide and humans.

Hadza Honey-Collecting Equipment

Harvesting wild honey does not always require equipment. The accessible hives of some species can be close to the ground or in a tree that is easily climbed. But *Apis mellifera adansonii* produces a small entry into its hives. The minimal equipment for harvesting the honey of this species would be a stick hammer for accessing the hive and a honey-extracting stick to extend the reach of the arm. Studies of foragers have made it clear, however, that acquiring large quantities of honey on a regular basis requires additional honey-collecting equipment. After locating and assessing a honey baobab, the Hadza reach the hive either by climbing accessible or neighboring trees or by using a branch ladder or peg ladder. Marlowe (2002:7) noted that German explorers in the 1890s produced the earliest written accounts of the Hadza, but he claimed these accounts were only second-hand descriptions. They were provided by guides and supplemented with observations of Hadza homes and the wooden climbing pegs in baobab trees for collecting honey. There are similar early reports of the San. Stow (1905:56) explained that despite the danger, the San "nevertheless scale [the baobab's] … castle-like walls by means of two rows of pegs driven deep into the bark to serve as a ladder." In a footnote, Stow also reported that he had "seen places where the … San, by a similar method, have climbed the face of a dizzy precipice where even a baboon could not have obtained a footing, to secure the much-coveted honey of a bees' nest." Old baobabs pitted with small holes and studded with small sticks that trail up the side of their massive trunk provide evidence of the importance of Hadza honey-collecting equipment.

The Hadza routinely climb the baobab by means of specially prepared hardwood climbing pegs that are 20–25 cm long. Men using peg ladders to climb the baobab is one of the iconic images of the Hadza, and it is not surprising to find many illustrations of this remarkable scene in publications and on the Internet. The Hadza use their axes to cut and sharpen suitable hardwood sticks into climbing pegs; they use them to hammer these pegs into the soft wood of the baobab's trunk; and they use them to break into the hives. Marlowe (2010:90) has also reported that Hadza men "use their axes to shave limbs into bows, as well as to pound the poison plant *panjube* into pulp. Sometimes they use them to sharpen women's digging sticks. They use them to crack open giraffe bones to get at the marrow. The main function of the ax, however, is to open up tree limbs or tree trunks to get at honey." It is noteworthy that the Hadza name for this most important component of their honey-collecting equipment is the "honey axe" (Peterson et al. 2013:122). Marlowe (2010:20) wrote:

> The ax may have been in the tool kit of hominins for almost 2 million years, given that Acheulean hand axes (associated with Homo erectus sites) probably were indeed used as axes, possibly to access honey (Allchin 1957; Mitchel 1955). I once watched a Hadza man who had gone on walkabout without his ax; he saw honey in several trees, and he could not resist an attempt to get some. He searched for and found a rock that happened to look very much like an Acheulean hand ax. After 10 minutes or so, he broke into the hive, albeit with much more difficulty than if he had had his ax. Woodburn (1970) mentions men using stones as well. I have also found large flaked stones in Hadzaland that look like they may be old stone axes.

In addition to locating honey baobab trees, assessing the available supplies of honey, climbing to the hives, and breaking into the hives, the Hadza also need a smoking fire (produced by their fire drill and fueled to produce smoke) to tranquilize the bees. In Benin, "Dried corollas [of the baobab flower] are burned and a heavy smoke, which repels insects, is developed and can be used to chase away bees while harvesting bush honey" (De Caluwé et al. 2009:438). The tools the Hadza require for collecting honey include containers for collecting and transporting honey; for storing it for individual and household consumption and kin and community sharing; and for barter or sale.

Falling from Honey Baobab Trees

The most common severe accidental injury to Hadza men is falling from baobab trees they have climbed, mostly for honey and fruit, but possibly also for water, fiber, game, and panoramic views of the surrounding landscape. With its smooth trunk and without handholds, accessible low branches, or climbable neighboring trees, a huge baobab is difficult and dangerous to climb. The Hadza use freshly cut green wood to make climbing pegs. These pegs soon become loose and easily dislodged from their sockets because of the baobab's soft wood, and dry or rotten pegs break easily. Hadza men risk crippling or deadly falls. Blurton Jones and Marlowe (2002:207) reported falls from baobab trees that resulted in two deaths and two cases of permanent disability. Similar deaths have been reported for the Kung (Elizabeth Marshall Thomas 1959:258–259). However, deadly falls from baobab trees are a small percentage of Hadza deaths from all causes. Of 75 deaths, Blurton Jones found that 1% were attributable to falls from baobab trees.

Another risk of falling from baobab trees involves the aggressiveness of African bees, as Sweeney (1969:79–97, 1974:54) has vividly reported from his own experience in the Sudan. Some accounts of Hadza honey collecting give the impression that the Hadza are indifferent to the stings of the African honeybee. For example, Cooper (1949:9) wrote: "They appear to be practically immune to the stings of the infuriated bees, as is the Ratel or Honey-badger. They employ little smoke to stupefy the bees, and in fact, take fewer precautions than any other native honey-hunters whom I have watched." Current research does not support this view. The Hadza know the temper of bees and they take precautions. Peterson et al. (2013:85) reported the Hadza as saying:

> All the flowers from the great diversity of plants in our environment contribute to honey production. We are able to recognize distinctive colors and flavors in honey especially from plants that are abundant and flower synchronously. All of us, men women and children know from experience that different flowers not only result in different flavors, but also affect the behavior of bees in different ways. For example when the *mukangako* (*Acacia tortilis*) blossoms profusely from January to mid February, it makes the honeybee very aggressive and you 'mess' with bees at you own peril.

To cope with the stings of bees, Peterson et al. (2013:127) reported, the Hadza boil the bark of the *katchiko* tree (*Strychnos potatorum*), which, in addition to other things, they use "for washing the body before harvesting honey to reduce bee sting pain." Bad climbing pegs and aggressive bees are not the only considerations, however. According to Marlowe (2010:160), "Old men are the most likely to fall out of tall baobab trees to their deaths, since they continue to try to collect honey into old age."

The Hadza Use of Honey

The Hadza have long been known for their love of honey, the consumption of which also includes bee larvae and pupae. "Perhaps the favourite dish [of the Hadza] is honey," wrote Bleek (1931:278), "considered a special delicacy and a treat for children." Honey, however, is more than a special delicacy or treat for children. It is even more than a sweetener, medicine, or health food as it is popularly considered today. Honey is a basic food in its own right for the Hadza. For example, Allport (2000:147) recalled a Hadza hunter in Arusha who was recovering from a puff adder bite he received while tracking for a private hunting safari: "I wasn't surprised when he told me that meat and honey were his favorite foods, but I was surprised by the quantity of honey that he said he consumed each morning. Three mugs, he told me, holding up his fingers to show me that each mug was the size of a small glass. That was his breakfast, a meal that usually lasted him until evening."

Bleek (1931:278) was right at the time to qualify her view that honey was the "favorite dish" of the Hadza with the word "perhaps," but current studies have confirmed that honey is indeed the favorite food of the Hadza, which they consume in large amounts. According to Mabulla (2003:4), "Bunn et al. (1988) and O'Connell et al. (1988, b) have, for example, noted Hadzabe hunters completely abandoning their hunting missions in favor of honey exploitation." Woodburn (1972) estimated that honey made up a significant part of the 20% of Hadza diet that is not composed of plant foods. The Hadza consume much of the honey they collect in its natural state. What they do not immediately consume they store only for a short time for later consumption and for exchange in the form of sharing and trade. They also consume honey combined with other ingredients, including baobab fruit pulp and seeds. According to Murray et al. (2001), honey and baobab fruit and seeds are the major non-meat foods of the Hadza throughout much of the year. Nutritionally, honey contributes an important percentage of calories to the Hadza diet. There are those who have argued that based on what we now know of its importance for contemporary foragers, honey has probably been an important factor in human evolution because of its nutritional contribution to the development of the human brain (Crittenden 2011; Marlowe 2010; Marlowe et al. 2014). Honey, like baobab fruit and seeds, is an important weaning food of the Hadza, and it is also used to treat cuts (Marlowe 2004:692). The latter use is in keeping with other therapeutic uses of honey, which occur widely among human beings and have received a great deal of attention in folk traditions and the literature.

Honey enters into Hadza songs and origin narratives and it is used as a ritual object. It should not be surprising that the honey-collecting process itself plays an important ritual role. In a story of the origin of the Hadza ancestral couple reported by Peterson et al. (2013:16–18), we learn that *Ts'ikayo* was accidentally left behind when her group moved camp. She was caught *by Duduk'we*, a fierce creature that ate people, who took her to his cave in a rock outcrop. He promised not to eat her, but to take care of her like a daughter. Fearful of the giant and distrusting his words, *Ts'ikayo* sought safety on the highest rock above the cave, which, at her request, grew taller than the giant could reach:

> But she was hungry and bored after so many days alone, and so when the honeyguide bird (*tik'ili*) called her, she came down from her rock and followed it. After a while the bird led her to a beehive and Ts'ikayo began harvesting the honey. When she pulled her hand out of the hive, she saw that something had spit on it. Surprised, she spat on her hand and put it in again in order to compare her spit with that from inside the hive. Comparing the spit, she cleaned her hand and did it again two more times – but the spit was still different. Unbeknownst to her, the honeyguide had flown into the hive and was spitting on her hand.
>
> She was busy with the hive when she heard a voice behind her, and turning around she found that the honeyguide had turned into a handsome young man.
>
> "Let me help you get the honey out of the tree," the young man offered.
>
> They took turns putting their hands into the hive unable to resist staring at each other. Together they harvested a lot of honey …

The honey-collecting process is also a ritual model in the Epeme Dance associated with the rite of passage for men and the rite of passage for women (Power 2015). With the Epeme Dance, male/female symbolic communication simulates the Hadza/honeyguide call-and-response relationship. With the rite of passage for young girls, Peterson et al. (2013:148) reported:

> When a mother feels that her daughter has come of age and is mature in body and mind, she sets in motion an initiation rite called *maitoko*. This usually occurs simultaneously for a small number of girls from any geographic area. The *maitoko* initiation can only take place when there is abundant food – berries, meat, and fat to anoint the girls. The initiates are adorned with beads and given specially decorated staffs (*narichanda*) made by their fathers but given to them by their mothers. For several days, the girls are encouraged to hit any young men in sight with these staffs.

> As part of the ritual the parents put honey or meat on the ground in between the line of maitoko girls and a line of boys who have gathered to see the girls. The boys, determined to get the food and to prove their prowess, are subject to severe beating during the ceremony … The ritual lasts a few days and when instruction is complete, the older women remove the beads and the girls are then said to have taken a step towards adulthood.

The Hadza also gather honey for its exchange value in gift-giving and trade. With respect to gifting, Peterson et al. (2013:161) wrote:

> Sons-in-law and mothers-in-law have a special relationship best described as an avoidance relationship. It is highly improper for a son-in-law to sit close to or to even talk to his mother-in-law. Even if they are in the same camp, their respective houses are carefully sited in terms of distance and position of entrance. Their houses must be close enough to show respect but far enough and facing away so as to avoid improper behavior. A son-in-law has great obligation to appease his mother-in-law by giving her gifts of meat, fat, honey, beads and tobacco.

Honey is often bartered with neighboring groups for alcohol, beads, tobacco, and other goods, and beeswax is sold to "Swahili merchants" (Woodburn 1972). The Hadza told Peterson et al. (2013:82) that in the past, honey collectors ate and shared most of the honey they harvested. They pointed out that although this custom continues, today they traded more honey than in the past. Peterson et al. (2013:82) presented a photograph that showed the entrance of a traditional Hadza branch-framed house, and taking up much of the space inside was a bicycle that was identified as having been purchased from honey sales. Although the Hadza do not make honey alcohol, the honey they collect does provide them with a traditional source of alcohol, albeit indirectly. Marlowe (2004:691) wrote: "They … trade honey to their agro-pastoralist neighbors who use it to make beer and give them some in return."

The Seasonality of Hadza Honey Collecting

One would think the importance of honey in the Hadza diet and in other areas of their practical and inspirational life would have resulted in a clear understanding of the seasonal nature of their honey-collecting activity, but this is not the case. There are widely differing views, including honey's year-round availability with unpredictable fluctuations, year-round availability with one or more peaks associated with either the wet or dry times of the year, and seasonal availability specifically associated with spring, summer, autumn, or winter. These differing views are not to be taken as merely contradictory, however; nor can they be dismissed simply as being indifferent to the nature of honey seasonality. What is most important is that the lack of general agreement represents a growing understanding of the seasonality of Hadza honey collecting, which is more complex than has been earlier assumed.

Woodburn (1968a:52) sidestepped this complexity. In his wet season/dry season framework, he characterized honey as being available year-round, but with unpredictable fluctuations. According to Woodburn, in the "wet season," the Hadza emphasized the gathering of roots and hunting small game, especially hyrax. In the "dry season," they emphasized the gathering of berries and the hunting of large game. It is significant that Woodburn made no reference to honey in his conception of Hadza seasonality. Instead, he noted elsewhere in his article that the "honey and grubs of seven species of wild bee are eaten [and that] … supplies of these vary widely from season to season and from year to year" (1968:50). There is no reason to doubt that annual and inter-annual fluctuations occur in the seasonality of Hadza honey consumption. However, most studies identify one or more seasonal peaks in the annual availability of honey.

The seasonal polarity approach dominates the study of the Hadza, and from this perspective, there are those who argue that the annual peak in honey collecting is the dry time generally or more specifically in the early part of the dry time, from the mid-autumn of May to the mid-winter of August. For example, Tomita (1966:168) reported that from the latter half of the mid-spring of November through the end of spring in December, the Hadza "start collecting honey," but supplies are "still small at this period." But in the early summer of January, "honey gathering starts," and this for Tomita is a significant development in the annual honey harvest. It is clear that for Tomita (1966:169), however, the "regular" honey season was in the mid-autumn of May and lasted through the end of the mid-winter of August. He wrote, "When May comes and the rains pelt the *madabi* fruit causing it to start deteriorating the regular honey gathering season comes." Along similar lines, Tomita (1966:170) reported elsewhere in his article that "From the start of the dry season honey gathering becomes possible. For this purpose [the Hadza] … move further into the [rocky baobab-clad] mountain area, where they can hunt more easily; a place a long way from other tribes, where they can carry out hunting in their original way." Cooper (1949:9) also captures this seasonal moment. As earlier noted, he made two visits to the Hadza, each lasting 10 days, and these visits occurred in the dry time at "the height of the baobab season." The important feature of the Hadza diet at this time of the year was "the usual baobab fruits and honey." It is noteworthy that the Economic and Social Research Foundation of Tanzania identifies the harvest time for honey as the "dry season" from the end of autumn in June to mid-winter in August. The fieldwork of Hawkes et al. (1997:555) supports this view. They noted that "By mid-May, the rains had diminished and the country had begun to dry out … Over the next two months, families foraged for honey, men hunted by encounter, and women dug tubers and collected *embiribe* and baobab."

Identifying the dry time as the period for gathering honey involves more than seasonal availability. Other factors that come into play are not always given the attention they deserve. A significant problem in efforts to account for the seasonality of Hadza honey collection is that researchers do not always make a clear distinction between the seasonal availability of honey and the seasonal collecting of honey. We are left with the false impression that for the Hadza, honey is always consumed to the extent of its seasonal availability. While these studies typically correlate the availability of honey with seasonal rainfall, they often overlook the mediating factors of seasonal flowering, honey-collecting traditions, the relative seasonal availability of preferred foods, and resource competition from other Hadza or outsiders. Consider the following account of Petterson et al. (2013:85), which reflects the complexity involved in determining the seasonality of Hadza honey collecting. These authors paraphrase the Hadza as saying:

> Following the 'old way', we Hadza tried to impose sanctions on gathering of honey during periods of the rainy season so that the bees are able to store a surplus. As the dry season sets in, elders from each locale send either a husband and wife or a small selected group of men to investigate the status of honey production to determine whether it is an appropriate time to open the honey gathering season. This delegation should return with a sample of honey on a piece of cut bark (*hek'wa*) so the elders can decide. Today, while still practiced, this tradition is under pressure due to outsiders who come in and harvest at any time, thereby undermining self-imposed sanctions for the great good.

Contrary to the views that identify the dry time as the honey harvest, a number of Hadza researchers point instead to the annual rains. According to Murray et al. (2001:3–4):

> In general, large animals, baobab (*Adansonia digitata*) fruit pulp and seeds, a few berry species, and a little honey are available during the dry season. Small animals, tubers, most fruits, baobab seeds and the majority of honey are available during the wet season (Woodburn 1968a; Vincent 1985a; Bunn et al. 1988; O'Connell and Hawkes 1988; Hawkes et al. 1989).

But the seasonal timing of honey harvesting is more than a matter of "the majority of honey" being "available during the wet season." Some authors who identify the honey harvest with the annual rains also associate the peak honey harvest with either summer or autumn, or both.

For Wekesa et al. (2005), the honey harvest is spring. In their "marketing study of the ecological resource products" of Tanzania, they reported that it was pointed out to them that "good honey harvests were made during rainy seasons" from the mid-spring of November to the start of summer in December "when there was enough bee forage." This view suggests honey was not plentiful in the March–April autumn rains. But Mabulla (2003:39) reported two harvest peaks in his study of the spatiotemporal distribution of plant foods and honey in the Eyasi Basin. He noted that honey was gathered from around the end of the mid-spring of November to the early part of the mid-summer of February and again from the latter part of the early autumn of April to around the middle of the mid-autumn of May. But his table tells us nothing about the availability of honey at other times of the year. Thus, even when two annual peaks of honey harvesting are identified, as in Mabulla's reporting, it does not mean there is recognition of a winter honey harvest.

A reasonable explanation for the different perspectives on honey seasonality in the literature emerges when we consider the seasonal ecology of honey production and the quantitative data presented by Marlowe (2010). The honey supply of bees depends largely on the seasonal pattern of flowering characteristic of the temperature and rainfall regimes of their environment. In the face of the freezing winters of temperate latitudes and the very dry winters of Africa's tropical savanna region, bees must store enough honey in summer, autumn, and early winter to survive the extreme conditions of the height of the dry time, from the end of winter in September to the early spring in October. Bees resume their intensive collection of nectar and pollen with the spring initiation of flowering in anticipation of the start of the annual rains that mark the return of the growing seasons. The honey supply from the spring flowering is partly the basis for the December spike in honey collecting recorded by Marlowe et al. (2014). The extensive flowering that characterizes summer means honey supplies would be ample in summer and autumn. The significant flowering in the initial part of the May–July dry time following the autumn rains partly accounts for the August spike recorded by Marlowe et al. (2014). This means bees' supply of honey would be most abundant in summer and autumn and especially in early winter in preparation for surviving the September–October dry time. This scenario is what Tomita (1966) reported and what Marlowe's (2010) data suggested.

According to Marlowe's (2010:108, 116–117) measurements of honey brought back to camp, most was collected during the annual rains from the end of spring in December to the mid-autumn of May. Within this period, significant quantities were collected in summer (identified as the "Early Wet" period lasting from the end of spring in December to the mid-summer of February). The highest amounts for the year were collected in autumn (identified as the "Late Wet," lasting from the end of summer in March to the mid-autumn of May). Smaller quantities of honey were collected in the June–November dry time. Within the dry time, the second-lowest amounts were collected in the "Early Dry" (identified as the end of autumn in June to the mid-winter in August), followed by the lowest amounts, in the "Late Dry" (identified as the end of winter in September until the mid-spring of November).

This means that for Marlowe (2010), autumn is the peak honey harvest of the Hadza. His account is particularly interesting because he showed that the peak seasonality of Hadza honey collecting was also related to the relative seasonal availability of other Hadza preferred foods. In a table representing "5 years of data in 24 camps spanning 11 years" in which the percentages of the Hadza top five foods brought into camp were measured in kilograms, Marlowe (2010:127) indicated that honey represented 8.1% of the total, compared with berries at 28.5%, meat at 27.1%, tubers at 24.7%, and baobab at 11.5%. What is particularly noteworthy is that in the data Marlowe presented for one camp in the early summer of January, we can see clearly the influence of the relative seasonal availability of preferred food on honey collecting. Honey, Marlowe indicates, represented only 1% of the top five foods the Hadza brought into camp, compared with berries at 67%, meat at 14%, baobab at 10%, and tubers at 8%. Reflecting on the foods brought into this one camp in the early summer of January, Marlowe (2010:126–127) made the important observation that "when *undushipi* [berries] was available ... berries completely dominate the foods brought back to camp during the berry season, and again because the vast majority of berries are eaten while picking them out of camp, the amount consumed is even greater than suggested by table 5.3."

This means the general view that the amount of honey gathered by the Hadza is precisely related to the available supply of honey is not accurate. In light of this, we have to consider the possibility that the relative seasonal availability of Hadza preferred foods also contributes to the intensive harvesting of honey during the autumn rains of March, April, and early May. Marlowe (2010:48) wrote, "Tubers are eaten much less often when foods like meat, honey, berries, and baobab are plentiful." He went on to note that "By switching to different foods, the Hadza are able to acquire a fairly level intake of daily kilocalories throughout the year (Marlowe and Berbesque 2009; Sherry and Marlowe 2007)." If what Marlowe says with regard to the relative seasonal availability of Hadza preferred foods is accurate, it could be argued that when the Hadza eat a lot of roots (and honey), as they do in autumn, it is because foods like meat, baobab, and berries are in limited supplies. This suggests that when more baobab, berries, and meat are taken in the "Early Dry" period from the end of autumn in June to mid-winter in August, it probably has to do not only with the seasonal availability of honey but also with the return of seasonal abundance which occurs during winter, spring, and summer in contrast to autumn.

The perspective on Hadza honey seasonality this work presents suggests the true seasonal polarity of Hadza life is not the "wet season/"dry season" contrast as widely conceived, but the contrast between the relative seasonal availability of their top five preferred foods between the drier time of the year represented by winter, spring, and summer and the wettest time of the year represented by autumn. This means the major seasonal transitions in Hadza life would be the mid-summer of February, giving way to the wettest time of the year from the end of summer in March to the early autumn of April, and the transition from the wettest time to the return of the drier time from the mid-autumn in May to the mid-summer of February. This drier time is associated with a resource concentration of baobab, meat, and berries. This is in contrast to the honey and roots of autumn when baobab, meat, and berries are less available. In the resource concentration of the dry time, camps are larger than the camps of the autumn rains when they are more dispersed. The time of resource concentration is the sociable time of the year for the Hadza in contrast to the autumn when they live more separated lives. It is also possible they live more separate lives in autumn because honey, their most-loved food that comes in small packets, has great exchange value and is especially important to husbands' wives. And life is easier in the dry time than in the rains, when people are more uncomfortable, must build more substantial dwellings, and have longer resource foray distances. The Hadza gather the most honey in autumn, the wettest time of the year, because the supply is plentiful and because their need is greatest at this time, given the limited availability of their other preferred foods. While there is no doubt that from an environmental point of view, the most dramatic seasonal changes in Hadza life are from the dry-to-wet and wet-to-dry times of the year, the critical difference in Hadza seasonal life is resource availability between the dry time of the year and the wettest time of the year.

Summary

Given the primate love of sugary fruit and honey's nutritiousness, it is likely that the baobab, Africa's honey tree, has been important in hominin evolution for a very long time. But honey is never free for the taking. The problem has always been climbing the baobab, neutralizing angry bees, and collecting and transporting honey for later consumption and for sharing and other kinds of exchange. Primate field studies have included observations of the use of plants and stone for weapons, shelter, and bedding; processing fruits, seeds, and nuts; collecting water and honey; catching invertebrates; and hunting mammals. In his discussion of Paleolithic technology and human evolution, Ambrose (2001:1749) presented a perspective long common in the history of anthropology: human evolution, including the capacity for cultural development, is correlated with technological innovation. And in the broad sense, technological innovation means handiness and the overall development of material culture. Ambrose wrote, "Assuming that the behavior of our closest living relative could also have been performed by our last common ancestor and its descendants, we can infer that the minimum level of technological capacity of hominids between 5 and 2.5 million years ago (Ma) was comparable to that of chimpanzees."

But this perspective is problematic. Chimps and the Australopithecines cannot be treated as evolutionary equivalents with respect to tool use. Great apes make and use tools, but following the chimp/human divergence, the Australopithecines – the earliest hominins whose bipedalism is here regarded as evidence of handiness and developed material culture – were probably more tool dependent than gorillas and chimpanzees. The extent of tool use was likely a significant contributing factor in the divergence between chimpanzees and the Australopithecines as handy hominins, just as it probably was in the divergence between the Australopithecines and the genus *Homo*. Hadza honey-gathering equipment clearly makes a difference in the amount of honey that can be collected and the ways it is distributed and used. But bees are not the only baobab-associated species that is important to human beings; there is also a great variety of birds. In contrast to monkeys, gorillas, and chimpanzees, handy hominins need equipment to make the best use of the baobab not only in collecting honey but also in harvesting fruit, processing fruit pulp and seeds, collecting tree water, and gathering nesting birds in spring.

Chapter 13
The Baobab and Birds

Birds, with their exquisite adaptations to land, water, and air, are among the most abundant vertebrates on earth (Allen 1961). One of the many benefits of the baobab for the Hadza is its well-known association with a great variety of birds, part of the remarkable biodiversity the tree supports. With its rivers, lakes, and marshes, and its many kinds of fruiting trees, the rich mosaic savanna of the eastern Rift Valley supports a diverse population of resident and migratory birds that have likely played a part in hominin evolution (Stidham 2005; Prassack 2011; Morelli et al. 2015; Blasco 2016; Negro et al. 2016). The significance of birds in hominin evolution is receiving increasing attention today; at the Eleventh Conference on Hunting and Gathering Societies in 2015, there was a section of oral presentations under the heading "Human-Bird Relationships in Hunting and Gathering Societies." Birds are indeed significant in the material culture and spiritual life of the Hadza and of other African savanna foragers, and the baobab/bird association is yet another contributing factor to the idea of the baobab as an ecological tree of life. As previously noted, it is this status as an ecological tree of life that makes the baobab a resource-rich environment for the Hadza, and would likely have made it a resource-rich environment for early hominins as well.

Tanzania is known for its many birds. Of some 10,400 species worldwide, 1116 occur in Tanzania. Thirty are endemic to Tanzania and sixteen to East Africa (iGo Terra 2020). Wickens and Lowe (2008) identified thirty-seven kinds of birds in Africa with baobab connections. The baobab's importance for birds is understandable. It provides a perch as a place for surveillance, eating, resting, preening, displaying, courting, and mating, as well as providing protection, food, water, roost, and nesting sites (Brown 1965; Owen 1974; Sweeney 1974; Watson 2007; Wickens and Lowe 2008; Hellekson 2009; Kopij 2019). For the Hadza, this makes the baobab, like figs, berry trees, and water sources, an important mediator of the Hadza relationship to the birds of their environment.

"There is no better place to watch … birds than under a baobab tree," wrote Leslie Brown (1965:168) in his discussion of the natural history of the Nyika, an extensive

J. Rashford, *Baobab*, https://doi.org/10.1007/978-3-031-26470-2_13

area of grassland and thorn thickets covering the dry eastern parts of Kenya and Tanzania. In 2008, Lyndsay Hellekson spent five days from June 19 to 23, observing a large baobab at the Epupa Falls in Namibia; it was forty-seven feet in circumference and an estimated twenty-one feet in height. She identified 248 holes, four of which held bird nests. She reported seeing twenty-four species of birds using the tree, mostly for perching, looking about, and signaling. None were "observed feeding in the tree, likely due to the fact that all but 2 fruits had been eaten already and the tree had already dropped its leaves for the dry season" (Hellekson 2008:6). Hellekson's brief survey is impressive, but limited with respect to the baobab's seasonal association with birds as she herself recognized in her reference to baobab seasonality. The time of the year when the baobab/bird association would be far more significant than she was able to record is spring, and during the annual rains generally. The baobab attracts many arthropods starting in the spring, when it puts out new leaves and begins flowering, and this is the time when many birds nest.

When considering the relationship between the baobab and birds, it is important to bear in mind that the specific avian community of each baobab is determined by a number of factors. They include, for example, the typically isolated position of the baobab in woodland and in tree-scattered grasslands and scrubland; the physical features of the tree itself, such as its age and size; the seasonal production and shedding of leaves, and its flowering and fruiting; the number and variety of holes in its trunk and branches, and the presence of water catchments; the environment in which the tree grows, particularly with respect its proximity to water and to human settlement; and the association with other plants, including its branch-resting canopy gardens, colonizing figs, parasitic mistletoes, understory plants, and neighboring herbs, vines, shrubs, and trees (Belsky et al. 1989; Imoro and Barnes 2013; Imoro et al. 2015). Watson (2007:113–115) has provided excellent illustrations of the baobab colonized by strangler figs, and the baobab/mistletoe connection has been discussed by Denis-Esdras et al. (2020a:46); these associations are particularly significant with respect to birds (Nickrent 2011). Strangler figs are keystone species that support a great variety of animals, especially birds, as do mistletoes.

The assumption underlying this chapter is that wherever baobab trees grow, whether in human or natural environments, they are significant contributors to the species richness and abundance of birds. In this context, the avifauna of settlement baobabs refers both to the trees concentrated in the landscape of Africa's residential communities, especially home environments, waysides, public spaces, cemeteries, and shrines; and to the park-like scattering of baobabs and other culturally significant trees in active and fallow fields and pastures. It is noteworthy that one reason some East African farmers give for not liking baobab trees in their farms is that the trees attract birds. When Maria Johansson (1999:14) asked "five elders in five villages in the Kondoa Eroded Area" of Irangi Hills, Tanzania, if there were "any negative effects" of having the baobab in their fields they said baobabs "take up a big area, the shade is not good for cash crops, [and] they hide birds which are harmful."

Because of its rich avifauna, we can assume that the baobab has played an important mediating role in the Hadza relation to the birds of their environment. This chapter looks at the contribution of the baobab to shaping the avian community and birding traditions of the Hadza, and its significance for theorizing hominin evolution. Following an introduction to birds in the environment and cultural traditions of the Hadza, the remaining sections of the chapter discuss birds and the making of the Hadza hunter, birds associated with the baobab, and the link between the baobab, birds, and seasonal hunger. Since there are no specific studies of the ways in which the baobab mediates the Hadza interaction with the birds of their environment, the approach taken here is to look at the Africa-wide baobab/bird relation as reported in the literature, noting those baobab-associated birds for which the Hadza have names in their own language, and for which Hadza uses have been reported.

Birds in the Environment and Cultural Tradition of the Hadza

An essential aspect of Hadza bird-hunting success is a detailed understanding of the birds of their savanna environment. According to Marlowe (2010:138, 127), the Hadza eat a "wide variety of birds." He illustrated this in a table titled "Number of species by type of food in Hadza diet." The table was based on five years of data from 24 Hadza camps over a period of eleven years. It showed that the greatest number of species by far were birds; of the 16 categories of Hadza foods presented, there were 741 species of birds, representing 84.4% of the 878 species of plants and animals on which the Hadza depend. Peterson et al. (2013:220–222) listed over 160 Hadza names for the birds of their environment; most were identified at the species level, but some at higher or lower taxa.

Birding traditions received little attention in early accounts of Hadza foraging. Woodburn (1968a) barely mentioned birds in his classic article on Hadza ecology. In reports predating Woodburn, most of what we learn about the significance of birds is incidental to discussions of other topics. Often, it was the sensational subjects – the eating of vultures, for example – that were thought worthy of elaborate commentary (e.g., Bagshawe 1925). In recent years, however, researchers have provided important information on the significance of birds in Hadza foraging, including insights on what the relationship between human beings and birds means in theorizing the development of the hunter-gatherer way of life of early hominins. The avian traditions of the Hadza include the gathering of chicks in the spring and the hunting of mature birds throughout the year; but as an in-camp source of meat, the contribution of birds as reported in recent evolutionary ecology studies appears to be relatively small compared with mammals. This, however, is a limited view. Hadza birding traditions involve more than birds as a direct source of meat. They also encompass the contributions birds make to acquiring other forms of meat – for example, the role of birds as indicator species, and in arrow fletching.

Birds enter into other areas of Hadza productive life as well, and into the most important areas of inspirational life. The Hadza depict birds in oral traditions, songs, and dance; birds are also significant in religious tradition, where they are featured in mythology and used as ritual objects and models. The contribution of Hadza birding traditions to diet and inspirational life also includes the ritual use of feathers, as in the Epeme Ceremony, which supports the necessary cooperation represented by the gendered economic activity. By custom, Hadza men link success in hunting with the use of raptor talons attached to their bows. Raptors have long been regarded as emblems of power in many parts of the world. This is particularly true of eagles; they play a widespread symbolic and ritual role in Hadza life, including, for example, the use of eagle feathers in hair pieces and headdresses. Hadza hunters make use of "cylindrical pieces of hide from animal skins or birds of prey leg skins (with talons) to decorate their bows and to keep splits in the wood from spreading." In addition to talons, baboon hair is used to decorate bows, a practice illustrated in several photographs presented by Peterson et al. (2013:40, 49). Raptor talons and arrow fletching are probably similarly believed to confer the power of the great birds on the Hadza hunters.

Birds and the Making of Hadza Hunters

The Hadza distinguish themselves as a "bow and arrow culture" (Peterson et al. 2013:35). This makes birds a critical target of their hunting equipment, which includes bird-harvesting sticks, baskets, fletching for arrows, and specialized arrows. For example, "The *//k'owa*, for shooting birds," wrote Bleek (1931:275), "is a solid arrow of tapering wood some 140 cm long with a pointed tip, to one side of which a small wooden barb [that] may be bound with sinew." Cooper (1949:12) presented two illustrations in his discussion of the size and power of the Hadza bow and the strength needed for its effective use; he noted that "The photographs of small boys with their toy bows used for shooting small birds … serve to illustrate how early they begin to develop in this art." Marlowe's (2010:157) account of the making of Hadza hunters points to the importance of birds in their forager adaptation:

> Boys get their first bows about 3 years old and thereafter spend hours every day in target practice, often shooting at a gourd on the ground … By 5 or 6 years of age, boys are good enough to kill birds and small rodents. It is through repeated practice and observation of older boys that they hone their skills. Boys almost never go hunting with their fathers, at least until they are grown (and even then it is rare). By the time boys are 8 or 9 years old, they go hunting for small animals, usually in twos. By 13 or 14 years of age, they usually go hunting by themselves, killing hyraxes, bush babies, did diks, and birds of all kinds. They may be gone nearly all day, longer than any other age-sex class. For example, males age 10–17 are gone an average of 7.3 hours a day. Most of what they acquire they eat while out foraging.

Today we know that some birds benefit not only from an association with human beings, but also from an association with the protective "nests of ants, bees, termites, or large birds of prey" (Brooke and Birkhead 1991:228). In addition to protection, the human environment also provides perch, food, nest site, and nest-building material. When the benefits are substantial, and the general reproductive success of birds interferes with human activities, these birds are labeled nuisance birds or pests to be deterred or eliminated (Conover and Chasko 1985). The major arguments against these birds are crop losses, health threats, aesthetic concerns, and damage to the built environment. What is theoretically important, however, is that some birds have likely always been attracted to the human environment (Dean and MacDonald 1981). Like farms, fields, villages, towns, and great contemporary urban centers (Jerolmack 2008; Tidemann and Gosler 2010; Elliott 2006; Deikumah et al. 2015; Støstad et al. 2017), forager camps probably also attract birds.

Hadza boys learn to stalk and shoot birds near camps, but it would be interesting to know if these near-camp birds are present merely by chance or because of benefit they derive from Hadza camps, especially protection and feeding opportunities from a camp's baobabs, figs, or berry patches, which are a magnet for birds. It should not be surprising then that Hadza boys who set out to shoot birds bring back baobab fruits, figs, and berries. In their study of tree preferences of wintering migratory birds and resident birds in West Africa (encompassing 183 species of trees and shrubs and 59 species of birds), Zwarts et al. (2015:131) found that "Migrants preferred thorny over non-thorny tree species, with highest densities of migrants found in thorny tree species except *Salvadora persica*, a woody non-thorny tree species which is only attractive when carrying berries." The Hadza are dependent on a variety of berries, as earlier noted, and they have told researchers (Peterson et al. 2013:111) "The most important are *undushibi* (*Cordia sinensis*) and *tafabe* (*Salvadora persica*) because they grow in dense stands with profuse berries. Between the months of October and March when these berries are in season, large camps with up to 80 people form in the vicinity of these bushes and life is easy." In their study of juvenile foraging among the Hadza and its implications for human life history, Crittenden et al. (2013:301) wrote:

> Of the total kilocalories young foragers brought back to camp in both regions, the majority (64%) comprised fruit, including baobab, figs, and berries. Fruit was followed by birds (19%), tubers (9%) and honey (7%). The lowest contributions came from small game meat (3%) and drupes and legumes (1%).

The Ostrich: Notable Bird of the Savanna

The largest existing birds today, including the African ostrich, Australian emu, and South American rhea, are birds of expansive plains. Although there are no reports of a direct link between the giant of all flowering trees and the giant of all birds, no account of Hadza life would be complete without reference to the distinctive ostrich (family Struthionidae, order Struthioniformes), which thrives in the environment of the baobab. Traditionally, the ostrich is the most important Hadza game bird and it was likely important in hominin history. Flocks of ostriches, a conspicuous feature of the African savanna, were a familiar feature of the Hadza environment when the Hadza were first contacted in the early twentieth century.

The ostrich is clearly of great importance in the practical and inspirational life of the Hadza and San. Lee (1979:28–29) identified three classes of Kung game birds. "Class 1" was "systematically hunted" birds, and Lee reported that the Kung targeted the ostrich, but for "eggs only." Class 2 was birds that were "occasionally hunted," which did not include the ostrich. And Class 3 was birds "eaten by some persons," and this included the ostrich, stork, and six other unspecified species. According to Yellen and Lee (1976:37), "Ostrich are still common and continue to provide the !Kung with a steady supply of ostrich eggshell and water containers and materials for making beads." Given the importance of the ostrich egg as a water container, it seems the ostrich is worth more alive than dead; it is conserved for the greater good of access to water in an arid land.

This does not seem to be the case among the Hadza, for whom this extraordinary bird was of far greater practical importance in the past than it is presently. Obst (1912) and Kohl-Larsen (1958) reported that the Hadza hunted the ostrich, and Bleek (1931:278) wrote "Game and birds are their staple food, the meat of giraffe and ostrich being preferred on account of its sweetness." In addition to their love of ostrich meat, the Hadza use ostrich sinew in making arrows. Ostrich feathers are placed on graves and used in making the headdress of the Epeme Dance, the most important ritual activity of the Hadza. According to Woodburn (1968a, b, 1982a, b), it is a male puberty rite conducted in sync with the dark phase of the lunar cycle and is intended to bring spirituality to bear on strengthening the cooperative unity of men and women, and in so doing, to promote the success of their productive and reproductive activities (Power and Watts 1997). In addition to headdress, Woodburn noted that in the past it was common to make shallow graves on top of which they placed Epeme feathers. It is indeed significant that according to Peterson et al. (2013:220, 54), the ostrich "is classified as an animal (kalimo) not a bird (tsela'alo)." They wrote:

> Ritual (*epeme*) animals such as eland, giraffe, impala, warthog, zebra, lion, leopard, and ostrich can only be butchered by initiated men who set aside special fatty cuts as *epeme* meat. Meat and gravy left over after the men's secret ceremony can be blessed. Once ritually blessed, this meat and fat can be eaten by males of any age but is strictly off limits for females. Ostrich eggs, the larger male oblong ones, are also subject to *epeme* rules.

Today there is no ethnographic or archaeological evidence of the Hadza using ostrich egg containers, especially for carrying water as documented for the San. Nor is there evidence of Hadza use of ostrich egg shell to make beads. Of the Hadza in the early twentieth century, Bleek (1931:276) reported "Both sexes sport any beads they can get, chiefly of European or Indian make. I saw no ostrich eggshell beads." The influences of a growing market economy and the restrictions placed on ostrich hunting by European colonial governments and the Tanzania government have impacted the Hadza; there is little doubt that the Hadza had greater access to the ostrich in the past.

Indicator Birds

Woodburn (2013:xv) paid tribute to the Hadza's extensive knowledge of their savanna environment in his introduction to the book by Peterson and coauthors (2013). This expertise was exemplified by the large number of birds the Hadza have named. Woodburn noted that the Hadza "not only distinguish and name a remarkable wide range of bird species but have an intimate knowledge of the behavior of the various species and particularly of those that they can put to use." It is noteworthy that the two important examples of Hadza bird use Woodburn presented were indicators of honey and dangerous animals.

Growing environmental concerns and restoration efforts have heightened attention to the role of birds as providers of environmental services, including plant pollination and seed dispersal, predation of arthropods that are troublesome to cultivation and livestock, and promotion of a healthy environment by helping decompose carcasses and clean up refuse. In addition, birds are indicators of such things as biodiversity, pollution, and successful environmental restoration (Schulze et al. 2004). They have probably served as important environmental indicators for a very long time. The value of land-indicating birds to oceanic Polynesian voyagers and early European explorers is well known. The concentrated feeding activities of ocean birds guide fishers to schools of fish, just as vultures guide scavengers to game. For the Hadza and cultures worldwide – and likely for early hominins as well – the variety of ways in which birds are important indicators inform understanding of seasonality, weather, time of day, food opportunities, danger, future outcomes, and spiritual phenomena.

The Honeyguide

In their summary of the literature on the association between the baobab and birds, Wickens and Lowe (2008:213) reported that "The greater or black-throated honeyguide (*Indicator indicator*) has acquired the habit of guiding humans to the nests of wild bees, which are often found in baobabs." But the connection between the honeyguide and human beings is not simply an "acquired habit" as suggested by Wickens and Lowe (2008:213); it is now widely viewed as a coevolving relation. The Greater Honeyguide (hereafter also identified as "the Honeyguide") receives sufficient benefits from its guiding behavior (including an opportunity for post-harvest scavenging or reward feeding, and exemption from being hunted) to have evolved the capacity for coevolutionary cooperation with human beings (Dean et al. 1990; Wood et al. 2014; Marlowe et al. 2014; Spottiswoode et al. 2016; Saha and Spottiswoode 2016; Giaimo 2016). Understandably, the honeyguide is both practically and inspirationally one of the Hadza's most important birds, since it is so highly prized as a food. As noted in the previous chapter, the Greater Honeyguide's unique guiding behavior is also inspirationally significant in Hadza aesthetics, oral traditions, and religious activity, where the relationship between the honeyguide and the Hadza symbolizes the relationship between men and women. It is noteworthy that the honey the Hadza obtain in the largest quantity and enjoy the most is that from hives of the African honey bee, which commonly nests in baobab trees (Guy 1972; Wood et al. 2014:541; Marlowe 2012; Marlowe et al. 2014:6). Because its unique coevolutionary cooperation with human beings, the Greater Honeyguide is the baobab-associated bird that has received the most attention in recent Hadza studies (Wood et al. 2014; Marlowe et al. 2014).

The honeyguide belongs to the Old World family Indicatoridae, which is divided into four genera represented by seventeen species, fifteen of which occur in Africa and two in Asia. Of the seventeen species, only the honeyguide has evolved the behavior of leading human beings to beehives in trees. Although not locally common, the small, solitary, cryptic honeyguide is found across much of East and Central Africa, particularly in woodlands, wooded grasslands, and bushlands. It also thrives in human environments. Williams and Arlott (1986:150) reported that the honeyguide is found in gardens and fields "where there are trees," and is "often encountered near native bee-hives hung in trees." The latter point has been documented by Martin Gruber (2018:182), who was told by honey collectors and honey producers in the Adamaoua Region of Cameroon that the honeyguide "had started leading humans to their own or other people's beehives instead of wild bee colonies." Williams and Arlott (1986:150) also reported that the Honeyguide "may be attracted by nailing lumps of bees' comb to tree trunks" (1986:149). The Portuguese missionary João dos Santos, observing the honeyguide feeding on the candle wax in his church and guiding people to beehives, was inspired to produce one of the earliest written accounts of the species (Dos Santos 1609:32).

Dean et al. (1990:100) noted that the coevolved cooperation between the Honeyguide and human beings has "developed to the extent that the birds will follow humans in motor vehicles and boats, and can be attracted by the sound of chopping wood, a specific human activity." We learn that "Greater Honeyguides are attracted to smoke and to human dwelling places," and that "They come into the gardens of small towns and settlements and attempt to guide humans if there are beehives nearby." In light of these observations and other studies, Dean et al. (1990:100) noted: "Greater Honeyguides persistently attempt to guide nonresponding humans (Friedmann 1955; WRJD, personal observations), suggesting that it is humans [and not the honey badger, baboon, or any other species for that matter] that provide the greatest stimulus for this [guiding] behavior."

Recent studies have dealt mostly with communication between the honeyguide and human beings, with little attention to the fact that the honeyguide "often" leads humans to baobab beehives. This is important, since the African honey bee, honeyguide, and humans are all part of the baobab's rich faunal ecology; all three derive a variety of benefits from the baobab, Africa's preeminent honey tree. The Honeyguide, as part of the baobab/bee complex, evolved both as an insectivore with a special appetite for baobab-acquired beeswax and as a brood parasite of baobab-associated birds (Spottiswoode et al. 2011). The most often reported host species of honeyguides are tree-hole nesting barbets and woodpeckers, which are phylogenetically close to honeyguides (Williams and Arlott 1986:149; Zimmerman et al. 1999:406; Wickens and Lowe 2008:214), as well as the ground-hole and cliff-hole nesting bee-eaters (Williams and Arlott 1986:149). Other hole-nesting host species of honeyguides identified by Diamond and Lovejoy (1985:289) "especially" included hoopoes and wood hoopoes, kingfishers, rollers, and starlings. One striking thing about the host species of honeyguides is their links to the baobab. Baobab birds build their nests at the top of the tree, resting on or hanging from its branches, and in its holes and hollow trunk. The birds identified above by Diamond and Lovejoy (1985:189) have long been known to nest in the holes and hollow trunk of the baobab; and, as brood parasites of baobab hole-nesting birds, honeyguide can itself be described as a bird that nests in the baobab.

The origin of the evolutionary view of human/honeyguide cooperation is generally attributed to Hoesch (1937). This view has subsequently received strong support from Isack and Reyer (1989), Dean et al. (1990), Wood et al. (2014), Marlowe et al. (2014), and Spottiswoode (2016). Wood et al. (2014), based on the influential work of Dean et al. (1990), concluded that "Since humans are the only species attested to partner with honeyguides, the would suggest that humans, or our ancestors, were the partners of proto-honeyguides as the habit evolved." Wood et al. (2014:544) began their theoretical explanation of the development of the hominin/honeyguide connection as a commensal relation; the honeyguide benefited, and humans were not harmed. But a more appropriate beginning would have been to consider the African honey bee, the baobab as a honey tree, the honeyguide as a resident bird of the baobab on the one hand, and humans, baobab residents who collect and consume honey and discard wax, on the other. Wood et al. (2014:544) theorized that the Greater Honeyguide dates back some 3 million years to about the time of the origin of the genus *Homo*. But this is not necessarily true of the honeyguides as a family; they evolved more than 3 million years ago. It is the Greater Honeyguide in particular that has been an evolutionary beneficiary of the human love of honey.

Vultures

There are fifteen extant species of Old-World vultures, belonging to the family Accipitridae. They are most often found in savanna and desert areas of Africa, southern Europe, and Asia (Ogden 1985). A ubiquitous feature of the savanna landscape, vultures are significant to the Hadza in ways that would also have been important for early hominins (O'Connell et al. 1988a, b; Morelli et al. 2015; Finlayson and Finlayson 2016). It is surprising that Wickens and Lowe (2008) did not capture the link between the baobab and vultures in their summary of the avian associates of the baobab, especially since the International Centre for Underutilised Crops (ICUC) noted this association in its 2002 factsheet on the baobab. Moreover, the most conspicuous aspect of the baobab's association with birds is the large stick nests of raptors, including vultures, the nests of storks and other water birds such as the hamerkop, and the distinctive nests of various weavers.

Discussions of the Hadza/vulture association have largely focused on food in the context of a single set of Hadza food choices across time and space. This has resulted in many ambiguities. For example, do the Hadza eat vultures? Or is this restricted to one Hadza region, or to some smaller subset of a regional population? Bagshawe (1925) said the Hadza eat "carrion birds," and so did Woodburn, who noted that some Hadza do not eat vultures now because of their neighbors. We know from early accounts, however, that vultures were important in Hadza culture. Their flight pattern and ground behavior signaled opportunities for scavenging (Marlowe 2010:119), and their feathers had important uses (Obst 1912). In his discussion of the metal-tipped poisoned arrow of the Hadza, Cooper (1949:11) reported "Guinea-fowl feathers may often be employed for fletching but those of the larger vultures are preferred." The Hadza generic name for vultures is *mudiga*, and Peterson et al. (2013:222) reported Hadza common names for the Lappet-faced Vulture (*Torgos tracheliotos*, *lokolongoko*), White-backed Vulture (*Gyps africanus*, *matui*), and Ruppell's Griffon Vulture (*Gyps rueppellii*, *hayaha*). Oschadleus (2016:15) offered the following report in his study of the nesting association between vultures and weavers:

> There are five African vultures that regularly breed in trees. Of these, four have at least one record of nesting with weavers and only the White-headed Vulture *Trigonoceps occipitalis* has no records found to date. Since the latter often nests in baobab *Adansonia* sp. (Mundy 1992), it is likely that associations may occur as buffalo-weavers and Red-headed Weavers regularly nest in baobabs (Weaver Watch 2017).

Dean (1974:111) reported seeing three nests of the Palmnut Vulture (*Gypohierax angolensis*) in baobab trees in his account of the distribution and breeding of the birds of Angola. According to Houston (1985:635), the Palmnut Vulture "rather resembles a sea eagle. It feeds on stranded fish and carrion, but its main food is the fruit of the oil palm and its distribution closely matches that of the palm plantations."

There are seven species of griffon vultures belonging to the genus *Gyps* that, according to Houston (1985:635), "feed almost exclusively from the carcasses of large mammals, and these species are usually the most numerous type of vulture in any region." In an article challenging the view of "most vulture authorities" of the time that considered the Ruppell's Griffon Vulture (*Gyps rueppellii*) "to be an exclusive cliff-nester," Rondeau et al. (2006:14–15) provided evidence of Ruppell's Griffon Vultures nesting in trees, including the baobab.

> The majority of tree-nesting reports found in the literature are relatively old, but there have been other more recent observations in West Africa. For example, in the 1980s, C. Barlow (pers. Comm.) observed this species nesting in borassus palms (*Borassus aethiopum*) along the Gambia River in loose colonies with African White-backed Vultures [*Gyps africanus*]. During the same period, F. Baillon (pers. Comm.) found Ruppell's Griffon Vultures [*Gyps rueppellii*] nesting in loose colonies on the top of large baobab trees … in the Saloum region of Senegal.

In addition to information received from personal communications, these authors also presented an illustrated account of three cases of Ruppell's Griffon Vultures nesting in different kinds of trees. They wrote: "The third nest, built on a large horizontal branch of a baobab, at about 9 m high … contained one chick of about 14 weeks old … Considering the amount of whitewash (vulture droppings) found in the vicinity of the nest, it is possible that it is not the first time it had been used by vultures."

The Baobab and Raptors

In addition to vultures, a number of other birds of prey (with their hooked beaks and sharp talons, and a diet wholly or chiefly of hunted animals or carrion) have also been identified with the baobab. Baobab trees located near water, food, and nest-building material (Long 1956), provide raptors with perch for hunting, eating, resting, roosting, and nesting. "Raptors" specifically refers to birds of the order Falconiformes, represented by the families Accipitridae (vultures, eagles, hawks, kites, and buzzards), Pandionidae (ospreys), Sagittariidae (Secretary Bird), Falconidae (falcons); birds of the order Strigiformes, represented by the families Tytonidae (barn owls) and Strigidae (typical owl); and birds of the order Cathartiformes, represented by the family Cathartidae (condors). With the obvious exception of the New World vulture or condor family (Cathartidae), each of these families has one or more members that have been associated with the baobab.

The five eagles (Accipitridae) with links to the baobab include the Wahlberg's Eagle (*Aquila wahlbergi*), Bateleur Eagle (*Terathopius ecaudatus*), Black-chested Snake Eagle (*Circaetus pectoralis*), Brown or Harrier Snake Eagle (*Circaetus cinereus*), and African Crown Eagle (*Stephanoaetus coronatus*). The most frequently mentioned, which was featured in a baobab documentary (Root and Root 1973), is the relatively slender Wahlberg Eagle, a widespread savanna raptor said to nest "sometimes" or "regularly" at the top of the baobab. Like hole-making woodpeckers and barbets, this eagle contributes to the avian diversity supported by the baobab (Owen 1974:8). By keeping snakes and predatory mammals at bay, the Wahlberg Eagle attracts birds such as rollers and weavers that are normally prey to nest in the tree. Additionally, the Wahlberg Eagle may benefit from the presence of smaller birds that serve as an alarm system and as sacrificial pawns, increasing the likelihood that the eagle's chicks will escape predation.

It is interesting that the eight species of eagles with Hadza names identified by Peterson et al. (2013: 220–221) did not include the Wahlberg's Eagle or Crown Eagle. However, Peterson and colleagues did include names for the Bateleur Eagle, Brown Snake Eagle, and Black-chested Snake Eagle, all of which have been identified with the baobab. According to Watson (2007:129), "Brown snake eagles are especially associated with baobabs, often in cultivated areas where the trees have been spared. They … move from perch to perch, but need to spend much longer on each, searching for the slow slither of a hunting snake, or for movements of the cryptically camouflaged puff adder, which is favored prey despite its size and toxicity."

In an interview about African Crowned Eagles (*Stephanoaetus coronatus*), Markusjais (2011:10), an outspoken voice for raptor conservation in Africa, said he was surprised to see photographs by a colleague in which this eagle was shown nesting "in baobabs on the side of hills in dry almost nyika terrain" (Virani 2011:11). In the interest of conservation, Thomsett noted, "Given the dry baobab nesting Crowned Eagles it begs that we not overlook similar habitats in the rest of Africa for these eagles."

The White-headed Buffalo Weaver is known to roost in baobab trees. According to Wickens and Lowe (2008:130), this weaver abandons its nest at the end of the annual rains. In Kenya and northern Tanzania, Zimmerman et al. (1999:309) report that the range of the Pygmy Falcon (*Poliohierax semitorquatus*, family Falconidae) "coincides with that of White-headed Buffalo Weavers whose unoccupied nests it uses."

Kestrels are diurnal raptors in the genus *Falcon* (family Falconidae), identified by the International Center for Underutilized Crops (2002) as nesting in the baobab. However, Peterson et al. (2013) do not report a common name for any of the five species of kestrels known in Kenya and northern Tanzania (Zimmerman et al. 1999:313–314).

The long-legged Secretary Bird (*Saggitarius serpentarius*, family Sagittariidae), a native of the African savanna, spends much of its time on the ground in grassland and woodland environments. It is well known not only for its highly distinctive appearance and foot-stamping foraging technique but also for its ability to prey on poisonous snakes. Zimmerman and co-authors (1999:292) describe it as widespread in southern Kenya and northern Tanzania, and the Hadza do have a name for this bird (Peterson et al. 2013:221). Wickens and Lowe (2008:212) note that the Secretary Bird "builds its grass-lined nest of twigs and turf in a baobab, whose proximity to any bat population [with which the baobab is often associated] must be an added attraction."

Kites, which belong to the family Accipitridae of the order Falconiformes, have also been identified with the baobab. Roualet (2016) posted a picture of "hundreds" of Scissor-tailed Kites (*Bhelictinia riocourii*) roosting in a large baobab tree on the site Internet Bird Collection. And the Yellow-billed Kite, *Milvus aegyptius*, is one of the twenty-one species of birds Kopij (2019:357) found nesting in the baobab trees of Outapi, Namibia. Peterson et al. (2013:220–222) listed Hadza names for the Yellow-billed Kite and the Black-shouldered Kite (*Elanus axillaris*), but not the Scissor-tailed Kite, a savanna bird that is widespread in Africa.

Owls, silent, nocturnal birds of prey with keen vision and hearing (belonging to the family Sagittariidae), are one of the most recognizable birds in the world, with deep cultural significance for human beings. In a quantitative assessment of the risks to birds traded for medicinal use, Williams et al. (2014) examined "the vulnerability of 354 bird species used for traditional medicine in 25 African countries, from 205 genera, 70 families, and 25 orders," and they report that "The Barn owl (*Tyto alba*) was the most widely sold species (seven countries)." As a time-of-day indicator, the Hadza call the Pearl-spotted Owlet (Glaucidium *perlatum*) the "good morning" bird because it "is very vocal at dawn" (Peterson et al. 2013:68).

Welman (1966) reported observing an owl that had taken up residence in a baobab close to his home in Sudan. While Owen (1974) identified the Barn Owl (*Tyto alba*) with the baobab, Wickens and Lowe (2008:212) identified the Giant or Verreaux's Eagle Owl (*Bubo lacteus*), which they described as having "a penchant for other birds' nests, which it either finds abandoned or forcibly takes over, such as the massive nest of sticks and mud of the aptly named hamerkop (*Scopus umbretta*) or that of the barn owl (*Tyto alba*)." In their study of the species composition of the prey of the Barn Owl (based on an analysis of regurgitated owl pellets collected from 1960 to 1961 in several parts of the Kruger National Park in South Africa), Coetzee (1963:119) indicated that one sample was collected from a hollow baobab. The Hadza have names for three species of owls (Peterson et al. 2013:220–221): Barn Owl, Marsh Owl (*Asio capensis*), and Pearl-spotted Owlet (*Glaucidium perlatum*). The Pearl-Spotted Owlet has also been associated with the baobab (Geraghty 2016).

Water Birds

The baobab attracts water birds when it grows close to water, especially when it is a waterside tree. "Water birds" is defined here simply as birds associated with water; in familiar descriptive terms, these include near-water birds, shorebirds, waders, waterfowl, seabirds, and in-water birds. Marlowe (2010:110–112) listed 19 species of birds in his table of Hadza foods. Fifteen of these were water birds, and the remaining four were relatively large terrestrial birds. It is not surprising that most of the birds Marlowe (2010:110–112) listed are water birds (Braun et al. 2010) since they are, because of their size, a big prize for hunters. Water birds are particularly important in the spring breeding season when they serve as rainfall indicators and a source of hunger-season foods. The Kidero Hills, in the heartland of the traditional territory of the Hadza, is ringed by aquatic environments that include the Barai River and its wetlands to the north, Lake Eyasi to the west, Yaeda Swamp to the south, and Lake Manyara to the east (Matagi 2004). In light of this, it is understandable that native and migratory aquatic birds are important to the Hadza, and that some of these aquatic birds might be routinely associated with near-water or waterside baobabs (Mlingwa: ND).

Pelicans

The baobab is home to birds that nest on its branches and in its cavities. Based on the personal observations of Wickens and citing Molien (1967), Wickens and Lowe (2008:212) reported that in West Africa, colonies of what is called the Great White Pelican or Rosy Pelican (*Pelecanus onocrotalus*) may sometimes nest on the branches of the baobab (see also Roberts 1940:16). The eight species of globally distributed pelicans are near-shore seabirds that are also found in inland aquatic environments. In "the latter part of the dry season" in the Sudan, Welman (1966:55) reported seeing flocks of migrating ibises and pelicans perched in a baobab close to his residence. The Great White Pelican and the Pink-backed Pelican (*Pelecanus rufescens*) are the two species that occur in Kenya and northern Tanzania (Zimmerman et al. 1999:274–275). Marlowe (2010:212) only included the Great White Pelican on his list of Hadza birds, but the list of Hadza bird names identified by Peterson et al.

(2013:220–222) indicates that the Hadza have a general name for pelicans, and specific names for the Great White Pelican and the Pink-backed Pelican.

Of the Pink-backed Pelican Long (1956:39) wrote: "If they are travelling to a new site or fishing ground and perhaps when on migration they fly in long lines or a large V. They always fly at a great altitude and it is a wonderful sight to see them mount higher and higher as they circle about." Birds are deeply inspiring for people everywhere. This is evident in the place birds occupy in the sky-earth oppositional symbolism characteristic of the religious cosmology that is fundamental to all cultures. Because birds fly, they share with mountains and trees a link to the sky as the spiritual realm, which, traditions hold, is a link to spirits, particularly the Supreme Being, contact spirits, and ancestors. For the Hadza, pelicans are indicators of future good fortune. According to Peterson et al. (2013:221), the "Hadza consider pelicans circling over a camp to be a blessing. They respond with a celebratory song, which pelicans respond to by circling until the singing stops."

Hamerkop

Hamerkop (*Scopus umbretta*) belongs to the order Pelecaniformes and is the sole representative of the family Scopidae. It is endemic to Africa and Madagascar and is widespread in Kenya and northern Tanzania. Hamerkop is a brown medium-size wading bird that feeds on the amphibians, fish, and crustaceans of the still or slow-moving shallow waters of lakes, marshes, rivers, estuaries, lagoons, and mangrove creeks (Searle et al. 1977; Williams and Arlott 1986:28; Borrow and Demey 2014:60). This bird's link to the baobab is conspicuous because it preferentially builds its massive chambered nest of sticks and mud in tree forks over or near water (Zimmerman et al. 1996:322), including those of the baobab (Wickens and Lowe 2008:212). In a second "expedition" to Angola to obtain "accessory materials" for a museum exhibit, Bowen (1932) collected the skins of 64 birds, representing 43 species. They included three Hamerkop specimens from the Hanya River. In his account, Bowen (1932:283) wrote that "Near by we saw two large globe-like nests of sticks and grass, about 6 feet in diameter placed about 25 feet high in baobab trees." Photographs of Hamerkop nests in baobab trees can be seen on commercial websites. The adaptable Hamerkop is not completely tree dependent in its nesting behavior, however. In the absence of suitable trees, especially trees close to water, the Hamerkop builds its nests on riverbanks and cliffs, on human structures such as walls, dams, and irrigation works, and even on the ground (Wilson and Wilson 1986).

Storks, Ibises, Egrets, and Herons

Storks, ibises, herons, and egrets are wading birds that occupy the same kinds of aquatic environments as the Hamerkop. The nineteen species of storks which occur worldwide are members of the Order Ciconiiformes, which also includes spoonbills and flamingos. Zimmerman et al. (1996:283–284) and colleagues identified eight species of storks in Kenya and northern Tanzania, but Wickens and Lowe (2008:129) only mentioned an association between the baobab and the Marabou Stork (*Leptoptilos crumeniferus*). The Marabou Stork and Black Stork (*Ciconia nigra*) are two of the five species of storks listed by Marlowe (2010:112). The species most associated with the baobab is the Marabou Stork; it is known to perch, roost, and nest on the branches of the baobab. In his study of the breeding colonies of large aquatic birds in the Port Herald District of what was then Nyasaland and is now Malawi, Charles Long (1956) indicated that in 1955 there were four Marabou nests in a baobab at the hospital of Port Herald, two in Cataika village north of Port Herald, and five in a baobab in Mbenje village south of Port Herald.

Peterson et al. (2013:221) also reported Hadza names for five species of storks that included the Marabou Stork, but unlike Marlowe (2010:112), he did not include the Black Stork. In their study of the importance of roosts for Black Storks that winter in West Africa, Chevallier et al. (2010:93) wrote:

> Of the 47 tree species in ten different families inventoried in the Nazinga reserve [in southern Burkina Faso, West Africa], Black Storks used twelve species for nocturnal roosting. African Birch *Anogeissus leiocarpus*, Wild Seringa *Burkea africana* and African Baobab *Adansonia digitata* were used most frequently as roosting sites. For example, we monitored the presence of birds continuously for *Adansonia digitata* (n=35), one of the most commonly used trees by roosting Black Storks. The trees were used by 610 Black Storks during October 2003 to March 2004, with a daily presence of 3–35 individuals (51–144 birds per month).

Storks are important to the Hadza (Oldfield 2010, 2017). Cooper (1949:9) reported:

> [The Hadza] … diet is not … confined to game animals and birds. They will shoot at almost anything with four legs, or with wings, that they come across. I once found a large group subsisting almost entirely – apart from the usual baobab fruits and honey – on the white European Storks [identified by Zimmerman and colleagues as simply White Stork, *Ciconia ciconia*], which fill the Yaida valley just after the rains have begun. They shoot the storks after they have gone to roost in the trees, often skewering two or three on one arrow. These birds are on the protected list in Tanganyika, but the numbers which these people kill can have little effect on the tens of thousands of storks which seem to fill the valley.

Ibises

The ibis is another wader among the nineteen birds listed by Marlowe (2010:167). Welman (1966) wrote of the great variety of animals associated with a baobab that he had observed over several months close to his residence in the Sudan. "I used the tree," he wrote, "as a kind of summerhouse and spent many hours there, lying in a deck-chair and doing little else but look." Welman (1966:155) noted that "In the latter part of the dry season the baobab's bare branches attracted few aerial visitors except the odd mournfully hooting hornbill or a flock of ibis or grey pelicans, who did not stay long." Welman does not identify the species involved, but there are four species of ibises in Kenya and northern Tanzania (Zimmerman et al. 1999:285). Of these, the Hadza have names for the Sacred Ibis (*Threskiomis aethiopicus*) and Hadada Ibis (*Bostrychia hagedash*) (Peterson et al. 2013:221).

Herons

In Kenya and northern Tanzania, Zimmerman et al. (1996:316–322) identified twelve species of the family Ardeidae that are called herons; among them was the Black-Crowned Night Heron, the last of the five aquatic birds on Marlowe's list (2010:112). Generally, herons frequent coastal and inland waters. A notable exception is the Black-crowned Heron, also identified as the Black-Headed Heron, which also frequents grasslands and pastures, where its prey includes rodents and large insects (Williams and Arlott 1986:26–28). This is the only heron for which Peterson et al. (2013:221) recorded a Hadza name. Zimmerman et al. (1996:317) described Black-Crowned Night Heron as active mainly at dusk and at night, acting alone or in small numbers and often roosting in "riverine trees."

Egrets

Other baobab-associated aquatic birds not mentioned by Marlowe include egrets and near-water birds, particularly the kingfishers and the African Fish Eagle and the Osprey. In their preliminary account of the birds of Tanzania's Songo Island, Werema and Nahonyo (2015) reported, "Of the species observed, the Little Egret *Egretta garzetta* and Cattle Egret *Bubulcus ibis* were found on active nests on a baobab tree." Zimmerman et al. (1999:278–279) identified six species of egret for Kenya and northern Tanzania, including the Little Egret (*Egretta g. garzetta*) and Cattle Egret (*Bubulcus i. ibis*); the Hadza have names for both of these birds as well as for the Great Egret (*Casmerodius albus melanorhynchos*) and Intermediate Egret (*Mesophoyx intermedia brachyrhyncha*) (Peterson et al. 2013:220).

Kingfishers, African Fish Eagles, and Ospreys

Kingfishers (of the family Alcedinidae) are known for foraging from the fringing trees and shrubs of rivers, lakes, and anthropogenic aquatic environments. Although Wickens and Lowe (2008:212) did not specify the species involved, they reported that kingfishers often "build their nests in holes and crevices" of baobab trees, and for most species of kingfishers, these baobabs would most likely be waterside and near-water trees (see also Owen 1974:8). Zimmerman et al. (1999:387–389) identified twelve species of kingfishers for Kenya and northern Tanzania, including the piscivorous genera *Alcedo*, *Megaceryle*, and *Ceryle*, and the genera *Ispidina* and

Halcyon, which seek terrestrial prey far from water. Peterson et al. (2013:221) only reported that the Hadza names for the Grey-headed Kingfisher (*Halcyon leucocephala*), Striped Kingfisher (*Halcyon chelicuti*), and Woodland Kingfisher (*Halcyon senegalensis*).

It is surprising that there is no mention in the literature of an association between baobabs growing near water and the African Fish Eagle (*Haliaeetus vocifer*) and Osprey (*Pandion haliaetus*). These near-water avian predators select prominent vantage points at large rivers, lakes, and lagoons, and it is likely that they, like the kingfishers, are associated with baobab trees. The African Fish Eagle is widespread in Africa, although Borrow and Demey (2014:82) describe it as "uncommon or rare" in some locations and "fairly common" in others. The Osprey, which also occurs throughout Africa, is described by Borrow and Demey (2014:82) as a "Locally common to uncommon Palearctic visitor," mainly in the period from the end of summer in September to mid-spring in April. Peterson et al. (2013:220) indicated that the Hadza do have a name for the African Fish Eagle, but no mention is made of the Osprey.

Game Birds

Lexically, "game bird" identifies any bird that is routinely hunted for food, commerce, or sport. Today such birds are mostly hunted for sport and are protected by game laws. Technically, the term is used to characterize the 250 chicken-like species of the order Galliformes represented by such familiar birds as guinea fowls, francolins, quails, and pheasants. According to Crowe (2016:1), "Africa is an evolutionary cradle of bird-kind for at least one of its major evolutionary branches, the gamebirds," and he notes that after ostrich-like birds, gallinaceous game birds (together with the duck/goose-like Anseriformes) "occupy the second oldest branch of the avian evolutionary tree." Game birds that have likely been important in human evolution are represented by fowls, which is to say, gallinaceous species such as those belonging to the families Numididae and Phasianidae; by water birds, especially ducks and geese; by the Columbidae family of the order Columbiformes, comprising pigeons and doves; by hornbills of the family Bucerotidae (Bucerotiformes); and by coursers and pratencoles of the family Glareolidae (Charadriiformes).

Marlowe (2010:110–112) identified nineteen species of Hadza game birds. In addition to the fifteen species of water birds, Marlowe (2010:110–112) highlighted four large terrestrial birds: two species of fowls (the Guineafowl *Numida meleagris* and Harlequin Quail *Coturnix delegorduei delegorguei*) and two species of coursers (the Violet-tipped Courser *Rhinoptilus chalcopterus* and Two-banded courser *Rhinoptilus africanus gracilis*). It is noteworthy that for contemporary African foragers, the same groups of large terrestrial game birds (sometimes represented by different species or varieties) are targeted (Ichikawa 1998). For example, Yellen and Lee (1976:37) noted that "only 8 species [of some eighty species] of birds are systematically hunted by the San for food: guinea fowl, francolin (two species), korhaan, kori bustard, sandgrouse, cape turtle dove, and the red-billed teal."

The Hadza have names for six species of francolins (which includes the Grey-breasted Spurfowl, *Francolinus rufopictus*), and it is surprising that Marlowe (2010) did not include francolins among the nineteen species he listed. Ndagala and Waane (1982:96) described francolins and guinea fowls as two of the seven animals the Hadza hunted the most, the others being dik-dik, impala, zebra, baboon, and wart-hog. Cooper (1949:9) reported:

> At several camps which I visited there were no signs of any recent hunting, except of birds. All the game-birds – Francolin, Bustard and Guinea-fowl – are extremely wary and hard to shoot in Kindiga-land. They all appear to know the range of an arrow, which usually causes them to get up well beyond the range of a shot-gun and the only effective weapon against them, as I have found, is the .22 rifle. Even then, one can seldom get a shot at under fifty yards or so.

While Peterson et al. (2013:220–222) indicated that the Hadza do have names for members of the family Otididae, including the Kori Bustard (*Ardeotis kori*), Black-bellied Bustard (*Eupodotis melanogaster*), Buff-crested Bustard (*Eupodotis gindiana*), and White-bellied Bustard (*Eupodotis senegalensis*), they did not indicate Hadza names for the Red-crested korhaan, also called Red-crested bustard, *Lophotis ruficrista*; or for the Southern Black Korhaan, also called the Black Bustard (*Afrotis afra*). Both of these korhaans occur in southern Africa, including Angola, Botswana, Mozambique, Namibia, South Africa, Swaziland, Zambia, and Zimbabwe. In the comparison of Hadza and San game birds, it is also noteworthy that the Hadza have a name for the Knob-billed Duck (*Sarkidiomis melanotos*), but not the Red-billed Teal (*Anas erythrorhyncha*), a highly gregarious non-migratory dabbling duck that is abundant in southern Africa and some parts of eastern Africa. Although the Hadza have the same terrestrial game birds as the San, the Hadza mosaic environment provides greater access to aquatic birds, as Marlowe's list makes clear. According to Lee (1979:29), among the San, the stork is only "eaten by some persons" rather than "systematically" or "occasionally hunted."

Guinea Fowls

Another of the nineteen species of birds identified by Marlowe (2010:112) as associated with the baobab is the guinea fowl (*Numida* spp.), one of the most important birds for the Hadza, both as food and as a source of the feathers for their arrows. The guinea fowl occurs in forest, woodland, and savanna environments. In Kenya and northern Tanzania, the group is represented by three large terrestrial species that are conspicuous for their dark-speckled plumage: the Crested Guinea Fowl (*Guttera pucherani*), Vulturine Guinea Fowl (*Acryllium vulturinum*), and Helmeted Guinea Fowl (*Numida meleagris*). The most widespread of these is the Helmeted Guinea Fowl (Zimmerman et al. 1999:319–320).

As abundant as game birds are in the forests, woodlands, and grasslands of Africa, and as much as they are hunted for food, commerce, and sport, one would think there would be many accounts, however casual, of the baobab's association with game birds. Some of the most familiar terrestrial game birds use trees for roosting and as an escape perch when threatened by predators, and although many game birds nest on the ground, some nest in trees. Of the many terrestrial game birds of Africa that belong to the order Galliformes, only the guinea fowl has been identified with the baobab. Wickens and Lowe (2008:214), who cited Mshigeni and Hangula (2001), reported that "In Namibia, guinea fowl (Numidinae) scrape their nests beneath the shady canopy of the baobab." It would be interesting to know if the association of guinea fowl with the baobab also holds true in the environment of the Hadza. Bagshawe (1925:124) wrote that apart from a fish trap made of reeds, "The only trap which they use is a running noose with which they catch guinea-fowl and other birds." It is possible that the cordage for making the noose was made from baobab fiber.

On their website, the Zazamalala Foundation (2019) provided a brief, but highly informative, account of the bird associates of the largest baobab (*Adansonia za*) of the Zazamalala forest in the Morondava area of west Madagascar. The article pointed to a relationship between the baobab (which hosted "more than 400 nests a year") and water birds, raptors, and large terrestrial game birds, including the guineafowl.

> Depending on the season, this huge tree hosts [not only] Squacco heron and Cattle egret, Green-backed heron, Large white egret, Black heron, Dimorphic egret, Little bittern, Black-crowned night-heron, Glossy ibis, but also rare and endangered species, such as Humboldt's heron, Madagascar pond-heron, and Madagascar sacred ibis. … Most of the year there is the happy sound of busy parents and hungry young, but tragedy is always near. Frequently young birds tumble down and sometimes with a dramatic result when they are injured and fall victim to one of the predators waiting in the vicinity, such as Indian civet cat and Narrow-stripped mongoose. Small scavengers and opportunists visit the area around the baobab every day, looking for dead birds, rotten eggs or the food that clumsy young birds spilled, sometimes entire tilapia or snakehead fish. Lots of cockroaches and beetles lurk in the leaf litter and underneath fallen branches, feeding on the bird excrements and of course, they attract tenrecs, Spiney-tail iguanids and other reptiles, as well as ground-dwelling birds, such as giant coua, Madagascar Magpie-robin, and Guineafowl. When the young birds begin with their flying lessons, also birds of prey appear, Black kites circling around the colony, Madagascar buzzard and Harrier-hawk patiently waiting in adjacent tree tops for fallen chicks. Our employees check the area regularly and often return with fallen birds that usually do well on a diet of fish, leaving us in good shape within a couple of weeks.

Doves

The Columbidae is a large group comprised of 45 genera represented by 310 species of which 16 occur in Tanzania (Zimmerman et al. 1996:422). Doves are not routinely identified with the baobab in the literature, but a few sightings have been reported. Hellekson (2009:9) saw four species of doves in the baobab she studied in

Namibia: the African Mourning Dove (*Streptopelia decipiens*), the Laughing Dove (*Streptopelia senegalensis*), the Cape Turtle Dove (*Streptopelia capicola*), also known as the Ringneck Dove and Half-collard Dove, and one "unknown" species Hellekson could not identify in the poor light of dusk. Kopij (2019:357), too, reported the presence of Laughing Doves nesting in one of the large urban baobab trees in the town of Outapi in northern Namibia. And in their observations of birds of the Soyo area in northwest Angola, Stavrou and Mills (2013:32) reported seeing Cape Turtle Doves "in a baobab on the coastal road." The African Mourning Dove, Laughing Dove, and Cape Turtle Dove or Ringneck Dove were all included in the 19 species of "pigeons and doves" identified for Kenya and northern Tanzania by Zimmerman et al. (1999:359–362). The Hadza have a generic name for doves, and they recognize five species including the African Mourning Dove, Laughing Dove, and Ringneck Dove (Peterson et al. 2013:220).

Birds that Nest on the Baobab

Mousebirds

The most noted birds of the baobab are those that nest in its holes or on its branches. In addition to raptors, water birds, and game birds, other branch-nesting birds include the mousebirds as well as weavers and a number of other passerine birds. The gregarious mousebirds, which form groups of about twenty individuals, occur today only in tropical Africa. The name mousebirds alludes to "Their habit of climbing and running about amongst branches with their long tails pointed downwards [which] gives them a very rodent-like appearance." They belong to the family Collidae in the order Colliformes. Zimmerman et al. (1999) described three species for Kenya and northern Tanzania. The Speckled Mousebird (*Colius striatus*) and the Blue-naped Mousebird (*Urocolius macrourus*) were part of the baobab-associated "native bushland" described by Werema (2017:36–37) for Godegode Village (in the Mpwapwa District of the Dodoma Region of east-central Tanzania); the study, titled "Preliminary comparison of birds inhabiting exotic Acacia and native bushland habitats in semi-arid east-central Tanzania," identified "native bushland habitats" as "areas of degraded native bushland … composed mainly of native *Acacia* spp., *Commiphora* spp. and baobab *Adansonia digitata* trees." Mousebirds build cup-shaped nests in trees, and Kopij (2019:357) found the Red-faced Mousebird (*Urocolius indicus*) breeding in the baobabs of Outapi, Namibia. Peterson et al. (2013:221) reported Hadza names for the Blue-naped Mousebird and the Speckled Mousebird, but not the Red-faced Mousebird, which is common in southern Africa; it "occurs in bush-country from southern Tanzania south through Zambia and Malawi to the Zambesi River" (Williams and Arlott 1986:122).

Weaverbirds

The conspicuous presence of weaver nests in baobab trees is often noted and photographically illustrated in published accounts. Eight species are identified with the baobab (Weaver Watch 2020). Of these, four regularly nest in the baobab and four only occasionally. One of the best-known of the four that regularly nest in the baobab is the Red-billed Buffalo Weaver (*Bubalornis niger*), which builds thorny stick nests on branches to fend off snakes. In their evaluation of the sustainability of baobab harvesting in the Gweta Area of Botswana, Rhodes and Setshogo (2012:139) reported:

> Baobabs appear to play a significant ecological role in that they provide the main nesting site for the red-billed buffalo weaver (*Bubalornis niger*) ... 34 of the 72 trees (47%), including some of the larger young ones, contained weaver nests. The nests tended to be constructed towards the ends of branches and were situated almost exclusively on the western side of the trees. The mean number of nest per tree was 5; the minimum and maximum were 1 and 13, respectively. No buffalo weaver nests were seen in trees other than baobabs.

The other three species of weavers that regularly nest in the baobab are the White-billed Buffalo Weaver (*Bubalornis albirostris*), which also builds large communal stick nests, the Red-headed Weaver (*Anaplectes rubriceps*), whose nests hang like baskets from the branches of the baobab, and the Sakalava Weaver (*Ploceus sakalava*), a colonial weaver endemic to Madagascar that is found in the drier western parts of the island. Of these baobab weavers, the two outstanding species are the Red-billed and White-billed Buffalo Weavers. According to the Weaver Watch (2020), They "are so named because when Andrew Smith first saw Red-billed Buffalo Weavers, they were associating with buffalo. They do not normally do so, however. Both Red-billed and White-billed Buffalo Weavers could more aptly be called Baobab Weavers, since they breed in baobab trees so often."

The four species of weavers that only occasionally nest in the baobab are the White-headed Buffalo Weaver (*Dinemellia dinemelli*), the Lesser Masked Weaver (*Ploceus intermedius*), Heuglin's Masked Weaver (*Ploceus heuglini*), and the Chestnut Weaver (*Ploceus rubiginosus* in Kenya). In addition to the species identified by the Weaver Watch (2020), Kopij (2019:357) also observed the Scaly-feathered Weavers (*Sporopipes squamifrons*) and Southern Masked Weavers (*Ploceus velatus*) nesting in the baobabs of Outapi, Namibia.

The WeaveResearch Unit did not mention the Red-billed Quelea (*Quelea quelea*). In Mali, Peters et al. (2012:591–592) examined the link between this baobab-associated weaver and the spread of plant viruses by birds. Several studies of the spread of the *Rice yellow mottle virus* (FYMV) in Mali were done between 1995 and 2005, and they reported:

> On two occasions ... nests of weaverbirds were found in baobab trees. At one location, a tree with approximately 80 weaverbird (*Quelea quelea*) nests was in a field with hardly any infection. However, infected plants were found in an area within a circle of 40–50 m in diameter around the tree. ... The second baobab tree with nesting weaverbirds was in a small field in which 10–15% of the plants were infected.

For African farmers, the Red-billed Quelea is one of the most destructive birds because of its impact on grain cultivation. In their review of the impacts of control measures against this weaver bird on other species, Mcwilliam and Cheke (2004:130) reported that "birds of prey, owls and passerines" have all been negatively impacted by the use of the organophosphate fenthion and explosives. These authors also noted that "Another control method, harvesting quelea as a source of protein, is benign and could contribute to local nutrition and economic needs." They concluded, however, that harvesting is unlikely to reduce quelea populations substantially." Mpala and colleagues (2015) present a different assessment in their article titled "Are quelea birds really a menace? Innovative use of indigenous knowledge systems in the harvesting and utilisation of quelea, *Quelea quelea lathamii.* in Hwange District of Matabeleland North Province [Zimbabwe]."

As in Zimbabwe, the Red-billed Weaver (*Quelea quelea*) is particularly important as a food source for the Hadza during the spring hunger season. In anticipation of the spring rains starting in October, and especially with their arrival, usually in November, the seasonal renewal of life comes into full view with flowering and the flush of new leaves. With this development comes an increase in the number of insects and other arthropod pollinators and herbivores. The baobab at this time of year, in addition to providing fruit pulp, seeds, and honey, is also a place to target birds that take advantage of the seasonal flush of small animals. This is especially true of nesting birds that start to breed around September, as reported by Mabulla (2003:49), and intra-African and Palearctic migratory birds that begin arriving around the same time.

Nestlings and migratory birds are important resources when other foods are scarce. Blurton Jones (2016:27), for example, reported that *Salvadora* berry trees "are also common nesting sites of weaver birds, sometimes in very extensive colonies, exploited by Hadza during the rainy season (seen by me in February 1989, and Marlowe 2010, p. 139; Crittenden 2009)." In her dissertation on allomaternal care and juvenile foraging among the Hadza, Crittenden (2009:97) was the first to report the importance of weaver birds as "a customary component of Hadza wet season diet." She documented that it was especially children from ten to thirteen years of age who gathered young birds, honey, berries, and tubers. A long stick was "sharpened and then used to pull individual nests down from the branches." Crittenden noted, "The hatchlings are exploited on a daily basis for only a short period of time, approximately five to eight weeks per year," but according to Marlowe (2010:139), "When the weaver birds, the red-billed Quelea (*Quelea quelea*), have their mass synchronized reproduction, Hadza eat almost nothing but the little chicks for 2–3 weeks." Blurton Jones (2016:17) found that in the camp where he stayed in February 1989, "11 children accompanied men and women on the very short walk to the [*Salvadora*] nesting trees, in periods of 124–672 min, collecting 140–3600 g of nestlings and eggs, for a presumably protein rich, mean return rate of 369.5 kcal/hr."

Crittenden (2009:98) also reported that children "target young birds that have recently left the nest," and she wrote:

> They prepare a stick (approximately 1–2 feet in length) with the sticky pulp of an inedible berry, called "rembo". The "rembo" is then used like a paste and spread on the stick, which is placed in a small body of water, typically a watering hole. When the young weaver birds fly to the water to drink, several birds land on the stick, get stuck in the paste, and are then captured by children waiting at the water's edge. The birds are prepared on a small fire tended by the older girls in the foraging party and consumed by the entire group.

In his account of the lack of substantial storage practices among the Hadza, Marlowe (2010:254) wrote:

> The Hadza do take advantage of bonanza resources such as the weaver birds (*Quelea quelea*) … But this does not lead to long-term storage or require any preparation other than finding an appropriate stick for yanking down the nests. They eat the chicks in short order, and a few days after the birds are all fledged, there are no more supplies to eat, even when they lay some out to dry. The same goes for the undushipi berries. In both cases, the proportion that was dried is eaten every day and gone within a week or less.

Studies that take seasonality into account suggest that the Hadza bring back more birds to camp during the spring than at any other time. Since spring is the hunger season, the consumption of birds at this time of the year, especially the nestlings of the spring breeding season, is one of the most important values of the baobab/bird association for the Hadza (Mabulla 2003:49).

Birds that Nest in Baobab Holes

The baobab's cavities, especially its hollow trunk, have been widely commented on in academic and popular accounts of the tree. The focus, however, has largely been on traditional uses, such as water container, shelter, storage, workspace, and tree crypt, and on incidental contemporary uses, such as dairy, bar, toilet, prison, bus stop, and storage barn. But in light of the discussion of the Greater Honeyguide as a brood parasite, especially of baobab hole nesters, the value of baobab cavities for birds deserves careful consideration. Ornithologists and other researchers have long documented the critical importance of nesting holes not only for many birds but also for a variety of other animals, most notably bees, reptiles, and rodents (e.g., Van Balen et al. (1982:1) and Joseph et al. (2011)). Taylor (2000:40), like many others, has noted the competition for tree holes. In their baobab documentary, Root and Root (1973) photographed the younger of two hornbill chicks being killed by bees in a hostile takeover of their nest. There is major competition for tree holes involving not only the parasitic Greater Honeyguide and the hornbill and bees, but also parrots, hornbills, and squirrels (Taylor 2000:40). According to von Haartman (1957:339), the greater success of hole-nesting birds compared with open nesters has been "confirmed by large bodies of data." In light of this, Haartman raised the question of whether "the classical theory about food as the ultimate cause limiting population density is not overdone." He concluded that "In the case of the hole-nesters it seems obvious that the number of holes, and not the amount of food, mostly acts as an ecological limiting factor, determining the maximum number of nesting pairs." With this in mind, it is significant that in her study of one of the large historic baobab of Upupa Fall, Namibia, Hellekson (2009:5) counted 248 holes. There are eighteen kinds of birds that nest in baobab holes. We have already touched on owls, kingfishers, and honeyguides; the other fifteen are parrots, lovebirds, turacos, swifts, spinetails, bee-eaters, rollers, hoopoes, wood hoopoes, hornbills, barbets, woodpeckers, swallows, white-eyes, and starlings.

Parrots and Lovebirds

"Parrots and their diminutives, lovebirds," wrote Watson (2007:127), "have a close affinity for baobabs. Wherever their ranges overlap the birds seem greatly attracted to the trees. They often nest in holes in the trunk, waddling clumsily around the branches, in distinct contrast to the screeching agility they show in the air." According to Zimmerman et al. (1996:427), there are 358 species of parrots worldwide of which 19 occur in Africa, and 6 in Tanzania. Nine of the African parrots are associated with the baobab, based on reports only from eastern and southern Africa. The Orange-bellied Parrot (*Poicephalus rufirventris*) is widespread in Kenya, but Zimmerman et al. (1999:363) did not indicate its presence in northeastern Tanzania. Nevertheless, Owen (1974:8) noted its baobab association. Zimmerman et al. (1996:32) included it among the eleven species of birds identified with "*Commiphora* bush with baobab trees, and Zimmerman et al. (1999:363) described it as "A characteristic parrot of dry bush areas, often near baobab trees." Peterson et al. (2013) did not report a Hadza name for this parrot.

The five species of Tanzania include the Red-fronted Parrot (Poicephalus *gulielmi massaicus*) in juniper forest, and the Grey Parrot (*Psittacus erithacus erithacus*), of which Zimmerman et al. (1999:362) identified over twenty years ago as "Now only in Kakamega Forest where probably fewer than 10 birds survive." The three remaining baobab-associated parrots are the Brown-headed Parrot (*Poicephalus cryptoxanthus*) of southeastern Africa, and the Orange-bellied Parrot (*Poicephalus rufirventris*) and Brown Parrot (or Meyer's Parrot, *Poicephalus meyeri*), both of which are savanna parrots.

Watson (2007:127) noted only that "Yellow-bellied parrots are said to eat baobab seeds, but otherwise the trees provide very little of the shoots, fruit and small seeds that make up the diets of all species of Poicephalus." Contrary to Watson's observations, the Meyer's Parrot is known to feed on the nectar of baobab flowers (Boyes and Perrin 2009, 2010). The Brown Parrot (*Poicephalus cryptoxanthus*) occurs widely in Kenya and in the eastern part of northern Tanzania and is the only species for which Peterson et al. (2013:221) provided a Hadza name. Chris Brown (2015:25) indicated that its nest "is usually a cavity in a large dead or living tree, either a natural hole or one made by a woodpecker or barbet (Hockey et al. 2005; Tarboton 2011)." He provided an illustrated account of a pair of Meyer's Parrots in a baobab "on the edge of the floodplain" of the Okavango River in the Mahango area of the Bwabwata National Park in Namibia.

> The bird drew our attention by its raucous screeching. It was soon apparent that there was a second bird calling in an un-melodious duet, but we could not locate it. After some time we traced the call to the vicinity of a scar some 9 m up and on the side of one of the large branches of the baobab. The scar was probably where a branch had broken off leaving a cavity. The cavity had been closed over with mud pellets by a swallow – typical of a Mosque Swallow *Hirundo senegalensis*, to create a nesting site.

In another account from southern Africa, Nina Geraghty (2016) posted a photograph taken by Vanessa Bristow of a Meyer's Parrot in a baobab tree on the Sentinel Ranch in Zimbabwe. In the Sudan, Sweeney (1974:62) reported seeing "several brown parrots feeding on broken baobab fruits on the ground."

According to Zimmerman et al. (1996:428), the Brown-headed Parrot "replaces Brown Parrot in the coastal lowlands," and they noted that it is "partial to baobab trees, but often roosts in coconut palms." In the second edition of the birds of Kenya and northern Tanzania, Zimmerman (1999:363) described it as "dry-country parrot partial to *Commiphora* bush with baobab trees." Werema too (2017:36) identified it as part of the baobab bushland he studied in Tanzania. However, in a study of the behavioral ecology of the Brown-headed Parrot in Southern Africa, Taylor (2000:40–41) wrote:

> Some authorities mention *Adansonia digitata* (Baobabs) as the principal tree species. This is incorrect. The species is not fussy, as long as the cavity is suitable. I have seen nests in *Entandrophragma caudatum* (Wooden Banana or Mountain Mahogany) *Colophospermum mopane* (Mopani), *Acacia nigrescens* (Knobthorn), *Afzelia quansensis* (Pod Mahogany), and various Combretum species, as well as Baobabs. The parrots have no preference whether the tree is dead or alive.

At two sites in Limpopo Province, South Africa, Symes and Perrin (2008) studied the habitat and associated bird assemblages of the Grey-headed Parrot (*Poicephalus fuscocollis suahelicus*), which reaches its southern limits in the Limpopo region. One site was Levubu, where the baobab population consisted of three garden specimens, and the other was Makuya. Although baobabs "were not common" at the Makuya site, these authors noted that they "were important in providing nesting sites for Grey-headed Parrots in the Area (Symes and Perrin 2004)." They concluded: "Most importantly to Grey-headed Parrots is the presence of certain seasonally abundant fruiting trees (e.g. *Parinari curatellifolia* at Levubu from August–October) and secure nest sites (e.g. *Adansonia digitata* at Makuya) during the breeding season (Symes and Perrin 2003, 2004)."

The Rüppell's Parrot of southwestern African, endemic to Namibia and Angola (Hellekson 2009:9), is also identified with the baobab.

Other baobab-associated members of the parrot family include the Yellow-collared Lovebird (*Agapornis personata*) and Fischer's Lovebird (*Agapornis fischeri*). Kopij (2019:357) identified the Rosy-faced Parrot, which is just another species of lovebirds, as nesting in baobab holes in Outapi, Namibia.

Rollers

Colorful rollers nest in baobab holes and use the tree as a hunting perch from which they can be seen "swooping to ground to catch large insects or small vertebrates" (Borrow and Demey 2014:270). Zimmerman et al. (1999:393–394) identified five species of rollers for Kenya and northern Tanzania. Of these, Watson (2007:128–129) said he "once watched a pair of lilac-breasted rollers [*Coracias caudata*] courting each other outside a small hole in the trunk of a Tanzania sea-edge baobab." And Kopij (2019:357) documented one case of the Lilac-breasted Roller nesting in the Outapi baobabs of Namibia. Peterson et al. (2013:221) indicated that the Hadza do have names for the Lilac Breasted Roller (*Coracias caudata*) and Rufous-crowned Roller (*Coracias naevia*).

Swifts

Zimmerman et al. (1996:454) reported 93 species of swifts worldwide, with 22 occurring in Africa, 12 of them in northern Tanzania. Swifts (of the family Apodidae) nest in the crevices of cliffs and caves, on buildings and bridges, and in "hollow trees" (Zimmerman et al. 1999:382). Wickens and Lowe (2008:212) noted only that "the gregarious swifts (*Apus* spp.)" are among the birds that "build their nests in holes and crevices of the [baobab's] trunk," but they do not identify the species involved. The appropriately named Palm Swift (*Cypsiurus parvus*) nests in a variety of palms, the most often referenced being the dum, Borassus, coconut, and oil palm (Williams and Arlott 1986:120). But Kopij (2019) also observed Palm Swifts nesting in the baobab trees of Outapi, Namibia. The Hadza have names for the African Palm Swift, the Little Swift (*Apus affinis*), and the Common Swift (*Apus apus* (Peterson et al. 2013:222).

Spinetails

Spinetails nest in the baobab's hollow trunk, and in this they are different from other baobab hole-nesting birds. In their field guide to the birds of East Africa, Williams and Arlott (1986:120–121) described the Mottled-throated Spinetail (*Telacanthura ussheri*) as "Local and uncommon in East and Central Africa." They noted that like the Boehm's Spinetail (*Neafrapus boehmi*), the Mottled-throated Spinetail was "often found in the vicinity of baobab trees in which it nests." In Kenya and northern Tanzania, Zimmerman et al. (1996:32) included the Mottled-throated Spinetail and Bohm's Spinetail among the birds associated with "*Commiphora* bush with baobab trees." Zimmerman and other colleagues (1999:455) have also indicated that the Mottled-throated Spinetail can be found "In pairs or small groups, foraging around or over trees, sometimes with Little Swifts [*Apus affines*], and typically in deciduous woodlands dominated by baobab trees." According to Wickens and Lowe (2008:130), "These birds acquired their common name from spines at the end of their tail feathers, not visible unless the bird is being held, which are used to prop the birds against the vertical inner walls of the hollow baobabs." These authors also note that "In South Africa, the Boehm's Spinetail is reported to breed and roost almost exclusively in baobabs and its continued survival may depend upon the conservation of the baobabs." According to Brooke (2018), "Conservation of Baobabs should ensure the local conservation of the Mottled Spinetail which is "rare" in South Africa (Brooke 1984b)."

In southern Africa, Brooke (2018) described the "principle habitat" of the Mottled-throated Spinetail as the "hot, low-lying valleys in Mopane veld in which Baobabs *Adansonia digitata* are common (Brooke 1971b; Benson and Winterbottom 1977)." The 1200-year-old Sagole Baobab (or Sagole Big Tree, as it is also called) is famous for being one of the largest baobabs in South Africa and is well known for its resident colony of Mottled-throated Spinetail (Pakenham 2004:87). Zimmerman et al. (1999:382) identified three species of spinetails in Kenya and northern Tanzania: the Sabine's, Mottled-throated, and Böhm's. The Sabine's Spinetail, they noted, is associated with forest and uplands, but the Mottled and Böhm's Spinetails are particularly common in coastal lowlands. Zimmerman et al. (1999) did not indicate the presence of spinetails in northern Tanzania, and Peterson et al. (2013:222) only provided Hadza common names for the Mottled-throated Spinetail.

Hoopoes

Hoopoes (*Upupa africana*), which are found in Africa, Europe, and Asia, are notable for their distinctive crown of feathers. The ground-feeding insectivorous African Hoopoe is a bird of savanna environments that nests in tree holes and crevices and prefers to feed in areas of short grass and bare soil. The Hoopoe also nests in the holes and crevices of cliffs, boulders, and the base of ant hills; it will even take over abandoned burrows. In the human environment, it has also taken up residence in walls, old buildings, and bird boxes (Searle et al. 1977; Fry 1985:289–290; Williams and Arlott 1986). Wickens and Lowe (2008:215) reported only that it "sometimes builds its nest in the crevices of a baobab trunk," but Kopij (2019:357) identified it among the twenty-one species of birds he found nesting in the old baobab trees of Outapi, Namibia; the Hoopoe was one of the six species identified as "strictly associated with large baobab trees." The Hoopoe was included in a comment that indicates the current cultural significance and economic and health-related uses of African birds; Shackleton et al. (2018:6) wrote of "low-density soft wood trees" like the baobab, fig (Moraceae), and marula (Anacardiaceae) as "keystone sheltering structures." In discussing these trees, they noted that "All of the hole-nesters (barbets, hoopoes, hornbills, woodpeckers) are commercially traded in Africa for traditional medicine (Williams et al. 2014)."

Hornbills

Hornbills are members of the Bucerotidae, a family of some 55 species of birds found in Africa, Asia, and Melanesia. Zimmerman et al. (1999:397) indicated that most hornbills nest in the cavities of cliffs and trees, and they are known for living in baobab holes (Root and Root 1973; Owen 1974:8; Watson 2007:127; Wickens and Lowe 2008:212–213). Root and Root (1973) showed the use of a glass-covered cutaway to film a remarkable sequence of a pair of hornbills nesting in a baobab. A similar sequence was produced by the filmmakers Deeble and Stone (2005) in their documentary on the sycamore fig (*Ficus sycomorus*). Like other savanna birds such as the hamerkop and weavers, hornbills face threats from predatory snakes, which has shaped their behavior, especially in the spring when they are nesting. The female uses mud and other materials brought by her mate to seal off the nest entrance, leaving only a narrow slit through which she receives food from her mate. She lays up to three eggs; often, only one chick survives to maturity (Kemp and Grimes 1985). Without identifying a particular species, Wickens and Lowe (2008:129) noted that hornbills nest in baobab cavities. However, Watson (2007:127) and Hellekson (2009:9) specifically mentioned the baobab's association with the Eastern Yellow-billed Hornbill (*Tockus flavirostris*), though Peterson et al. (2013) did not report a Hadza name for this species.

The Southern Ground Hornbill is widespread in tropical Africa, with deep cultural significance throughout its range. Coetzee et al. (2014:1) indicated that the main themes of their research involved the Southern Ground Hornbill as "(a) a bringer or signifier of death/destruction/loss/deprivation, with the bird commonly being regarded as a bad omen or evil spirits and announcer of calamities; (b) a protector against evil spirits, against lightning and against drought; (c) an enabler/causer of altered perceptions, which include remote viewing, foreseeing the future, and creating an illusion; and (d) a timekeeper that announces the beginning and end of a working day and of seasonal changes." It is noteworthy that the Hadza share some of these beliefs associated with the hornbill, which point to the importance of birds as indicator species.

Zimmerman et al. (1999:400) identified twelve species of hornbills for Kenya and northern Tanzania. Of these, the Hadza have names for four (Peterson et al. 2013:68), including the Von Der Decken's Hornbill and the Ground Hornbill, which are identified as rain indicators. The call of the Van Der Decken's hornbill "means the rains are close," while that of the ground hornbill "is considered a harbinger of rain and good time ahead." According to Peterson et al. (2013:68), the Ground Hornbill (which is also called the Southern Ground Hornbill) is one of the few birds not eaten by the Hadza. Current research suggests that other hornbills are easily accessible to the Hadza. Hawkes et al. (1991:84) reported:

> Small animals were rarely taken by Hadza hunters during the period covered by our fieldwork, and returns measured by mass were extremely low. When caught, they were usually consumed before the hunter returned to camp. This means that our record of small animal captures comes from focal follows of foraging men. During 45 follows in 1985–1986, on

which a total of 75 hunter days were monitored, men shot at small game (mainly guinea fowl (*Guttera* spp.) and francolin (*Francolinus* spp.) about once or twice a day, but actually obtained only 14 individuals. Ten of these were immature hornbills, snatched from the nest as the hunter walked by.

Barbets

Barbets comprise eighty-four species in eleven genera found mostly in the tropics of the Americas, Africa, and Asia. Like woodpeckers, barbets play a part in the avian diversity the baobab supports. In their field guide to the birds of Western Africa, Borrow and Demey (2014:282) described barbets as "nesting in self-excavated tree holes." These birds may be especially attracted to the baobab. The Piciformes (2020) website noted that "because the bill of the barbet is not as 'chisel-like' as a woodpecker they often dig their nesting holes into softer wood or plant materials." The site also reports that "an African species digs into a dirt bank." Neither Watson (2007:128) nor Wickens and Lowe (2008:212) specified the kind of barbets that nest in the baobab. However, Peterson et al. (2013:220) listed names for three species of barbets. Two belong to the genus *Trachyphonus*, including *Trachyphonus erythrocephalus*, which "nest and roost in holes in river banks," and *Trachyphonus usambiro*, which "nest and roost in holes in the ground" with communal roosts having "as many as ten birds." The third species was the widespread White-headed Barbet, *Lybuis leucocephalus*; like the majority of barbets and their related tinkerbirds, this species roosts and nests "in cavities excavated in soft wood" (Zimmerman et al. 1999:400). Kopij (2019:357) identified the Black-collard Barbet (*Lybius torquatus*) and Acacia Pied Barbet (*Tricholaema leucomelas*) as nesting in the baobab trees of Outapi, Namibia.

Woodpeckers

The baobab's cavities, especially its hollow trunk, have been widely commented on in academic and popular accounts. But the focus has largely been on the cavities' traditional uses for containing water, shelter, storage, workspace, and tree crypt, and its incidental contemporary uses such for such things as dairy, bar, toilet, prison, bus stop, and storage barn. But in light of the discussion of the Greater Honeyguide as a brood parasite, especially as a baobab hole-nester, the value of baobab cavities for birds deserves consideration. Along with the Greater Honey Guide, the other eleven kinds of birds that have been identified as baobab hole-nesters are woodpeckers, barbets, parrots, love birds, hornbills, hoopoes, rollers, starlings, swifts, Swallows, and owls.

Woodpeckers, like barbets, differ from most other baobab hole-nesting birds in their ability to produce their own holes (Watson 2007:128). When these tree excavators abandon their holes, the holes are often taken over by other birds, small reptiles, or mammals. In this sense, the woodpecker can be described as the home-builder for other cavity-nesting species. Woodpeckers are identified in current research as a keystone species and, as such, an important environmental indicator of biodiversity (Mikusiński et al. 2001; Joseph et al. 2011). Zimmerman et al. (1999:408–412) identified fifteen species of woodpeckers for Kenya and northern Tanzania, but there are few reports of woodpeckers being associated with the baobab. Hellekson (2009:8) observed an "unknown woodpecker" in the tree she studied at Epupa Falls, Namibia. In a blog on the Baobab Foundation website, Nina Geraghty (2016) posted photographs by Vanessa Bristow of a pair of Bennett's woodpeckers in a baobab in Zimbabwe. Another picture of "Grey Woodpeckers feeding in a baobab" taken by Bram Piot was presented on the Africana Bird Club website (2017).

Peterson et al. (2013:222) indicated that the Hadza call the Nubian Woodpecker *kulupuko*, which is also their name for woodpeckers in general, but the names for other specific kinds of woodpeckers were not given. Peterson et al. (2013:69) also reported a belief "identical to that of the Dorobo and Maasai" that the location of the Nubian Woodpecker's call indicates good or bad future outcomes. "If [the Nubian Woodpecker] … calls in front it means danger (e.g. buffalo), on the left, bad luck (e.g. bush camp you are going to have hunger), on the right it means good luck (e.g. arrow will fly true), from behind one can ignore."

Swallows

David Bryant (1985:571) noted that "In many aspects of morphology, habits, and their insect diet taken on the wing, they [i.e., Swallows, mostly of the family Hirundinidae belonging to the Passeriformes, suborder Oscines] are convergent with the swifts (Apodidae)." Swallows that nest in baobab holes include the Mosque Swallow (*Hirundo senegalensis*), a Palearctic migrant; the Lesser Striped Swallow (*Hirundo abyssinica*), an intra-African migrant (Brown 2015); and the Wire-tailed Swallow (*Hirundo smithii*), which occurs throughout Africa (Kopij 2019). The association between the Mosque Swallow and the baobab has long been noted in the literature (Sclater and Moreau 1933; Owen 1974; Williams and Arlott (1986:262); Wickens and Lowe 2008; Borrow and Demey 2014; Werema 2017; Kopij 2019). According to Zimmerman et al. (1996:515), this widespread swallow of Kenyan, Tanzania, and elsewhere builds its "gourd-shaped mud nest" in hollow trees, "especially" the baobab. Peterson et al. (2013:222) noted Hadza names for swallows include the Mosque Swallow, Barn Swallow (*Hirundo rustica*), and Rock Martin Swallow (*Ptyonoprogne fuligula*).

Starlings

Starlings, which belong to the family Sturnidae (suborder, Ocines), are small to medium-sized passerine birds native to the Old World. The family of some 25 genera comprising 110 species is well represented in Africa (Feare 1985; Zimmerman et al. 1996:454). It should be noted here that it is not just hole-excavating birds that augment the avian diversity and overall biodiversity that the baobab supports. Nest-building birds produce a similar effect when their abandoned nests are taken over by other birds such as starlings, as well as by arthropods, reptiles, and small mammals. According to Feare (1985:563), "Most species [of starlings] nest in holes, especially in trees, but holes in cliffs and buildings may also be used, as may holes bored by other species like woodpeckers and barbets." Starlings are among the small birds that have been reported to nest in the baobab because of the protective presence of the Wahlberg's Eagle, and because, like the Superb Starling (*Lamprotornis superbus*) in particular, they take over the abandoned nests of what Watson (2007:129) identifies as the buffalo weavers, and Wickens and Lowe (2008:213) as the white-head buffalo weavers. The widespread, gregarious Superb Starling is described by Williams and Arlott (1986:392) as "fearless" of human beings; they note this has made it a common bird in "the vicinity of human dwellings."

The Ashy Starling (*Cosmopsarus univolor*) is the species most often identified with the baobab in the literature (Brown 1965:169; Owen 1974:8; Watson 2007:129; Wickens and Lowe 2008:212). It is a Tanzania endemic that Williams and Arlott (1986:393) described as "locally common in bush country, acacia woodland and open park-like country with baobab trees." Wickens and Lowe (2008:212) identified the Red-winged Starling (*Onchognathus morio*) as nesting in the baobab. And, of the twenty-four species of starlings described for Kenya and northern Tanzania, Zimmerman et al. (1996:32) identified the Golden-breasted Starling (*Cosmopsarus regius*) as associated with "*Commiphora* bush with baobab trees" in southeastern Kenya. Hellekson (2009:9) reported sightings of the Meve's Starling (*Lamprotornis mevesii*) and the Pale-winged Starling (*Onychognathus nbouroup*) in the baobab tree she studied at the Upupa Falls in Namibia. Peterson et al. (2013:221) provided Hadza names for four species of starlings, including the baobab-associated Ashy Starling and Superb Starling, but not for the Golden-breasted or Red-wing starlings.

Birds and the Baobab's Flower

Although bats have long been identified as the principal pollinators of the night-blooming baobab, other floral feeders include bush babies, birds, bees, wasps, ants, and moths (Baum 1995a:322–323; 1995b:443–444). In addition to accommodating bats, Wickens and Lowe (2008:168) noted, "It is likely that the nocturnal opening also reduces pollen and nectar losses to diurnal animals such as bees and sunbirds (Baum 1995a)." Sunbirds, green wood-hoopoes, and canaries are known nectar-seeking visitors to the baobab's flower (Watson 2007:123–124). The Yellow-fronted Canary (*Serinus mozambicus*) is typically a ground feeder, but Wickens and Lowe (2008:213) noted that it "will feed on buds and flowers in the tree tops and is often associated with baobabs in the coastal districts of East Africa." The White-bellied Canary (*Crithagra dorsostriatus*), which is actually a species of finch in the family Fringillidae, was sighted in the baobab-associated "native bushland" of east-central Tanzania (Werema 2017:37). And the Black-throated Canary (*Serinus atrogularis*) was one of the 29 birds Kopij (2019) identified as breading in the baobabs of Outapi, Namibia.

The best-known representatives of nectar-feeding birds are the sunbirds of Africa and southern Asia; the hummingbirds and honeycreepers of the New World and the honeycreepers of Hawaiian; and the honeyeaters and flowerpeckers found mostly in Australasia (Nicolson and Fleming 2003). Sunbirds comprise a distinctive family (Nectariniidae) of nectar-feeding birds with pointed bills that are long, slender, and curved. Zimmerman et al. (1996:638) indicated that there were 120 species worldwide, with 74 in Africa and 26 in Tanzania. With the exception of the hummingbirds of the New World (where there are no native baobabs), sunbirds are frequent visitors to *Adansonia digitata* in Africa, and to several *Adansonia* species in Madagascar (Wickens and Lowe 2008:215), and honeyeaters are frequent visitors to *Adansonia gregorii* in Australia. Zimmerman et al. (1996:32) mentioned the presence of the Hunter's Sunbird (*Chalcomitra hunteri*) in the "*Commiphora* bush with baobab trees" near Mtito Andei in southeastern Kenya. At Godegode Village in the Dodoma Region of east-central Tanzania, Werema (2017:32, 37) recorded four species of sunbirds in "areas of degraded native bushland … composed mainly of native *Acacia* spp., *Commiphora* spp. And baobab *Adansonia digitata* trees." Three of these – the Western Violatel-backed Sunbird (*Anthreptes longuemerei*), Variable Sunbird (*Cinnyris venustus*), and Beautiful Sunbird (*Cinnyris pulchella*) – were only found in the "native bushland" and not in the "exotic *Acacia* bushland."

The Scarlet-chested Sunbird (*Chalcomitra senegalensis*) is one of the most widespread and abundant savanna member of the Nectariniidae. It is the only species found in both the native bushland and in the exotic *Acacia* bushland, and the only species Watson (2007:123–124) identified by name in his discussion of the baobab and sunbirds. According to Williams and Arlott (1986:330–331), the Scarlet-chested Sunbird is common in East Africa not only in forests margins and riverine acacias, but also in woodlands and tree-dotted grasslands and bushlands. They also describe it as a "common and conspicuous species" of "cultivation" that "often visits gardens." According to Zimmerman et al. (1996:643), the Scarlet-chested Sunbird "feeds low in garden flowers or at tall flowering trees." In West Africa Searle et al. (1977:233) described this showy sunbird as "often near human habitation and frequently close to the nest of a wasp." Sunbird flowers are generally tubular, and, according to Gill (1985:568), "typical sunbird flowers in Africa include species of *Erythrina*, *Spathodia*, *Symphonia*, *Loranthus*, *Aloe*, *Leonotis*, *Kniphofia*, Strelitizia, *Lobelia*, *Protea*, *Tecomaria*, *Haileria*, *Burchellia* and *Schotia*."

Sunbirds appears to be seasonally specialized in their feeding pattern. It is likely then that another reason for the sunbird's attraction to the baobab is not only its flowering during the annual rains but also the abundance of insects and other arthropods during the growing seasons from spring through autumn. Sunbirds are more nectarivorous in the dry time of year, when they feed on the flowers of the end of autumn, winter, and early spring, and they are more insectivorous during the annual rains from early spring in April to early autumn in October. Sunbirds breed during the annual rains and feed their young on insects (Pettet 1977). The start of the baobab flowering period in early spring and through the annual rains provides sunbirds with both nectar and arthropods.

In addition to nectar and arthropods, there is a third reason why sunbirds might be attracted to the baobab. It is possible that they also benefit from the floral nectar and fruit of the hemiparasitic mistletoes of the family Loranthaceae that they themselves have transported to the baobab (Nickrent 2011). Zimmerman et al. (1996:638) noted that sunbirds "tend to be solitary or in pairs, but regular seasonal movements in response to flowering can result in concentrations at such favoured plants as aloe, the orange-flowered mint *Leonotis*, *Erythrina* (coral-bean) trees, and parasite misteltoes." There are a number of studies of mistletoes of the family Loranthaceae being a serious pest of cocoa, coffee, and rubber plantations (e.g., Denis-Esdras et al. 2020b). It is interesting to note that the Scarlet-chested Sunbird in particular is considered a pest because it spreads the mistletoe in cocoa plantations (Cheke and Clive 2008:71–79). In their ethnobotanical study of the medicinal use of Loranthaceae species in the Ivory coast, Denis-Esdras et al. (2020a:46) gave an idea of the extent to which the sunbird is associated with loranthaceous mistletoes, and as a consequence, with the baobab. Under the section heading, "Famous host species of Loranthaceae," they wrote:

> Twenty-seven (27) known host species for which Loranthaceae species are actively sought were cited by traditional healers and herbalists in Aboisso and Maféré (Table 3). The Frequency of Citations (FC) of the plants cited by the respondents ranged from 0.29 to 0.70. … The specific host species with the highest FC value is *Mangifera indica* (56.72%) Fig. 6a. It is followed by: *Adansonia digitata* (46.27%), *Manihot esculenta* (46.27%) (Fig. 6), *Ceiba pentandra* (44.78%), *Bombax buonopozense* (43.28%) and *Alstonia boonei* (43.28%).

Along with the floral nectar of the baobab and mistletoe, baobab arthropods, and the mistletoe fruit, a final factor in the baobab/sunbird relation that should be noted here is that some sunbirds have been reported to roosts in hemiparasitic mistletoes, possibly in mistletoes that have parasitized the baobab. Peterson et al. (2013:221) recorded Hadza names for three species of sunbirds: the Scarlet-chested Sunbird, Eastern Violet-backed Sunbird (*Anthreptes orientalis*), and Variable Sunbird (*Cinnyris venustus*). No Hadza names were reported for any of the canaries.

Passerine Birds

In addition to weavers, starlings, swallows, and canaries, a number of other baobab-associated birds of the order Passeriformes (Gill et al. 2018) have been identified in the literature. As earlier noted, Marlowe's (2010:127) table of the species of plants and animals that make up the sixteen categories of Hadza foods indicated that 741 (or 84.4%) of the 878 were birds. Subtracting raptors, water birds, and game birds from the targeted population of 741 species suggests that the great majority of birds taken by the Hadza are passerines – that is, "perching birds," a common name appropriately applied to the order Passeriformes – and "song birds," to the suborder Ocines. It is not surprising that many passerines are associated with the baobab. Passerines constitute the largest order of birds with some 57,000 species representing seventy-four families, "compared with only about 4069 species of all other birds" (Britannica Online Encyclopedia 2020). Of this "largest order of birds and the dominant avian group on Earth today" (Gill et al. 2023), Hall and Moreau (1970) identified 962 species as breeding in Africa. With the exception of the most conspicuous members of the order, few passerines have been identified with the baobab. This is probably because the Passeriformes is extraordinarily diverse order not only with regard to genera and species, but also with regard to significant differences even within species – differences between males and females, juveniles and adults, and breeding and non-breeding individuals.

The baobab-associated passerines that have not been previously identified include the following: Red-eyed Bulbul, *Pycnonotus nigricans* (Hellekson 2009; Werema 2017; Kopij 2019); Short-toed Rock Thrush, *Manticola brevipes* (Hellekson 2009); Spotted Palm Thrush, *Cichladusa guttata* (Werema 2017); Familiar Chat, *Cercomela familiaris* (Hellekson 2009; Werema 2017); White-browed Robin Chat, (Werema 2017); African Grey Flycatcher, *Melaenornis microrhynchus* (Werema 2017); White-browed Scrub Robin, *Cercotrichas leucophrys* (Zimmerman et al. 1996; Werema 2017); wabler (Welman 1966); Black-chested Prinia, *Prinia flavicans* (Hellekson 2009; Werema 2017); Fork-tailed Drongo, *Dicrurus adsimilis* (Guy 1972; Wickens and Lowe 2008; Werema 2017); Pied Crow, *Corvus albus* (Werema 2017; Kopij 2019); Southern Grey-headed Sparrow, *Passer diffusus* (Hellekson 2009); Swahili Sparrow, *Passer suahelicus* (Werema 2017); Red-billed Firefinch, *Lagonosticta senegala* (Werema 2017); Blue Waxbill, *Uraeginthus angolensis* (Hellekson 2009; Kopij 2019); Cinnamon-breasted Bunting (Hellekson 2009:9); and Somali Golden-breasted Bunting, *Emberiza poliopleura* (Zimmerman et al. 1996).

Hellekson (2009:9) sighted 25 bird species in her study documenting the high biodiversity associated with a single baobab in Namibia. Fifteen were passerines, of which eleven were identified by scientific name, and four she was unable to identify because they "landed very briefly." She noted that the Cinnamon-breasted Rock Bunting (*Emberiza tahapisi*) and the Rosy-faced Lovebird (*Agapornis roseicollis*) were by far the most frequently sighted in the baobab, or were seen flying overhead. The Somali Golden-breasted Bunting (*Emberiza poliopleura*) was one of the birds Zimmerman et al. (1996:32) identified with the "*Commiphora* bush with baobab trees" in southeastern Kenya. Peterson et al. (2013:220) indicated that the Hadza have names for the Cinnamon-breasted Rock Bunting and the Golden-breasted Bunting (*Emberiza flaviventris*); they did not report a Hadza name for the Somali Golden-breasted Bunting, which Zimmerman et al. (1996:703) described as recorded in Kenya, but only "sporadically at Olorgesailie in the Rift Valley." The

Hadza have names for most of the passerines identified here (see Peterson et al. 2013:220–222), although not all the species identified are represented in their environment. For example, Peterson et al. (2013:220–221) reported Hadza names for the Dark-capped Bulbul (*Pictnonotus tricolor*), but not the Red-eyed Bulbul (*Pictnonotus nigricans*), which is found in the savanna environment of southwestern Africa.

Kopij (2019) and Passerine Birds Nesting in the Old Baobabs of Outapi, Namibia

In his study of the population density and structure of birds breeding in the urban habitat of the town Outapi in northern Namibia that was dominated by large baobab trees, Kopij (2019:357) identified 29 species of birds of which 15 were linked to the baobab; and of these, 6 were identified as "species strictly associated with large baobab trees": the Pied Crow (*Corvus albus*), Red-eyed Bulbul (*Pycnonotus nigricans*), Black-throated Canary (*Serinus atrogularis*), African Hoopoe (*Upupa africana*), Black-collared Barbet (*Lybius torquatus*), and Rosy-faced Parrot (*Agapornis roseicollis*). Of the 15 baobab associates, 10 were passerine birds, and it is noteworthy that of the 6 "strictly associated" with Outapi's baobab trees, 3 were passerines: the Pied Crow (*Corvus albus*), Red-eye Bulbul (*Pycnonotus nigricans*), and Black-throated Canary (*Serinus atrogularis*). Kopij (2019:357) identified the House Sparrow (*Passer domesticus*), Blue Waxbill (*Uraeginthus angolensis*), and African Palm Swift (*Cypsiurus parvus*) as the dominant species in his study, and it is surprising that the "alien House Sparrow was by far the most numerous species comprising 48.4% of all breeding birds," and that it was the only species that did not nest in Outapi's baobab trees. Kopij suggest the reason for this is the presence of the Pied Crow.

The Pied Crow is an especially noteworthy bird of Kopij's study. Crows, largely ground-feeding omnivores renowned for their resourcefulness, are one of the most interesting groups of passerine birds with links to the baobab. Crows and allied taxa of the family Corvidae include 118 species worldwide with 15 in Africa, 7 in Kenya, and 5 in northern Tanzania, one of which was introduced (Zimmerman et al. 1996:626). The Pied Crow (*Corbus albus*) is one of the most successful of the crows, well known for its ability to thrive in a variety of environments. Zimmerman et al. (1996:212) described its distribution in Kenya and northern Tanzania as "Widespread up to 30000 m, including all urban areas." The Pied Crow is identified as a bird of "open country" that also frequents riparian woods, marshes, and lakeshores. But it is particularly noted for its presence in the human environment, especially around habitations, in cultivated fields and at dump sites. The Pied Crow is the only crow included on the list of Hadza bird names presented by Peterson et al. (2013:220). Kopij (2019:359) wrote in his account of the Pied Crow's influence on the other birds of the Outapi baobab trees:

The Rock Pigeon, usually so numerous in southern African Towns (Kopij 2015), was uncommon in Outapi, while the Speckled Pigeon *Columba guinea* was not recorded at all. Both suitable nesting sites and food resources are commonly available for these species in the town. So low densities of doves and pigeons could have been caused by the exceptionally high population density of the Pied Crow. In addition to breeding birds, a large non-breeding grouping of the Pied Crow occupies the town throughout the year. The non-breeding grouping has been estimated at about 100–150 birds. These were probably immature birds, as Pied Crows reach sexual maturity at age of 2–4 years (Kilham 1989). The baobabs are their favoured roosting sites (own observ.). The Pied Crow is omnivorous, feeding often on fruits, grain products and carcasses if these are freely available (Dean 2005). … The Pied Crow is also known to prey on fruit bats (Dean 2005), which commonly roost on the baobabs (own observ.). As crows are nest predators (Kilham 1989; Dean 2005), some bird species, especially doves (their nests are relatively easy to locate and access for the crows) may be vulnerable to high nest predation. As a result, they may breed in much lower population density than under conditions where crow population density is much lower.

Other Potential Baobab Associates

The baobab is a dominant, though often solitary, element of savanna woodlands and tree-dotted grasslands and scrublands, and each vegetation type is assumed to have its distinctive population of birds with many overlapping species between them. This chapter has focused on birds specifically identified with the baobab, but there are studies documenting birds identified with baobab-associated vegetation, rather than with just the baobab. For example, Werema's (2017:34) reported in his comparison of birds inhabiting exotic *Acacia* bushland and baobab-associated native bushland in semi-arid east-central Tanzania that "The prediction that native bushlands would have higher bird abundance and species richness compared to exotic *Acacia* bushland was confirmed." All 71 species were found in the "baobab-associated native bushland," but of the 71 species, only 25 were observed in the exotic Acacia bushland. A number of species on Werema's list (2017:36–37), which includes an additional 19 passerines that have not been discussed in this chapter, are possible associates of the baobab.

Similarly, Zimmerman and colleagues (1996:32) identified twelve birds typically associated with the "*Commiphora* bush with baobab trees" near Mtito Andei, in southeastern Kenya. Ten of these have already been discussed: the African Orange-bellied Parrot, Red-billed Hornbill, Bohm's and Mottled spinetails, White-crowned Shrike, Golden-breasted Starling, White-crowned Shrike, Hunter's Sunbird, Lesser Masked Weaver, and Golden-breasted bunting. The remaining two birds for which the Hadza do have names (Peterson et al. 2013:220–222) are the Go-away Bird (*Corythaixoides personata*) and the White-browed Scrub Robin (*Cercotrichas leucophrys*). Werema (2017:36) included the White-browed Scrub Robin on his list as present in native bushland and in exotic *Acacia* bushland, but not the Go-away Bird. Although there are no reports, it is likely that these birds, and a number of other birds identified by Werema (2017:36–37), are also associated with the baobab, even if only for a safe perch.

Summary

However small, it is important to consider the amount of meat birds contribute to the forager way of life, both out of camp and in camp. But to limit our attention to the consumption of birds is to overlook the overall importance of birds in the practical and inspirational life of savanna foragers like the Hadza. From this broader perspective, we can assume that birding in all its complexity would have been an influential part of hominin evolution from early on. The baobab is a place where a great variety of birds converge, and the discussion presented in this chapter suggests the baobab-bird association is important for the Hadza, as it would also have been for early hominins. This is especially true of the overall contribution that birds make to the foraging success of the Hadza. As has been argued, birds are not only a direct source of food in the form of mature birds, eggs, and chicks, they are also important in food acquisition generally, as with scavenging from raptors, hunting equipment, and indicator birds. But birds are only one contribution that animals make to the baobab as an ecological tree of life, one of the most important resource-rich environments of the African savanna. Other baobab-associated animals also play a part in the Hadza hunter-gatherer adaptation.

Chapter 14
The Hadza Diet and the Baobab as a Source of Other Animal Foods

The baobab is a resource environment for many species for whom it provides such things as protection, food, residence, and breeding site. In their excellent documentary titled *Baobab, Portrait of a Tree*, which focused on the biodiversity of single old baobab, Root and Root (1973) filmed the remarkable behavior of some of the many arthropods, reptiles, birds, and mammals that were part of the tree's rich faunal ecology (Sweeney 1969, 1974; Owen 1974; Root and Root 1973; Wickens 1982; Du Puy 1996; Wickens and Lowe 2008; Hellekson 2009:6; Niassy et al. 2011). Although Africans obtain a variety of food from the baobab, its nutritious fruit pulp, seeds, and leaves are the most often noted in the literature (Rashford 2018). But for foragers like the Hadza, the baobab is also a source of animal foods. These include honey bees and birds as earlier indicated, as well as a number of small and large mammals that are targeted for food and other resources (Tomita 1966; Shipman 1986, 2010); Hawkes et al. 1991; Schoeninger 2001b), Marlowe 2010; Peterson et al. 2013). Although largely overlooked in the literature, the dietary contribution of the baobab/animal association is an important aspect of the tree's significance as an ecologically based multipurpose tree of life for human beings.

This chapter examines the baobab's contribution to the Hadza consumption of animal foods. Given this focus, Hadza's dependence on animal foods has to be understood as more than simply the hunting of large game, the overwhelming focus of researchers in recent years. Big-game hunting has been viewed as an expression of prestige hunting connected to sexual selection rather than to household provisioning. In the discussion of Hadza hunting and meat-eating, small game such as birds and small mammals have largely been set aside. But if every baobab indeed supports its own diverse community of plants and animals, then targeting the baobab as a resource destination must also include not only its birds and small mammals but also its many invertebrates as well as its large leaf, flower, and fruit-eating mammals. Since the goal here is to understand the ecology of the baobab and its contribution of animal foods to the Hadza diet by species rather than by calories, the focus is on both small and large animals.

J. Rashford, *Baobab*, https://doi.org/10.1007/978-3-031-26470-2_14

The baobab contributes to three of the five most important categories of Hadza foods, which are baobab fruit, honey, and meat. It is noteworthy indeed that even though Marlowe (2010:127) indicated that animals made up only 35% of the 838 plants and animals the Hadza brought back to camp, they made up 92.9% of all the species on which the Hadza depended; plants were a mere 7.10%. Birds (84.4%) and mammals (6.4%) were 90.8% of the animals acquired, and the remaining 2.1% were listed as invertebrate (0.1%), insect (0.1%), honeybee (0.8%), reptile (0.2%), eggs (0.1%), and domestic (0.8%).

Marlowe (2009:608) examined the sex differences in food preference of the Hadza in a sample that include 94 individuals, and the basic category of Hadza foods included honey, meat, berries, baobab, and tubers. He reported that "Both males and females ranked honey first, baobab third, and tubers last." He discovered that "females ranked berries second and meat fourth, whereas males ranked meat second and berries fourth." Baobab fruit, honey, and meat are the preferred foods of the Hadza and this means the baobab contributes to three of their five preferred foods. The baobab's multifood contribution is especially significant in autumn when husbands and wives forage together or in small groups. At this time of the year, men target honey and small game, and they contribute to the collecting of baobab fruits and to the collecting of baobab seeds from baboon dung. Women gather berries or roots in the absence of baobab fruit. This means that the baobab's contribution to Hadza meat consumption (Marlowe 2006:366) is likely to be significant, especially when we consider the importance of small animals, which are mostly eaten out of camp, in the household of hunters, or shared in the smaller camps of the autumn rains. It is not known at this point how much of the Hadza in-camp animal food is baobab-associated, but as a meat source the baobab would be more significant if the focus was on small animals eaten out-of-camp. Nevertheless, the literature indicates an association between baobabs and meat-eating that deserves scrutiny.

The three ways in which the baobab contributes to the consumption of animal foods throughout its range are the consumption of bee larvae, birds, other game that inhabit or visit the tree and its surrounding environment, and the use of the baobab's bark fiber and fruit pod to make traps. The absence of studies of the baobab's contribution to Hadza animal foods means the same approach used to describe the baobab/bird connection in the previous chapter is followed in this broader discussion of the baobab/animal connection. That is, this chapter focuses on animals associated with the baobab throughout its African range that are present in the environment of the Hadza. The discussion offers only the suggestion that Hadza's access to these animals could be partly mediated by the baobab.

Animal Foods

Hominin consumption of big game is thought to have developed over the last two million years in association with *Homo erectus*. Given this perspective, we are left to wonder about the taking of small animals during the five million years prior to

Homo erectus. If savanna baboons and chimps are able to hunt small game effectively without hunting equipment, there seems little reason to doubt that the evolving handy biped would have been able to do the same. In thinking about the baobab/animal connection and its possible importance in hominin evolution, the focus is on small and large game as the two key aspects of a single story. Half of the 16 "types of food" Marlowe (2010:127) identified for the Hadza were from animal-sourced categories labeled as domestic, mammals, honeys, birds, reptiles, invertebrates, insects, and eggs.

Invertebrates

The great diversity of life is represented not by vertebrates, but by invertebrates, including microbes, annelids, molluscs, and arthropods, and it is this diversity of invertebrates that vertebrates are dependent on. In popular discourse, the invertebrate/vertebrate distinction is often discussed in ways that suggest we are talking about equivalents, when in fact vertebrates represent only one diverse and highly successful branch of the phylogenetic tree, while invertebrates, as a paraphyletic group, represent the greater diversity of microbes, insects, annelids, and molluscs. Arthropods and other invertebrates are often overlooked in popular and scholarly accounts of the ecology and uses of the baobab. Yet, herbaceous and predatory arthropods are prey for resident and visiting reptiles, birds, and small mammals targeted by human beings.

Arthropods

Arthropods, especially insects, make up over 80% of the order Arthropoda, and they are important in the overall ecology of the baobab. The baobab is a temporary or permanent settlement site for many arthropods, a group that includes crustaceans, insects, myriapods, and spiders. Their attraction to the baobab is a consequence of such things as its enormous size; its forks, folds, crevices, and holes; its water catchments; its leaf production and flowering through the annual rains; and its fruits that ripen during the dry time of the year from mid-autumn in May until early Spring in October. The limited attention to arthropods in theorizing hominin evolution is partly due to the fact that insects and other arthropods are small, of numerous kinds that are difficult to distinguish in the field and even in the lab, and are generally not considered edible in the European tradition. As earlier indicated, honey bees are the most noted of the numerous arthropods that have been identified with the baobab. This is understandable, given the importance of honey in the diet of human beings wherever it can be had. The value of honey, as we have noted, is not only the eating of bee larvae and honey, however, but also the use of honey and beeswax as trade goods and honey as medicine. However, apart from honey bees and baobab-associated insects that transmit viruses to crop plants, arthropods as a large and important group have mostly been overlooked in studies of the baobab's ecology and of the impact of its ecology on the human uses of the tree.

With the notable exception of bee larvae, and of winged termites and locusts (Marlowe 2010:118), the Hadza largely reject arthropods as a source of food. In southern Africa, however, the caterpillar, commonly known as the mopane worm (*Gonimbrasia belina* Westwood), is widely eaten as a delicacy. Although chiefly associated with the mopane tree (*Colophospermum mopane*) from which it derives its name, this large caterpillar (commonly called the emperor moth) is also found on baobab trees, where it feeds on the leaves, and is much loved by the Kung. Today it is sold in markets and is an important source of nutrition for many rural people (Gabaza et al. 2018).

Molluscs

Molluscs, a major group among the gastropods, are a large class of animals with soft unsegmented bodies that are adapted to aquatic or damp environments. They include snails, slugs, mussels, and octopuses, and they comprise the second-largest group of invertebrates after arthropods. Marlowe (2010:110–112), in listing 97 plant and animal foods of the Hadza, included the category of "snail" in addition to plants (30), mammals (38), reptiles (1), and insects (1). And under the heading of "snail," for which he gave "N/alaka" as the Hadza name, he provided only the English common name "land snail." Peterson et al. (2013:222) also recorded "n/alaka" as the Hadza name for "land snail," and they identified this "land Snail" as *Achatina fulica*. This species is probably one of Africa's most well-known molluscs, the Giant African Snail or the Giant African Land Snail (*Achatina fulica*); it is now considered a serious pest in many parts of the world where it has been introduced. The Global Invasive Species database includes it among the 100 most invasive species in the world.

The herbivorous African Land Snail feeds on numerous agricultural crops and garden ornamentals. In their study focused on the relationship between Madagascar's baobabs and their pollinators, predators, and seed dispersers, Andriantsaralaza et al. (2016) reported their results showed that "land snails of the Achatinidae family may also exert a predatory action on young baobab plants, consequently impeding their regeneration." One wonders if the land snail is ever attracted to the aquatic or damp micro-environments of the baobab's water catchment sites. Unfortunately, "as a host," wrote Wickens and Lowe (2008:218), "the baobab has been greatly neglected and could be the subject of much fruitful entomological research." So could other invertebrates, like molluscs. The African Land Snail is discussed here because of its possible association with the baobab, its important cultural significance for the peoples of Africa, and its value for the Hadza as a source of food and shell containers (Harrison et al. 1997; Miller et al. 2018).

Fish

We know from many reports in the literature that baobab trees in proximity to large bodies of water have provided important resources for fishing communities. Most often noted are bark fiber for cordage and for weaving fishnets, and baobab fruit shells used as fishnet floats and boat bailers. The Rift Valley is the most striking geomorphological feature of the eastern region of the great African plateau. With respect to biodiversity, one important feature of the Rift Valley is its many fresh and saline lakes, both permanent and seasonal, that distinguish this part of the continent from all other major regions. East Africa is indeed the lake region of Africa. The wooded hills that are at the center of the Hadza's Rift Valley mosaic savanna contribute to the small rivers that give rise to Lake Eyasi and the Yaeda wetlands, and they make fishing an option for the Hadza. However, fishing is not most common during the rains when these areas have the most water, but at the height of the dry time, when they are on the verge of drying up.

The Hadza eat fish, which Bagshawe (1925:121) identified as "among the common articles of diet." He also noted that they made a "reed fish trap." Tomita (1966) reported that the Mangola Hadza catch eels and fish by hand in Lake Eyasi. Neither Marlowe (2010) nor Peterson et al. (2013) identified fish among the targeted species of the Hadza. And despite the accounts of Bagshawe and Tomita, the Hadza do not seem to relish fish as a routine aspect of their diet. They take fish at the height of the dry time from the end of winter in September to the start of Spring in October – the period of seasonal food scarcity – when they would be most in need of hunger-time and hunger-only foods, and when it would be the easiest time of year to catch fish. The dietary importance of fish, swarming termites, and locusts for the Hadza is their availability during the period of seasonal hunger; and as earlier indicated and further explained in the next chapter, the same is true of eating spring-nesting quelea birds. These are all hunger-time foods that are easily acquired at the time they are targeted.

Amphibians

Tadpoles have been found in the water collected in the baobab's hollow trunk, as noted in the discussion of the baobab's water catchments. According to Cooper (1949:9), the Hadza do not eat snakes, frogs, and toads, and Marlowe (2010:110–112) did not include amphibians among the 97 Hadza food sources he identified. Peterson et al. (2013:222) indicated only the Hadza generic name for large frogs, which they identified as the African Bullfrog (*Pyxicephalus adspersus*), and the Hadza generic name "for small frogs and toads."

Reptiles

Lizards

The baobab is an attractive resource environment for reptiles, including lizards, snakes, and tortoises, and possible also for terrapins when it grows close to the margin of wetlands (Sweeney 1974; Wickens and Lowe 2008:216–217; Hellekson 2009). Lizards and snakes, the most successful modern reptiles, are routinely identified with the baobab. On their list of Hadza names for reptiles, Peterson et al. (2013:222) included nine lizards, eleven snakes, two tortoises, and one terrapin. With the exception of the terrapin, which is adapted to wetlands, all the others have baobab connections, especially lizards and snakes.

Even though Hellekson's (2009) study was only done over the five-day (June 19 to June 23) transition from autumn to winter, she reported seeing "a wide diversity of reptiles" in the baobab she studied at Upupa Falls, Namibia, and in the leaf litter beneath the tree. "The presence of the geckos," Hellekson (2009:6) explained, "was the primary reason this particular study [of baobab biodiversity] was chosen." The geckos included the Barnard's Day Gecko (*Rhoptropus barnardi*), Turner's Thick-toed Gecko (*Pachydactylus turneri*), and Boulton's Day Gecko (*Rhoptropus boultoni boultoni*), which were seen in the tree; and the Bradfield's Day Gecko (*Lygodactylus bradfieldii*), Spotted Sandveld Lizard (*Nucras intertexta*), and Ovambo Tree Skink (*Mabuya binotata*), which were seen moving through the leaf litter. The Common Ground Agamas (*Agama aculeate aculeate*) and Namibian Rock Agamas (*Agama planiceps*) "were seen within 500 meters of the study area. The first two were found along the banks of the river, but one Rock Agama was seen in very close proximity to the [study] site in similar habitat." Peterson et al. (2013:222) reported Hadza names for the Turner's Thick-toed Gecko and Dwarf Gecko (*Lygodacatylus mani* or *similar*), but not the Dwarf Gecko and Red-headed Rock Agama (*Agama agama*) recorded by Hellekson.

Snakes

Wickens and Lowe (2008:216) noted that "especially pythons (*Python* spp.), boomslangs (*Dispholidus typus*), mamba (*Dendroaspis* sp.) and Egyptian cobra (*Naja Haje*) seek refuge in the hollows of the [baobab] tree." In his account of the natural history of the Nuba Mountains of Sudan, Sweeney (1974:65) recorded his encounters with borrowing vipers (*Typhlops schlegelii*) and the Somali Snake (*Micrelaps boettgeri*) under baobab trees. Peterson et al. (2013:222) recorded Hadza names for eleven kinds of snakes, but they did not record a Hadza name for the boomslang, which Wickens and Lowe (2008:216) identified as linked to the baobab. According to Dellatola (1983:27), "The deadly boomslang spreads death and destruction among the birds which nest in baobab trees."

Tortoises and Terrapins

As earlier noted, Hadza names for reptiles (Peterson et al. 2013:222) include those for lizards, snakes, tortoises, and terrapins. According to Bagshawe (1925:121), the Hadza ate lizards and snakes, but this view has not been supported by more recent studies. With respect to lizards, however, Cooper (1949:9), like Woodburn (1968a:52) and Marlowe (2010:102), noted that when it comes to food, the Hadza "drew the line" at "snakes, frogs and toads," but he added, "I believe they sometimes eat lizards." Tomita (1966:160), writing of the Mangola Hadza, said lizards were not eaten, and this is also what the Hadza told Peterson et al. (2013:38). But Woodburn (1968a:52) mentioned only that the eastern Hadza did not eat monitor lizards. There is general agreement in the literature that the Hadza do not eat amphibians and most reptiles, the only exception being land turtles.

Sweeney (1974:66) reported seeing the Common Hinged Tortoise (*Kinixys belliana*) under a baobab in the Nuba Mountains of Sudan, but he tells us nothing of its association with baobab trees, as his discussion was essentially concerned with the variability of this species. Petignat and Jasper (2012) reported that today "young baobab trees are scarce in Madagascar's dry forests and thickets," and they wrote:

> Likely candidates for the baobab disperser include large-bodied extinct lemurs in the genera *Archaeolemur* and *Pachylemur*, which occurred in southern and western dry forests until the arrival of humans, and survived until at least 1,000 years ago. Studies indicate these species were principally frugivorous. Other possible candidates are the two extinct giant tortoises in the genus *Aldabrachelys*, which one occurred in large numbers in Madagascar's dry forests. The closely related Aldabran Giant Tortoise (*A. gigantea*) has been shown to readily consume baobab seeds without destroying them. The seeds then show enhanced germination once they have passed through the reptile's gut.

Although Cooper (1949:9) did not identify the kind of tortoise involved, he wrote of the Hadza: "I have … seen them cook – or merely scorch – such diverse fare as a Squirrel, a Tortoise and a Kite." Peterson et al. (2013:222) recorded Hadza names for the Leopard Tortoise (*Geochelone pardalis*) and Pancake Tortoise (*Malacochersus tornieri*), and for the Helmeted Terrapin (*Pelomedusa subrufa*), also commonly known as the Marsh Terrapin, Crocodile Turtle, and the African Side-necked Turtle. But Woodburn (1968a:52) reported that the Hadza did not eat terrapin, and Marlowe (2010:110–112) did not include it on his list of Hadza food species. Marlowe (2010:63) only identified the Leopard Tortoise (*Geochelone pardalis*) as targeted by the Hadza and he was definite in stating:

> In fact, [the Hadza] … eat no reptiles except for tortoise (*k'oloko*; *Geochelone pardalis*). Only women and children eat tortoises and the large land snail (*nlalaka*; *Achatinidae Burtoa nilotica*). When a man is out foraging and finds a tortoise, he is not inclined to think about taking it back to camp, but if I am with him he may ask if I want to take it back. The women and children are glad to get it and promptly cook it and eat it. The men say that if they were to eat the tortoise, it would ruin their arrow poison. Apparently, some men eat tortoises (I have seen one), and younger boys regularly eat them with the women and girls.

For Marlowe (2010:127), reptiles represented a mere 0.2% of the 878 species of plants and animals upon which the Hadza depended, and this consisted only of the Leopard Tortoise (*Geochelone pardalis*).

Mammals

Rodents

Rodents are often identified with the baobab. They are one of the most successful orders of mammals worldwide with several hundred species in Africa, including mice, rats, squirrels, gerbils, shrews, spring hares, and porcupines. Because birds and rodents are noteworthy components of the small animals taken by the Hadza, the link between the baobab and these two major groups of animals is an important factor in Hadza foraging. Of the twelve kinds of animals Welman (1966:150) identified in the baobab tree he observed in the Sudan, six were birds and three were rodents (including a green squirrel identified as *Aethosciurus*, "a dozen or more gerbils," and a "family part of diminutive elephant-shews"); the remaining three were a mongoose and a genet, and bats. Some rodents are known to feed on the fruit of the baobab and are regarded as possible dispersal agents for the species. According to Owen (1974:8), "Squirrels, rats, and monkeys carry the [baobab's] fruit some distance from the parent trees and, if the seeds are deposited on suitable soil, germination may take place."

Cooper (1949:9) indicated that the Hadza hunted squirrels. Peterson et al. (2013:35) recorded the Hadza as saying "boys learn to hunt by stalking and shooting reptiles, birds and rodents near camp," and he made specific reference to field mice. The author of a blog titled "Measuring Trees with the Hadzabe" wrote: "So, here we are! A team of 6 from Arusha and a team of variable size from the local Hadza community. The aim being to locate points I'd randomly identified in advance using a GPS, get there and measure all the trees, whilst training everyone to do it themselves too." The author presented a photograph of himself standing in front of a baobab tree, and with him was a Hadza man with his head inside a small hole at the base of the tree. In the caption to this photograph, the author wrote: "Brief stop in a Baobab on the off-chance of mice … since our random point took us past that way." According to this author, "every Baobab has a mouse hole cut in it, and/or a ladder of pegs to climb for fruits and honey." The stick ladder, which is chiefly used for gathering honey, is a conspicuous aspect of the Hadza use of the baobab; it has been commented on and illustrated from the earliest accounts of the Hadza. But the same cannot be said of mice holes cut into the base of the baobab. Nevertheless, to the extent that they do exist, these mice holes probably contribute to the development of the baobab's hollow trunk, which enhances the baobab's status as an ecologically based multipurpose tree of life. It is interesting to note here that one use of the baobab fruit pod in East Africa is to make a rat trap (Wickens and Lowe 2008:123). Eating rodents, although not emphasized in the scientific literature, would especially take place out of camp. In terms of baobab regeneration (taken up in Chap. 24), hunting rodents inadvertently protects baobab seeds; in the same way, the Hadza may be protecting young baobab trees by hunting herbivores that would otherwise feed on the tree.

Bats

There are more than 1200 species of bats worldwide (Wilson 2020), of which some 200 occur in Africa (African Wildlife Foundation 2021). Next to rodents, they are the second most species-rich mammals in the world. Even though there are no reports of the Hadza eating bats, the baobab/bat association is discussed here because bats are recognized as the chief pollinator of the baobab, and they roost in its hollow trunk; because bats contribute to the tree's biodiversity by serving as prey for snakes, birds, and other mammals; and because bats and the baobab are inspirationally significant for human beings wherever they occur.

The baobab's association with bats is well documented (Van der Pijl 1936; Jaeger 1945; Harris and Baker 1959; Coe and Isaac 1969). The fact that in its relatively open canopy the baobab produces large, pendant, glossy-white flowers on long stalks that are nocturnal, odiferous, and nectar-rich is generally attributed to the tree's evolutionary association with bats. Citing a number of authors, Wickens and Lowe (2008:207) identified nine species of baobab-associated fruit bats. Some feed on the pollen and nectar of the baobab's flower, and in so doing they serve as pollinators, and some roost in the baobab's hollow trunk. In terms of baobab-associated biodiversity, bats are also prey for other baobab-associated species, such as the Pied Crow (Kopij 2019:359) and snakes. Peterson et al. (2013:219) indicated that the Hadza have names for bats in general. They also distinguish by general names between fruit bats and insectivorous bats. The yellow-winged bat was the only species specifically noted, though they also referred to a Hadza name translated as "bat," and described as a "solitary small bat [that] lives in trees (named after scops owl due to call)."

Giraffes

Baobab herbivores include invertebrates, chiefly insects, and a variety of mammals of all sizes; the latter include the giraffe, the most extraordinary tree browser of them all, and the elephant, the ultimate Baobab predator. Giraffes are known to favor feeding among the leaf-bearing branches of thorny acacias, but they also prey on the foliage of the baobab when the opportunity arises. This is particularly so with respect to young baobab trees. Of those baobab that survive destructive herbivores when they are young, browsing giraffes and elephants may well be an important contributor to the generally tall nature of savanna baobabs, compared with the squatter forms that are relatively common in the New World, India, and in other places where the trees are not subject to major browsing mammals.

Elephants

The elephant is one of the best-known faunal associates of the baobab (Owen 1974:8; Wickens 1982:188; Wickens and Lowe 2008:204–206), but it is more famous for the destructive consequences of its well-documented baobab feeding practices than it is for its contribution to the dispersal of baobab seeds (Ridley 1930; Barnes 1985; Teel 1985; Swanepoel and Swanepoel 2008; Swanepoel 1993; Campbell 1995; Kupika et al. 2014). Elephants target the baobab most of all during the annual May–October dry time, particularly at its height from the mid-winter of August to the early spring of October, when seasonal runoff and pools have dried up and plants are just beginning to put out leaves in anticipation of the spring rains and the return of the growing seasons. Most baobab herbivores eat only the baobab's fruit or its leaves, or they sip its sap or floral nectar. Elephants, however, eat the baobab bark, wood, and leafy branches in addition to its fruit and leaves. At times, they have been reported to eat almost the entire tree (Kupika et al. 2014). Excellent photographs of baobab-associated animals are now commonplace on the Internet, including iconic images of elephants hollowing out the baobab's trunk for food, water, and minerals. It is not just elephants destroying mature baobabs, however. Citing Guy (1982) and Venter and Venter (1996), Sanchez (2010:31) noted that "seedlings are believed to be eaten by elephants and other herbivores," a point also made by Wickens and Lowe (2008:335), who earlier wrote that elephants "are held responsible for the absence of the younger age classes of *A. digitata* in parts of East and southern Africa." These authors went on to note that "Desertification, climate change, fire and agricultural development have also been blamed for lack of [baobab] regeneration."

When they have the opportunity, the Hadza do eat the meat of elephants, of which "they are inordinately fond" (Cooper 1949:9), but according to Woodburn (1968a), they do not hunt elephants. Marlowe (2010) did not include the elephant on his list of Hadza food species. The Hadza explained to Peterson et al. (2013:38) that "while there are still a small number of elephants on our land, we don't hunt them. Even when elephants were plentiful, we didn't hunt them because our poison isn't strong enough to kill an elephant."

Rhinoceros

The black rhinoceros is not routinely included among the dispersers of the baobab tree. However, based on personal communication from the ecologist Malcolm Coe, Wickens and Lowe (2008:75) reported that the "seriously endangered black rhino

(*Diceros bicornis*) also eats baobab pods." We can assume then that baobab seeds are spread in its dung. The rhinoceros was not included on Marlowe (2010) list of Hadza animal foods. The Hadza told Peterson et al. (2013:38) that "large animals, such as giraffe, eland, buffalo and zebra, which were commonly hunted in the past, are now rarely shot as their numbers have dwindled. None of the younger generation has even see a rhino and many have never seen a buffalo."

Primates

The nine baobab-associated primates of Africa's woodlands and wooded and open grasslands identified by Wickens and Lowe (2008:207) include two species of bush babies of the superfamily Lorisoidea (family, Galagidae); the baboon and vervet, blue, green, and colobus monkeys of the family Cercopithecidae; and the savanna chimp and human beings among the great apes (family, Hominidae). With respect to baobab-associated animals, primates are a key part of the tree's residential complex, its flower-visiting and pollinating complex, and its fruit-eating and seed-dispersal complex. The baobab is a resource tree for the four savanna primates that are part of the diet of the Hadza, and both Marlowe (2010) and Peterson et al. (2013) included them on their lists of Hadza animal foods. These four are the Lesser Bush Baby (*Galago senegalensis*) and Greater Bush Baby (*Otolemur crassicaudatus*), the vervet monkey (*Chlorocebus pygerythrus*), and the baboon (*Papio*). They represented 10.5% of the 38 mammals targeted by the Hadza that were listed by Marlowe (2010:111), and they were 5.8% of the 68 mammals identified by Peterson et al. (2013:219–214).

Bush babies are among the small animals hunted by Hadza boys (Marlowe 2010). These highly active arboreal and mostly nocturnal primates roost in baobab holes. In the absence of such holes, they also build leafy nests in the canopy of the baobab and other trees. They are especially attracted to the baobab for its flowers, pollen, and nectar (Coe and Isaac 1969); they are considered minor pollinators of the baobab. They also feed on the baobab's many arthropods, on the eggs, nestlings, and fledglings of its birds, and on its small reptiles and mammals (Root and Root 1973). Even though Peterson et al. (2013:219) listed Hadza names for the Olive Baboon (*Papio anubis*) and Vervet Monkey, and also for the two species of bush babies, they did not include bush babies among the fourteen species of animals most hunted by the Hadza today.

Apart from human beings, the baboon is the most important primate associate of the baobab. It is most likely the species responsible for the baobab being called the monkey bread tree (Owen 1974:9–10), although not all "monkey names" for the baobab can be attributed to "monkeys" eating the fruit (Rashford 1994a). The Olive Baboon (*Papio anubis*) is common in the Hadza environment, while the Yellow Baboon (Papio cynocephalus) is common in the Ruaha Park of southeastern Tanzania. In an account of the "core dry-season diet" of the Yellow Baboon (identified as *Papio hamadryas cyocephalus*), Pochron (2000:347) reported that by weight, "the diet consisted primarily of baobabs (62.3%), tubers (21.0%) and corms (11.0%)." According to Mabulla (2003:49), baboon meat "is highly valued by Hadzabe foragers for its protein and medicinal uses. It is believed that baboon meat, when eaten and the fat applied on the body protects the Hadzabe from diseases and evil spirits (Mahiya, pers. Comm.). The occurrence of baboon bones at Mumba Rock Shelter (Mehlman 1989) shows the persistence of baboon exploitation in the Eyasi Basin."

Like other ground-dwelling monkeys, the baboon has a more varied diet than arboreal primates. Their plant foods include ground edibles such as roots, rhizomes, bulbs, tubers, and corms; leaves, grasses, and herbs; stems and bark; and floral parts including petals, pollen, and nectar. In addition to these plant foods (including baobab flower and fruit and the root of small baobab saplings), the omnivorous baboon eats invertebrates; the eggs, nestlings, and fledglings of birds; and small reptiles and mammals. It is noteworthy that the baboon was among the eight animals that the Hadza said are most hunted today (Peterson et al. 2013:38). The other seven were comprised of three ruminants including the dik dik, impala, and kudu, as well as the warthog, bush pig, vervet monkey, and hyrax.

The vervet monkey eats baobab leaves as well as the baobab's flower, pollen, and floral nectar. In an account that points to the hunger-season activity of a troop of vervet monkeys, Watson (2007:124) wrote:

> With the countryside often so parched and barren when baobabs start blooming, their flowers may constitute food of last resort. In October 2000, after six dry months, a few baobabs in Kenya's Meru National Park were showing early white blooms. One of these had been invaded by a troop of vervet monkeys which seemed desperate to get at the flowers … Only later did I find out what had probably attracted the monkeys: the globs of nectar they could lick from the underside of the sepals, after pulling off the rest of the flower.

In considering the baobab/primate association, it is important to note the connection between the savanna chimpanzee and the baobab, even though chimpanzees are not part of the environmental range of the Hadza today. The chimpanzees of the savanna of West Africa and eastern Tanzania include the baobab among the trees in which they sometimes nest (Wickens and Lowe 2008:208). Like the vervet monkey, impalas, and other ruminants, they eat baobab flowers, fruit pulp, and seeds. The chimpanzee is well known for using stone and wooden hammers and anvils to access the fruit pulp of the baobab's woody indehiscent fruit (Ambrose 2001:1749; Marchant and McGrew 2005; Lindshield et al. 2021). Interestingly, the baboon, unlike the chimpanzee, eats both ripe and unripe baobab fruits. The immature seeds of the unripe fruits of summer and early autumn are of course destroyed in the process, and in this sense, the baboon is a baobab predator. However, the seeds of the ripe fruits eaten from the mid-autumn through the early spring pass through the baboon's digestive tract and are dispersed in their dung. In this sense, the baboon is one of the baobab's most important dispersal agents. There are no reports of the baboon eating baobab seeds collected from their own dung, as occurs with the savanna chimpanzee (Wickens and Lowe 2008:208). The savanna chimpanzee is also known for targeting baobab honey bees using wooden tools to access their hives.

Teleki (1975:159) highlighted the dietary similarity of chimpanzees as "collector-predator" and the San as "hunter-gatherers." He wrote:

> [M]any human food-getting techniques, and especially those used in collecting wild plants and small land fauna, do not differ substantially from those performed by other primates. At what point, then, can we draw a meaningful line between collector-predator and gatherer-hunter patterns? Combination of the lower size limit in large prey killed by humans (c. 15 kg) and the upper size limit in small prey killed by baboons and chimpanzees (c. 10 kg) yields a range of 10–15 kg. as a marker for this subsistence shift from predator to hunter ... But we must not, at the same time, disregard the fact that much of the game taken by human hunters and by nonhuman primate predators falls within the same size range, and that monkeys, apes and men are all omnivores practicing basically similar subsistence patterns overlaid by some differences in habits, skills and techniques.

The problem with this view is that it minimizes the difference between human beings and chimpanzees and other primates at a time when differences were often dogmatically emphasized. But in fact, there is a real difference that should be highlighted. Chimpanzees collect the food they eat, but they are not collectors in the sense of gathering, transporting, and exchanging food. That is, they do eat baobab fruit they pick from the tree or pick up from the ground, but there is no sexual division of labor involving a temporary residential settlement in the form of a camp centered on exchange and reproduction. And in this sense, the savanna chimp is not truly a fruit collector. But as omnivorous primates, both chimpanzees and human foragers are plant and animal predators, and for human beings, this means a greater reliance on herbivory than animal foods in tropical and temperate environments. The essence of the difference in plant and animal predation between human beings and chimps and other primates is human handiness enhanced by material culture in the form of equipment, and this means much greater access to roots and seeds, and to honey, birds, and mammals.

Savanna Predators

Predatory snakes and raptors are well-known associates of the baobab, but the same cannot be said for mammalian predators. There are some 40–60 species of mammalian carnivores in Africa. The continent's great expanse of wooded and open grassland is home to large mixed groups of zebras, antelopes (notably, wildebeests and gazelles, especially the Thompson Gazelle), ostriches, and baboons. These communities are routinely trailed by predators, including the lion, cheetah, and leopard, and by canids, especially the wild dog and hyena. They are also trailed by scavengers, notably the hyena and jackal among mammals, and vultures and the marabou stalk among birds. The less prominent carnivores are also cats (striped polecat, caracal, civet) and canids (jackal, fox), as well as the African striped easel, honey badger, and mongoose.

Of the baobab's links to the big savanna cats, Wickens and Lowe (2008:208) noted only that "The hollows are much used during the day by sheltering leopards (*Panthera pardus*)." However, in sparsely wooded grasslands, lions are known to rest in the shade of the baobab, and recent studies have shown that in the face of intraspecific competition in the form of predator aggression and confrontational scavenging, leopards succeed because of their adaptation to trees, including the baobab. The leopard is the most tree-dependent of the big savanna cats. Van Valkenburgh (1985:106) noted that the importance of tree-climbing for the leopard involved not only a place to stash food, but also to eat, shelter, rest, conduct surveillance, and escape.

> Because of the tree-climbing abilities, leopards are not as vulnerable as wild dogs and cheetahs. Nevertheless, they are often the targets of attempted kleptoparasitism. In Kruger Park, spotted hyenas discovered about half of 55 leopard kills but were unable to steal them if the carcass was in a tree (Bailey 1993). Lions have been known to steal leopard kills that were not placed high enough in trees (Turnbull-Kemp 1967; Schaller 1972), and wild dogs are able to chase leopards of kills on the ground (Pienaar 1969; Schaller 1972; Creel and Creel 1996). Thus, it appears that the tree-caching ability of leopards is critical to their success. African leopards occasionally engage in intraguild predation on hyenas, cheetahs, and jackals and have themselves been the victims of spotted hyenas. (Pienaar 1969; Caro 1987; Bailey 1993)

Peterson et al. (2013) recorded a Hadza name for the leopard and so did Marlowe (2010).

Ruminants

Ruminants are among the mammals identified by Marlowe (2010:110–112) and Peterson et al. (2013:38) that are linked to Hadza baobab-based hunting. Of the roughly 200 species of ruminants belonging to the suborder Ruminantia, 186 occur in Africa (Groves 2021). The six families that comprise this group are the most consequential animals for human beings worldwide. This is particularly true of the family Bovidae, represented in Africa by cattle, wild and domesticated goats, sheep, buffaloes, giraffes, and numerous wild antelopes. As previously noted, Marlowe (2010:110–112) described the Hadza as dependent on 878 species of plants and animals, and of these, he identified 97 species by common name, Hadza name and scientific name. We can assume that these 97 species – 10 tubers, 1 vegetable, 10 berries, 7 fruit, 2 fruit/nut, 7 honeys, 1 insect, 38 mammals, 1 reptile, 1 land snail, and 19 birds – are the most important "foods eaten" by the Hadza. And compared with carnivores (7) and primates (4), it is clear that birds (19), together with ruminants (15) and non-ruminant mammals (12), are the most important food species of the Hadza. Of the mammals, no group was more important than ruminants, especially antelopes. In contrast to Marlowe (2010), the Hadza told Peterson et al. (2013:38):

> Dik dik, impala, kudu, warthog, bush pig, baboon, vervet monkey and hyrax are the animals most frequently hunted these days. Large animals such as giraffe, eland, buffalo and zebra are now rarely shot as their number have dwindled. None of the younger generation has ever seen a rhino and many have never seen a buffalo. While there are still a small number of elephants on our land, we don't hunt them. Even when elephants were plentiful, we didn't hunt them because our poison isn't strong enough to kill an elephant.

Fallen Baobab Flowers as Animal Food and Hadza Hunting Strategies

Six hunting strategies are discussed in theories of hominin evolution that are based on Hadza foraging and meat-eating. They are encounter hunting, targeted hunting, stealth hunting, ambush hunting, flush hunting, and pursuit hunting. Three of these are routinely discussed Hadza hunting strategies. All Hadza researchers recognize that they are ambush hunters, best known for nighttime hunting that takes advantage of animals along game trails leading to water holes. Woodburn (1968a:51), however, emphasized their status as stealth hunters. Hawkes et al. (1991:244) and Marlowe (2010:118) did not follow up on this idea, highlighting instead the idea of encounter hunting (which Marlowe also identified as "walk-about" hunting). In addition to Hadza ambush, stealth, and encounter hunting, there has also been a discussion in the literature of pursuit hunting. And in this discussion, pursuit hunting is often contrasted with ambush hunting, wrongly suggesting that they are the two mutually exclusive forms of animal predation. Often overlooked in the discussions of predation strategies are flush hunting and targeted hunting; both occur widely in nature and are well represented in the ethnographic literature.

Baobab ambush hunting would especially target large game. The baobab's flower attracts a variety of species when it opens on the tree and when it falls to the ground; it is a significant contributor to the baobab's status as an ecological and multipurpose tree of life. During the annual flowering from spring through early autumn, the ground beneath the baobab is littered with spent flowers. There is no evidence of the intentional use of the picked or fallen flowers to feed livestock, as occurs with baobab leaves in various parts of Africa. Nevertheless, Palmer and Pitman (1961:232) reported that "cattle are said to love the flowers like pudding," and Wickens and Lowe (2008:78) noted that "Domestic animals, including camels, and wild herbivores eagerly browse accessible leaves and fallen leaves and flowers, especially in the drier regions." Burton-Page (1969:333) saw deer feeding on them in India but he knew "of no instance of them being collected as cattle fodder (as with *Bombax*), for example." Lamprey (1963) noted that impala eat the fallen flowers.

Not only do the abscised flowers attract animals, which makes the baobab an ambush site, the flowers are also collected for use as bait for hunting and catching insects. De Caluwé et al. (2009:438) documented the use of the flower to make a termite trap. The abscised "corollas are collected, dried, and then mixed with cattle feces and placed in a bowl. Subsequently, this bowl is placed upside-down close to a termite hill and left for 24 h." The trap is used to provide food for poultry and it also serves as "an example of local veterinary medicine to reduce poultry mortality rates."

Watson (2007:123–124) has given us a glimpse of what the flower could have meant to ancestral foragers of the African savannah, but it is a contradictory view of the ecology status of the flower. He claimed that "The baobab's fallen flowers seem to be one of nature's least utilized by-products, instantly redundant once they hit the ground, and ignored by most creatures as a source of food," and yet he wrote:

> Munyao Mwakavi is a Wakamba driver, old enough to have some grey hairs and also to remember a much wilder environment than exists today. As a youth he would build a platform in a tree next to a flowering baobab, and wait with his bow and arrow for bushbuck to emerge in the evening from their daytime cover and start grazing on fallen flowers. Sometimes Munyao would even scatter handfuls of flowers below his perch, to give himself a better chance of shooting the animals cleanly. Now the bushbuck have gone, to be replaced by goats and sheep, and as soon as he lets these out every morning, they too scamper over to the nearest baobab to see if any flowers fell in the night.

The ecological associates of the baobab flower make it far more important in human evolution than is evident in Watson's perspective. At the September–October height of the May–October dry time, Hadza foragers ambush large game along well-worn tracks to permanent sources of water. In autumn, they ambush animals that prey upon the ripening crops of their farming neighbors which they are hired to protect. And during the November–April rains, they ambush animals that visit the baobab to feed on its fallen flowers (Peterson et al. 2013:35).

Other Baobab-Associated Mammals

Honey Badger

Other baobab-associated mammals with Hadza names (Peterson et al. 2013:219) that are mentioned in the literature include the honey badger (Sweeney 1969; Wickens and Lowe 2008:2008), which should be considered part of the baobab's flowering complex. Although it is known for its love of honey (as indicated by its name), the diet of the omnivorous Honey Badger routinely includes invertebrates, frogs, tortoises, lizards, snakes, birds, rodents, and eggs, as well as fruit and ground edibles such as roots and bulbs.

Wild Pigs

The widespread warthogs and bushpigs of tropical Africa are represented by five indigenous species: including the Common Warthog (*Phacochoerus africanus*), Desert Warthog (*Phacochoerus aethiopicus*), Red River Hog (*Potamochoerus porcus*), Bushpig (*Potamochoerus larvatus*), and Giant Forest Hog (*Hylochoerus meinertzhageni*). Marlowe (2010) included a single species of "warthog" among the thirty-eight mammals targeted by the Hadza, and he identified this species as the Desert Warthog, which is especially common in northern Kenya and the Horn of Africa, including Somalia, Ethiopia, and Djibouti.

In contrast to Marlowe (2010), the only warthog Peterson et al. (2013:219) identified in their list of Hadza names was the Common Warthog (*Phacochoerus africanus*). According to Penrith (2021:4), this species is known to shelter and nest in "burrows, often in old termite mounds and may use burrows of other species e.g. ant bears, as well as road culverts and hollow baobab trees."

Bushpigs are widespread in Africa, but *Potamochoerus larvatus* is the only species for which Peterson et al. (2013:219) presented a Hadza name. Wickens and Lowe (2008:208) identified the Bushpig as also occurring in Madagascar, where it is known to seek out baobab fruit. They speculated, based on Perrier de la Bathie (1924), that the Bushpig probably eats baobab fruit in Africa just as it does in Madagascar.

Baobab Materials for Hunting Animals

The use of baobab bark-fiber cordage to make snares is common throughout its range where hunting occurs (Marks 1976:80). Bagshawe (1925) reported that the Hadza made a "running noose snare," but he did not indicate the materials from which it was made. Current research suggests that the use of snares is not common among the Hadza today, but as (Marlowe 2002:263) points out, they are able to use snares if they choose to do so.

The Baobab and the Taking of Small Versus Large Animals

Theoretically, the idea of Hadza "small game" specifically means molluscs in the form of the African Land Snail, reptiles (specifically the tortoise), and numerous small birds and mammals, the latter including rodents (rats and squirrels), primates (bush babies, vervet monkeys, and baboons), ruminants (dik dik, impala, kudu), warthog, bush pig, and hyrax. Small game is particularly significant for the Hadza today, since they told Peterson et al. (2013:38):

> Large animals such as giraffe, eland, buffalo and zebra, which were commonly hunted in the past, are now rarely shot as their numbers have dwindled. None of the younger generation has even seen a rhino and many have never seen a buffalo. While there are still a small number of elephants on our land, we don't hunt them. Even when elephants were plentiful, we didn't hunt them because our poison isn't strong enough to kill an elephant.

For the Hadza, the baobab's significance as a source of meat turns in part on whether they in fact pass over small game for larger game or whether they take all game they encounter. On this point, there is a fundamental disagreement among the most important evolutionary ecologists who have studied the Hadza. On the one hand, Hawkes et al. (1991, 2001a, b) presented an either/or scenario based on the idea that men focus on hunting large game almost to the exclusion of small animals. They argued that the Hadza do so to gain a reputation as successful hunters, which comes from camp-wide meat distribution. This enhances their opportunities for mating and strengthens their mate-bonded households and in so doing provides the best chance for reproductive success.

If the view of Hawkes et al. (1991, 2001a, b) is correct, then the baobab's links to meat-eating would be insignificant, because the tree is especially associated with small animals. On the other hand, Marlowe (2010) argued that Hadza men take large and small game whenever they can; and small game would be especially beneficial to hunters and their households and kin, since there is less of an obligation to share. Woodburn (1968a) reported that small animals were mostly taken during the November–April rains. According to Marlowe (2010:216), boys under eighteen take a lot of small game. They do so because they are unable to take larger animals, and because small game is essential for self-provisioning and for delayed reciprocity with co-hunters and friends. "I have found," wrote Marlowe (2010:218), that "Hadza men willing to shoot at almost any mammal or bird they could hit," and he argued that adult male hunters, by virtue of being better hunters, take more small prey.

Hadza men are opportunistic foragers. They take large and small game whenever they can and collect baobab fruits when they encounter them on the ground, or while gathering honey or other baobab resources in the tree. But it is significant that this does not always involve a tradeoff when targeting resource-rich environments such as the baobab. With the baobab as an ecologically determined multipurpose tree of life, Hadza men have the option, among other things, to rest in its shade, collect water, gather baobab fruit and honey, and hunt resident and visiting animal. The baobab as a multisource food tree is relevant to Marlowe's (2010:116) discussion of tradeoffs in pursuit of honey verses game verses baobab fruit:

> When men are out on walkabout, they always have their bows and arrows, but they often stop at trees to check on beehives. They can tell by watching the bees whether there is much honey ready to harvest. A man often puts his ear to a tree trunk to listen to the bees. After monitoring trees in the area around their camp, they may decide that it is time to raid hives, and then they carry their axes (in addition to their bows). If they should see game worth pursuit, they abandon the search for honey and pursue the animal until they kill it or it gets away.
>
> Honey collecting does not pose a substantial tradeoff with hunting; the Hadza (especially men) are very opportunistic foragers, which makes it difficult to code their forays into honey vs. game vs. baobab search time the way optimal foraging models suggest can be done.

In addition to Hadza opportunism, the baobab's status as a multipurpose tree could help explain why there is no substantial tradeoff between targeting honey, game, and baobab fruits, and this could account for the difficulty in coding in accordance with optimal foraging models.

If baobab-sourced meat consisted of small animals, then it would mostly be consumed out-of-camp, or brought back to camp for household and kin provisioning. If, however, it involved large game feeding on fallen baobab leaves, flowers, or fruits, or taking advantage of its shade, or its bush cover, then baobab meat would be part of the 19% of in-camp meat consumption. The perspective of Hawkes et al. (1991, 2001a, b) suggests men focus on hunting large game to the exclusion of small game to gain a reputation as successful hunters. The evidence supporting this argument is that the children of better hunters do not seem to benefit from the fathers' successful hunting of large game. But Hawkes et al. (2001a:687) encountered a contradiction, of which they write:

> Hunters were much more successful in the late dry of 1985 than at any other time. In this season, which accounted for 33% of the observation days of the time-allocation sample, hunters took > 60% of the total amount of meat acquired during that period (table 1). Perhaps small additional portions had negligible value when meat was generally available – in fact, this relatively heavy meat-eating season was the only one in which children generally lost weight. The children of better hunters may have gained more weight (or lost less) in the other seasons because even small additional amounts made a large difference when meat was usually scarce. When the late dry season is excluded from analysis (fig. 1), there is a positive relationship between father's overall hunting success rate and children's weight changes.

Hawkes and colleagues (2001a:688) noted: "Still, in the honey seasons, as in the rest of the year outside the dry season, the children of better hunters gained more weight … Since this variation appears to be unrelated to differences in the amount of food provided by fathers, something else may link men's long-term hunting success rates to children's weight changes."

As Bell (2001:695–696) pointed out, Hawkes and colleagues did not emphasize seasonality in their perspective, and this was the basis for the ambiguity they encountered. The Hadza children lose weight at the height of the dry time in September–October when fruits are scarce, and this includes the children of the better hunters. Even though at this time of the year Hadza hunters are most successful in targeting large animals by nighttime ambush hunting along game trails, Hadza children still lose weight. This fact points to the importance of fruit in the diet of the Hadza as a fruits-and-roots people. But Hadza men take large and small game, and small game benefits hunters and their households and kin more than large game. Hawkes et al. (2001a:686) wrote:

> Although Hadza hunters generally ignored small animals, experimental data (Hawkes et al. 1991) indicate minimum return rates of about 0.25 kg/hunter-day for encounter hunting small game and 0.78 kg/hunter-day for snaring. Small animals taken as prey draw little attention and so are easily kept by the hunter or, if brought home, by his household (Woodburn 1968a; Barnard and Woodburn 1988). If the meat of small animals is mostly consumed by household members, then the whole acquisition rate for these prey is the marginal gain that a man could earn for household meat consumption by taking small animals…
>
> This means that each time a hunter encounters a guinea fowl and passes by in order to continue earning 0.08 kg/hr searching for big game, he is reducing his household income. Even if neighbors made claims on small animals, a man who pursued them and kept less than half of the lowest-return small prey would still earn a greater nutritional benefit for his own household than he would get from specializing in big game.

In the rainy period, however, especially in the spring hunger season and during the summer and autumn fruiting seasons, men who are successful hunters of small game, and honey collectors for their family and kin households, are the ones that help their children most. All Hadza do better during the fruiting season than at the end of the dry time of September and October. But it is the biological children of the best male foragers who are better off during this time, as men target small animals, honey, and baobab fruits (Woodburn 1968a, Hawkes et al. 1991:182). This suggests, contrary to Hawkes et al. (1991, 2001a, b), that the baobab is indeed an important meat source for the Hadza.

Given the omnivorous diet of most primates and the fact that chimps regularly hunt based on local environmental conditions and cultural tradition, it is a reasonable assumption that small game hunting would have been important over the course of hominin evolution. Prior to the development of manufactured stone tools some 2.5 million years ago, the only large game that would have been regularly taken, we can theorize, would have been from scavenging. According to Marlowe, with whom I agree, the focus of Hadza men is on provisioning, and this view is also supported by Wood's (2006) test of foraging goals among the Hadza. This view, however, is not incompatible with the idea that the efforts of some men are intensified by the desire for hunter prestige and its greater sexual access, including mating opportunities and strong pair-bonding.

Bell (2001:696), in a commentary on the perspective presented by Hawkes et al. (2001a:696), asked: "So, why do better hunters have healthier children?" And he wrote, "The answer to this question requires a fuller appreciation of the rewards of being known as a good hunter," particularly with respect to the annual rains when the Hadza break up into small family camps.

> For many hunter-gatherers there are times of great plenty and times of great hunger. At these extremes there is a tendency for available supplies to be distributed broadly (Ingold 1980:146–147). However, during periods of modestly reduced availability, there is a tendency for the camp to disperse into smaller units, often single-family units, foraging for smaller patches of gathered foods and smaller prey. As Ingold makes plain, better hunters are desired as leaders of these smaller groups, given their prestige for competence.
>
> The families of better hunters are clearly advantaged during periods of dispersal [associated with the rains and with fruit plants] not only because of the superior abilities of the particular hunter but also because men of prestige are better able to select companions whose presence would be most advantageous, creating thereby more effective teams. In the African context, the focus of attention at such times would be on gathering, and we would expect the wives (and other family members) of better hunters to be advantaged by optimal aggregations in their gathering activity.
>
> However, seasonal variations in the levels of social aggregation are not prominent in the discussion by Hawkes et al. We are told that "in some seasons they search specifically for honey, often *in nuclear-family parties*" (my emphasis). This tells me that there is an unexplored variable here. Can that be where the true culprit lies?

I agree with Bell. The seasonality of Hadza life has not been adequately treated, as earlier argued, and this, too, obscures the overall significance of the baobab as a source of animal foods in Hadza life.

A particularly puzzling aspect of Hadza life is the fluctuating weight of Hadza children in relation to the contribution of good hunters and the seasonality of berries, honey, and small game. Marlowe (2010:108) reported that berries "mostly come into season toward the end of the dry [time] … and continue throughout the first half of the rains," and that "honey then becomes plentiful." The "first half of the rains" is spring; this is the seasonal hunger time during which baobab fruit is still available although in diminishing amounts. Honey becomes "plentiful" after the "first half of the rains," meaning, in summer and autumn. At this time of the year, Marlowe (2010:108) noted that baobab fruit "is least plentiful." According to Marlowe (2010:108), "when honey becomes plentiful, husband and wives often go foraging together with the man looking for honey and the woman digging tubers or collecting baobab." Perhaps families living in small camps during the autumn rains when they are most dependent on small game are trying to avoid delayed reciprocal entitlement, which has been described in the ethnographic literature as coercive, aggressive, or demanding. At the height of the dry time, all children lose weight because most fruits are unavailable. In the wet time of the year, when fruits are abundant, all children gain weight. But the children of better male foragers are better provided for at the family level and in small camps with honey, meat, and baobab. In addition, better male foragers marry better female foragers, who target baobab fruit or berries, and if neither is available, roots.

Summary

This preliminary discussion of the baobab as a mediating factor in Hadza access to animal foods suggests the tree is more important in Hadza meat-eating than is generally recognized. It is especially noteworthy that this use of the baobab would be particularly valued during the spring hunger season, and during the autumn rains when the Hadza are especially dependent on small game.

Chapter 15
The Baobab as a Hunger-Time Tree of Life

Hunger has long been associated with seasonal changes as well as such things as drought, natural disasters, ruinous economic circumstances, social unrest, and warfare. This chapter describes the baobab as a diverse food source for all seasons; it can aptly be described as Africa's hunger-time tree of life, one that is as important today for the Hadza and other savanna dwellers as it would have been for early hominins. The baobab's value as a hunger-time food source is especially appreciated during the annually recurring period of food scarcity which, depending on latitude, local ecology, and mode of subsistence, most often occurs between the end of winter and the beginning of summer (Speth 1987). More broadly, the baobab provides hunger relief in many parts of Africa because it is a multisource food tree whose edible parts are available year-round or for several months and whose leaves and fruit can be preserved for use throughout the year, especially in the dry time. The tree produces leaves with the return of the annual rains in spring, sheds them at the beginning of the dry time in autumn, and is leafless in the winter dry season. The naturally dry ripe fruits are stored as they are. But leaf storage requires collecting and cleaning the leaves, drying them, then pounding, grinding, and sieving them to make a powder (Raji and Adeyemi 2018) for household use or sale in the dry time.

Given the baobab's well-known reputation as a food source in times of hunger, it is not surprising that Freedman (2018) included it in his list of famine foods. According to Awori (1989), the baobab "has often been used as a life sustaining plant," and he noted that it "is often the only thriving food plant during … famines especially when they are accompanied by prolonged drought." As Wright and Kerfoot (1966:52) pointed out, "Much of the tree can be eaten in one form or another," and in "times of scarcity it becomes a staple." Similarly, Sidibe and Williams (2002:37) identified baobab leaves, fruit pulp, and seeds as "wild-gathered foods," noting that they "play a significant role in the preparation of traditional dishes and as sources of food during times of scarcity and famine (Sai 1969)." In his study of Nigerian household food strategies in response to seasonality and famine, Longhurst (1986:27) emphasized the importance of secondary crops. These he divided into gathered crops, crops mixed into fields of staples, cultivated vegetables

J. Rashford, *Baobab*, https://doi.org/10.1007/978-3-031-26470-2_15

in home gardens near the compound, and non-staple root crops grown as a contingency reserve. Gathered crops included "wild vegetables occasionally cultivated, such as species of *Cassia* and *Loranthus*, [and] leaves of the baobab (*Adansonia digitata*)." It is also significant that the field manual on baobab cultivation produced for farmers and extension workers by the Southampton Centre for Underutilised Crops (SCUC 2006:2) reported that "The processing and sale of baobab products, especially in urban areas, offers a secondary means of income ... especially in times of drought and famine."

As Africa's hunger-season and famine-time lifeline, the baobab has likely been a vital component of human adaptation to the African savanna for a very long time, but this seasonal importance of the tree is one of the least explored aspects of its many uses (Rashford 1987b). In identifying the baobab as hunger-time tree of life, this chapter distinguishes between hunger-time and hunger-only foods, and it discusses the value of the baobab as a hunger-time food source throughout the year. This lays the foundation for understanding the significance of the baobab as a hunger-time and hunger-season food tree for the Hadza; also discussed is the tree's similar value in West Africa, Sudan, East Africa, and Southern Africa. The final section of the chapter discusses the seasonal significance of roots in Hadza diet, especially in relation to the baobab and seasonal hunger.

Distinguishing Hunger-Time Foods from Hunger-Only Foods

There is a widespread tendency in the literature to characterize seasonal hunger as a post-harvest or pre-harvest phenomenon associated with farmers and herders. This is a limited perspective, however, since it is ultimately the seasonal growth and reproduction of wild and tended plants and animals that is responsible for the hunger period of both food producers and foragers (Rashford 2002). Although the September–December hunger season incorporates the end of winter and the beginning of summer, it is identified here simply as the spring hunger season, a period that is also characterized in the literature as the lean season, early rainy season, or planting season (Richards 1939; Fortes and Fortes 1936; Miracle 1961; Ogbu 1973; Jenike 1988; Huss-Ashmore and Goodman 1989; Sahn 1989; Ferro-Luzzi 1990a, b). Generally, there is continuity in the availability of food from summer through autumn and winter. The natural seasonal break in the annual flow of food is spring following the winter harvest and preceding the summer harvest. In tropical climates with unimodal rainfall that occurs largely in summer, the period of seasonal hunger can extend from the spring hunger season through some part of the summer.

It is important to identify not only the kinds of baobab foods that are available during the hunger season and in times of famine, but also to determine whether these foods are part of a regular diet or exclusively eaten in desperate circumstances, which raises the idea of fallback foods. From its introduction in the 1970s, the concept of fallback foods has emerged as an important theoretical issue in the works of primatologists and paleoanthropologists, but as Constantino and Wright (2009:599) wrote in their introduction to a symposium issue on the subject published by the *American Journal of Physical Anthropology*:

> Despite the recent introduction of fallback foods into the primatological lexicon, there has been little consensus as to how they should be defined and how inquiry into their importance in primate evolution should be operationalized. This was the primary reason behind a symposium organized for the 77th meeting of the American Association of Physical Anthropologists held in Columbus, Ohio in April 2008. The symposium had three objectives: define fallback foods, discuss methods for identifying and measuring fallback foods and fallback dietary strategies, and explore the role of fallback foods in the ecology and evolution of primates through the presentation of recent field studies.

Current definitions of fallback foods are generally based on the dichotomy of preferred foods and fallback foods, which is also expressed as preferred foods (more preferred foods or regular foods) on the one hand and less-preferred foods on the other. The problem with this dichotomy in whatever form is that it is inherently ambiguous. It implies that hunger-season foods are not regular foods (which is synonymous with preferred foods), but fallback foods in the specific sense of hunger-only foods – foods ordinarily ignored but chosen in times of desperation.

To fully understand why the baobab can justly be regarded as the hunger-season tree of life, it is necessary to distinguish seasonal hunger-time foods and famine-time foods on the one hand from seasonal hunger-only and famine-only foods on the other, and to do so recognizing that the concept of fallback foods has been essentially defined in terms of seasonal hunger-only and famine-only foods. Hunger-time foods are eaten in times of hunger because that is when they are available. They are staples during the annually recurring time of scarcity resulting from the influence of the alternating wet and dry periods on the growth and reproduction of plants and their dependent animal populations. For many groups, the baobab's fruit pulp, seeds, leaves, honey, and game are seasonal hunger-time and famine-time foods. In Gambia, for example, Annegers (1973:254) found that in the spring hunger season, "boiled leaves and baobab seed were commonly the only ingredients in meals." Mertz and colleagues (2001:285) provided insight into the significance of the spring harvest of baobab leaves in their study of the consumption and marketing of wild and cultivated vegetables in Burkina Faso. They found that "*Adansonia digitata* and *Annona senegalensis* follow the same pattern as Corchorus spp." Of Corchorus they wrote, "The peak season is clearly in May–June, the early part of the rainy season, when the desirable fresh new leaves are most abundant and cultivated vegetables are in short supply." But we also know from the literature that baobab's leaves, fruit pulp, and seeds are hunger-only foods for some Africans, as suggested by Riley and Brokensha (1988:197). Writing of the Mbeere of the plains south of Mount Kenya, they report that "young leaves in times of famine are used as vegetable relish when they appear at the start of the rainy season." This suggests that there are some groups for whom the new leaves of spring can be considered a seasonal hunger-only food. The Hadza have not been reported to eat baobab leaves, but they are most reliant on the baobab's fruit pulp and seeds during the peak harvest of the winter dry season. This reliance continues through the spring, when ripe fruits are still relatively abundant at the height of the dry time of September–October and dwindle in the early autumn, when fruits are least available for the year. For the Hadza, the baobab fruit pulp and seeds are clearly hunger-time foods.

By contrast, seasonal hunger-only foods and famine-only foods are foods that are only eaten when nothing else is available, and famine can occur at any time of the year for a variety of reasons earlier noted. In the face of starvation, the baobab foods of last resort for most groups seem to be the baobab's germinating seed, root, bark, bud, shoot, and flower (Layser 2001; Kaboré et al. 2011; Sharma and Savant 2012:12). Root and bark can be gathered year-round, but buds, shoots, leaves, and flowers become available during the annual rains from the mid-spring of November through the early autumn of April. There are no reports of the Hadza use of these largely hunger-only foods. Nor do the Hadza make use of the edible taproot of the baobab, which can be had throughout the year. However, baobab honey, which is prized throughout the year, starts to become plentiful in spring and is most available during the autumn rains. Because the baobab provides both kinds of hunger foods throughout the year, it is a crucial link in the annual food chain that tides people over the hungry time associated with the end of the old annual growing period and the beginning of the new. It is a lifeline in times of food crisis.

The Baobab as a Hunger-Time Food Source Throughout the Year

In his influential article "Seasonal Hunger: A Vague Concept and Unexplored Problem," Marvin Miracle (1961) argued that the often-reported existence of seasonal hunger in Africa did not in fact occur. For Miracle, seasonal hunger, which was precisely defined in biological terms, was said to occur only when peoples' diets included fewer calories than needed at a particular time of the annual cycle even though their overall yearly intake was adequate. Ogbu (1973:317) challenged this perspective:

> Most of our evidence comes from anthropologists or others with close acquaintance with particular rural populations. These were neither trained in nutrition nor capable of calculating the daily calorie intake of the people they worked among. What they did report is that people at these seasons bore out the fact that since they ate less food, shifted to food they disliked, they became querulous and showed other signs of stress, and appeared to lose weight … There is abundant literature attesting to the fact that rural people in many parts of Africa are not able to meet their own nutritional standards throughout the year.

In support of his position, Ogbu (1973:317) cited the works of Fortes and Fortes (1936), Lynn (1937), Richards (1939), Haswell (1953), Platt (1954), Johnston (1958), Phillips (1959), Scudder (1962), Allan (1965), May (1965), and de Schlippe (1956). Other important anthropological studies not mentioned by Ogbu (1973:317) include Evans-Pritchard (1940) and Thomas (1959). A number of studies have also been published since Ogbu's article (e.g., Anderson et al. 2017) including works on the San (Marshall 1976, Lee and DeVore 1976, Lee 1979, Shostak 1981). Today, the topic of seasonal hunger is especially important among primatologists and paleoanthropologists, although the focus of their concern has come to be identified in the literature as "fallback foods" (Pennisi 1999, Laden and Wrangham 2005, Constantino and Wright 2009, Altmann 2009, Wrangham et al. 2009).

West Africa

In a survey of the Sahel to determine which trees were important to the peoples of Niger, Mali, Burkina Faso, and Senegal and why, scientists from the Canadian-funded Semi-Arid Lowlands of West Africa program, together with national agricultural research organizations, learned that the most valued trees were those that provided food security and were a source of income. "Baobab appears in the list in each country" and "tops the list in three countries" (Anonymous 2004a).

Sturtevant (citing Monteiro [1875]) referred to the use of baobab seeds "for food in times of scarcity" (1972:24), and Uphof (1968:8) wrote: "Pounded seeds are used in some regions as famine food." From general statements like these, it is not clear whether baobab seeds are considered a desirable food, or whether they are only eaten in times of scarcity because that is when they are available or because there is no alternative. There is a great deal of variability across Africa in the food value of baobab seeds and it is probably the case that many Africans eat baobab seeds not as a seasonal hunger-only and famine-only food, but as a seasonal hunger-time and famine-time food; for these groups, baobab seeds are in fact a staple. For example, Vaid (1978a:40) was more precise in noting the "seeds which are pounded and eaten as food in times of scarcity or regularly mixed with meal of millet," and Dalziel (1937:14) noted in *The Useful Plants of West Tropical Africa* that baobab seeds are "much used as food stuff," except in East Hausa, where "the pounded seeds serve as famine food, called charon." It is important to recognize, then, there are "some regions" where baobab seeds are "regularly" eaten as a staple, and what Dalziel says of West Africa might well be true of Africa in general – baobab seeds are a preferred food for most people, although there are exceptions like the East Hausa.

The occurrence of seasonal hunger in the northern savanna region of Ghana was already recognized in the early twentieth century, but British colonial officers stereotypically attributed it to what they considered the backward nature of traditional farming. Starting in the 1930s, studies showed that a period of seasonal hunger did occur, but not all people suffered because of it. Individual wealth and kin connections made a difference, as Destombes (2006:190) made clear in his review of the literature of this period. He identified seasonal hunger in northern Ghana as the time of the year when people were living largely on "soup" made of the leaves of grasses, herbs, and the baobab. In what he described as a traditional military state of Ghana's northeastern savanna, comprised at the time of 60 square miles and a population of 14,000, Hunter (1967:171) wrote in his article "Seasonal Hunger in a Part of the West African Savanna: A Survey of Bodyweights in Nangodi, North-East Ghana":

> Few families, particularly the poorer ones in terms of livestock and manure (no livestock means no manure, hence smaller crops), manage to survive the hungry season without distress. When granaries are exhausted recourse is had to wild fruits and nuts collected in the bush, to shea nuts [*Butyrospermum parkii*], dawa dawa [*Parkia filicoidea*] and baobab seed, and to greenstuffs, the latter in enormous quantities towards the end of the hungry season. These are abundant early in the rains but 'cabbage soup' (the nearest 'western' equivalent) will not starve off the pangs of hunger for long. A few groundnuts may be chewed but these are scarce and need to be reserved for seed.

Baobab seed is not a hunger-only food for all Ghanaians, however, and it is not the only hunger-season food the tree provides. In northern Ghana, fresh and dried leaves are a staple for groups that also eat the seeds. In one of the few articles that dealt specifically with the baobab and seasonal hunger, Kranjac-Berisavljevic et al. (2009:85) documented "the role of the baobab tree in securing Gruni households against hunger." They reported that "during food shortage" the Gruni (Frafra), one of the largest ethnic groups of Ghana's northeastern savanna region, ate "the tender fruits, leaves (dry and fresh), seed, and the pulp from the mature fruit of the baobab tree." They also noted that baobab products were "sold in local markets to help people cope during the hungry season."

There are similar reports from other areas of the West African savanna. Annegers (1973:254) noted, for example, that in Gambia during the spring, boiled leaves and baobab seeds were commonly the only ingredients in meals. "In Benin," wrote Assogbadjo et al. (2005), "the fruit maturation period extends from [the end of autumn in] December to [the beginning of spring in] March, approximately 6 to 7 months after the beginning of flowering," and they reported that this "also corresponds to the season of food shortage." In a paper titled "Household Food Strategies in Response to Seasonality and Famine," focused on northern Nigeria, Longhurst (1986:27) discussed the importance of secondary crops which he divided into gathered crops, crops mixed into fields of staples, cultivated vegetables in home gardens near the compound, and non-staple root crops grown as a contingency reserve. In his comments on gathered crops, Longhurst noted that they included "wild vegetables occasionally cultivated, such as species of *Cassia* and *Lorathus*, [and] leaves of the baobab (*Adansonia digitata*)." The baobab's fruit harvest is at its height in the winter dry season and is already in decline by the early spring of October with the end of durational and dry-time differentially delayed fruit fall. It is least available in the annual rains. Unsurprisingly, during the spring hunger season, the baobab's leaves and seeds are used as hunger-time or hunger-only foods.

Sudan

In their account of the contribution of non-timber forest products to reducing poverty in the Nuba Mountains of South Kordofan, Sudan, Tahir, and Gebauer (2004:3) reported that some 70% of their respondents identified 40 species of plants used for food, and among the 20 most frequently tagged was the baobab. These authors wrote, "The respondents mentioned that the fruits of all the above-mentioned species … are consumed during time of food shortages."

East Africa

According to Weiss (1979:40), baobab seeds are eaten in Angola and East Africa "when food is short." In their 1988 account of the Mbeere who live on the plains south of Mount Kenya, Riley and Brokensha (1988:197) were told the "young leaves in times of famine are used as vegetable relish when they appear at the start of the rainy season." Writing of the baobab as a hunger-season food on the same plains, Joe Ombuor, in an August 4, 2004, article in *The Nation*, Kenya's major daily newspaper, said:

> The sun beats down mercilessly from the cloudless skies. For miles around all one can see is drying crops and vegetation. The local residents are weak from hunger, with many of them having exhausted whatever little food they had. Such is the situation in the plains below Mount Kenya, where the seasonal rivers have dried up following a long, dry spell. A tour across the newly created Makuyu, Mbeere, Tharaka and Mwingi Districts brings one face to face with a looming disaster. In Mwingi, District Commissioner Joash Miyoma leaves doubt about just what lies ahead for the area residents. We could starve pretty soon without urgent food support, he says. The famine in Tharaka is so serious that many children are skipping school to look for food ... Women are seeking ways of feeding their children, even using fruit from the baobab tree; they grind the dry pith into flour and use it to make porridge. (Ombuor 2004)

A similar representation of the baobab as a hunger-season tree of life involves other groups in Kenya. Ukambani, the traditional homeland of the Akamba, comprises the central highlands of Kenya, which lie east of the Rift Valley and south of Mount Kenya and are dissected by the Athi River and its tributaries to the west of the Yetta Plateau and by the Tira River and its tributaries to the east of the Yetta Plateau. Ukambani is today largely contained within the borders of the Machakos and Kitui districts and is an area that has long been associated with annual food scarcity and with famine. It is noteworthy that the baobab as a hunger-season food is not a recent phenomenon for the Akamba. Akamba elders say their original home was to the south of Kenya, from which they migrated north, passing Kilimanjaro, to settle in the Taita Hills in southeastern Kenya. There they experienced famine and turned to wild fruits, and because they regarded the baobab as nourishing, it was in high demand. One group left to look for baobab fruits and did not return; they became the Kambas. Those left behind were called Taitas, from the Swahili word Mtaitwa. The Kambas split into three groups: those who moved eastward to what is today the Kitui district, those who moved to the northeast to what is today the Makueni district, and those who moved to Kilungu. In July 2004, *The Nation* newspaper reported:

> The current [summer] famine in Ukambani [where "85 per cent of the farmers are now buying food"] may be blamed on the failure of the short and long rains, with some areas receiving less than normal and others none at all. Experts say that this sordid situation has made families feed on low quality maize, occasionally leading to aflatoxin (mould) poisoning, which has claimed more than 100 lives in Makueni and Kitui districts. The people who cannot afford food supplements eat wild fruits like baobab and tamarindus.

Since 1983, Kenya Energy and Environmental Organizations (KENGO) have been studying the traditional knowledge of indigenous trees "to generate basic data … for use at the community level to enhance their propagation, conservation and utilization." As a part of this effort, Arum (1989) reported on the baobab as one of these species, and he too made specific reference to Ukambani:

> The baobab has not been observed in association with [definite planting] … though many people leave it in their shambas [agricultural fields]. For example in the Southern parts of Machakos and Kitui Districts of Kenya and in Shinyanga region of Tanzania it is often found growing in cropland as one of the trees left in shambas. Some people believe that having baobab trees on farmland is a good omen that will result in bountiful harvests. Clearing baobabs from the farm or selling land on which baobabs stand will result in bad luck. When in season, a medium sized tree can produce up to 200 Kg of fruits and since the fruits play an important role in local diets, especially during famines, the baobab is sometimes considered as a crop.

Southern Africa

The baobab has also been reported as a hunger-season food source in Angola (Weiss 1979:40) and Zimbabwe. In the case of Zimbabwe, The *Daily News of Zimbabwe* published an article titled "Zimbabwe: Starving Villagers Claim Ownership to Wild Fruits" that highlighted the life-saving importance of the baobab in a time of drought-induced famine.

> Have you ever thought of owning a baobab tree, which grows wild in the bush? The drought and starvation in the country has made such ownership necessary among desperate villagers of St. Mary's in Hwange District. Many of them now own a variety of wild fruit trees, the most important being the famous baobab … used to make sour porridge for human consumption. Villagers in St. Mary's said they often go for weeks without eating sadza as their crops failed last season due to the drought … One villager, Methuseli Sibanda, said: Now there is nothing at all left in our granaries. So we have learnt to survive on the baobab fruit. The villagers said most of the wild fruit trees … are now owned individually as families scramble to get the most out of them. The area around the tree is swept clean by the owner to indicate the tree belongs to someone. It is now an offence for one to pick up wild fruit from any tree, as the picker can be hauled before the chief or even be attacked if caught red-handed, said another villager, Martin Tshuma. He said gone are the days when a traveler could stop and freely pick wild fruits from the bush … When I was young, wild fruit belong to God and the ancestors but today you find people owning wild trees. This means that the food situation is very desperate, he said.

For the Kung of the Kalahari, the summer rains usher in a period of plenty which lasts until the mid-winter in August. After this comes the "spring dry season," which Lee (1979:104) identified as "the least attractive time of the year."

> The early spring begins in late August [which is actually towards the end of mid-winter] with a rapid increase in daily temperatures and ends in October or early November with the onset of the first rains. This is the least attractive time of year. Although humidity remains low, the days are exceedingly hot, with highs from 33 to 430C (92 to 1100F) in the shade. Work is difficult, and the better foods may be available only at distances from camp. It is in

> this season that the !Kung make use of the widest variety of plant foods. Fibrous roots, ignored at other times of the year, may be dug and eaten without enthusiasm. The men say the heat makes the animals less wary and their movements more predictable, so they are easier to stalk and kill ... But by and large, spring is a time of waiting and eagerly scanning the horizon for the first sign of rain.

In winter, especially at the height of the dry time which includes the end of winter in September and the early spring of October, the Kung seek out the baobab for its ripe fruits. Lee (1979:261) reported that the water hole called "!Kubi gets many visitors from August to October during the height of the baobab season."

The Hadza and Seasonal Hunger

As earlier noted, the Hadza view of the significant seasonal moments of their savanna environment is based on temperature and rainfall, with the major periods of the year identified as being cool, hot, or rainy. It is cool from June through August; then comes the hottest time of the year from the end of winter in September through the early spring of October, identified as at the height of the dry time. The dry time is followed by the annual rains from the mid-spring of November to the early autumn of April, after which drier cooler conditions return from the mid-autumn of May onward. The spring rains bring about dramatic changes in the landscape. The hard, dry, dusty earth quickly turns muddy, as seasonal runoff and pools begin to appear. The most active period of plant growth – the germination of seeds and the budding, shooting, releafing, flowering, and fruiting of herbs, shrubs, and trees – transforms the brown landscape of the dry time to the greens of the rains. With the active growth and reproduction of plants comes an upsurge in the activity of insects, including swarming termites, noisy cycads, and annoying mosquitoes; it is the time of frogs and catfish and the avian breeding season.

The shift from the May-October dry time to the start of the annual rains is also a shift from the enhanced opportunity to take large animals to a greater reliance on small game, which includes the benefit of the breeding season abundance, especially in the taking of nesting birds. During the November–April rains, large animals are taken less frequently than during the September–October height of the dry time because large animals are more scattered, better hidden by the lush vegetation of the growing seasons, and more difficult to track. The most important plant foods of the spring hunger season are berries. These include tafabe (*Salvadora persica*), which fruits from early spring in October to mid-spring in December; *Cordia sinensis* and *Cordia gharaf*, which fruit during the rains from the mid-spring of November through the summer; and *Dobera lorathifolia*, which fruits at the same time as the *Cordia* berries. Other food plants of the spring hunger season include the fruit-bearing tamarind (*Tamarindus indica*), the *Hyphaene ventricosa* palm which are in full season in the spring, and the baobab, whose year-round harvest is approaching its lowest availability for the year. With the November–December start of the annual rains comes the period when honey and roots approach their peak harvest in autumn. The largest amount of roots for the year are taken by the Hadza during the autumn rains, but the second largest amount of roots for the year is taken in spring.

Tomita on the Hadza and Seasonal Hunger

Tomita (1966:169–170), who spent a year with the Mangola Hadza, reported that the "food the Hadzapi can gather in the rainy season grows far more abundantly than in the dry season. They depended mainly on food gathering in the rainy season but when the dry season comes they must do more hunting than they did in the rainy season." Although he did not include the baobab in his explicit view of the seasonality of Hadza foods, Tomita (1966:169) was fully aware that the baobab was an important food source for the Hadza, and he recognized a baobab-fruiting season. He reported that fruits started ripening from the end of summer in March and that the hard shell protected the fruit from breaking open when it fell to the ground and kept it from rotting. This, he thought, was the reason baobab fruit could be gathered throughout the year. However, within the framework of this seasonal polarity perspective, Tomita (1966:169) reported that the hunger time for the Hadza was the dry time from the end of autumn in June to early spring in October, especially the height of the dry time in September and October. Unlike Woodburn, Tomita (1966:169) did include the seasonal gathering of honey and the ripening of the third and largest crop of berries for the year, which occurs around the end of summer in March through mid-autumn in May.

> This is the time when they forget hunger and spend the most peaceful period of the year. When May comes and the rains pelt the *madabi* fruit causing it to start deteriorating the regular honey gathering season comes. From the end of the rain season until the next rainy season when the *tafape* again starts mellowing the Hadzapi are obliged to undergo the most critical time of the year. This period is the dry season of Mangola, where they daily endure strong winds. From June to October of the dry season, at least until June, July and August honey gathering is possible but by the time August [is] over all the honey has been collected. Then, the Hadzapi have only roots and stalks to gather for food. Some of the agricultural people in Mangola also have to go low on food supplies in this period.

The dry-time honey gathering coincided with the baobab harvest, and it is clear from the passage below that Tomita (1966:170) overlooked the importance of baobab-clad hills (Vincent 1985a, b) in Hadza camping during the dry time.

> From the start of the dry season honey gathering becomes possible. For this purpose they move further into the mountain area where they can hunt more easily; a place a long way from other tribes, where they can carry out hunting in the original way.

Woodburn on the Hadza and Seasonal Hunger

Although both Tomita (1966) and Woodburn (1968a) worked within the restricted wet season/dry season framework, they provided very different perspectives on the seasonality of Hadza life. The significant difference is the emphasis placed on the seasonal round in contrast to the wet/dry oscillation. Tomita was clearly more attentive to the seasonal round, but Woodburn's work has had a greater impact on the field of hunter-gatherer studies and on anthropology in general.

As earlier noted, Woodburn's (1968a) influential article on Hadza ecology was part of a radical shift in the way anthropologists viewed contemporary hunter-gatherers and their relevance for theorizing the adaptation of prehistoric foragers. His study of the Hadza and his theoretical reflections on the nature of hunting-and-gathering societies, especially the attention he gave to the difference between immediate and delayed-return hunter-gatherers, was an important contribution to this new view. Far from living a desperate life, modern foragers came to be viewed as the original affluent society (Sahlins 1968). What Woodburn (1968a, b:52) emphasized was a "wet season"/"dry season" oscillation as representing the perspective of the Hadza in which the activities "which are characteristic of the wet season tend to be avoided during the dry season and vice versa." And in this view, what is significant is that "In the wet season the emphasis is on root gathering and hunting small game, especially hyrax; in the dry season the emphasis is on gathering berries and hunting large animals." In contrast to widespread reports of seasonal hunger among farmers, Woodburn (1968a, b:52) reported that the Hadza never experience a scarcity of food, even in times of drought. All the preferred foods of the Hadza – which include honey, meat, baobab, berries, and roots – have been identified as available year-round, and this is the view of hunter-gatherer affluence that Woodburn (1968a, b:52) emphasized.

> The range of foods in the bush is so great, if one knows what these are and how to obtain them, that if weather conditions should cause the failure of some type of root or berry, or the migration of some of the game, some other type of food is always available. For a Hadza to die of hunger, or even to fail to satisfy his hunger for more than a day or two, is almost inconceivable.

It is noteworthy, however, that with respect to berries being principally obtained in the dry time, more recent studies identify the rains from mid-spring in November to mid-autumn in May as the main berry season. This means berries are available year-round, with *Cordia* species being common in the wet time and *Grewia* species in the dry time. It is also noteworthy that Woodburn did not include honey or baobab in his seasonal polarization, writing instead, "The honey and grubs of seven species of wild bee are eaten; supplies of these vary widely from season to season and from year to year" (1968:50). As earlier indicated, honey is available year-round, but the peak honey harvest is in the rains. Baobab fruit, he wrote is "always abundant even at the height of the dry season" (1970:50). The height of the "dry season" is September–October, which is the end of winter to early spring, but the use of the word "even" is misleading since the baobab, as earlier noted, is available through the long dry time from the end of autumn in June to the early spring of October, and September–October is the end of the tree's main fruiting period which peaks not in September–October, but in the winter dry season.

These issues notwithstanding, Woodburn's account of the Hadza and his contributions to the idea of hunter-gatherer affluence became widely accepted. Jelliffe et al. (1962:22), for example, began their account on the health of Hadza children with the following understanding: "The Hadza, as described more fully by anthropologist Woodburn (1968a), are a prime example of an original affluent hunter-gatherer society, one in which basic subsistence requirements could be satisfied with as little as two hours labor a day per adult and no seasonal shortage could be found." The hunger season, whether the Hadza experienced it or not, did not come into focus in the perspective of Woodburn and Jelliffe and colleagues, and their approach obscured the specific nature of Hadza adaptation to the spring hunger season.

Marlowe on the Hadza and Seasonal Hunger

Marlowe (2010:108) seem to suggest that the dry time of the year is the period when the Hadza should be seasonally stressed, and in doing so, he overlooked the year-round importance of baobab foods. "One might think," he wrote, "the dry season would be a lean season, but people do not lose weight during the dry season, mainly because at that time men take more meat through ambush hunting at water holes." In effect, Marlowe presented a view of Hadza year-round food abundance that is consistent with that presented by Woodburn. But while Woodburn saw the baobab fruit as available year-round, Marlowe (2010:108) identified it as "least plentiful" in the summer, which is when honey "becomes plentiful."

The Hadza During the Period of Seasonal Hunger

Data presented by Hawkes et al. (1997) indicated that the Hadza experienced the hunger season, which manifest itself in physical changes, but they did not routinely suffer because of it. They lost weight, but not enough to have a significant impact on their overall wellness, for along with roots as their most dependable food, they also had the baobab, an important hunger-season food source. Hawkes et al. (1997:557) wrote:

> Almost all of the variation in children's weight changes is associated with seasons 1 [the late September-October dry time] and 2 [the mid-spring of November]. During season 1, children lost weight. With the onset of the wet season and the related increase in their own foraging efforts (as well as those of their mothers and grandmothers), they made larger weight gains than at any other time of the year.

The baobab clearly plays an important role in the Hadza ability to thrive during the spring hunger season. According to Marlowe and Berbesque (2009:756), were it not for the baobab the consequence of the hunger season would be much more significant.

> Women's %BF declined when more baobab was taken [in the May-October dry time, especially the September-October height of the dry time]. Given the baobab is high in vitamin C and the seeds are high in fat, this must be because baobab is more available at times when other important foods (berries and meat) are less available... [A]ll other foods except for tubers are negatively correlated with baobab. Women's %BF would be much lower at such times but for the benefits of baobab.

Given the increasing scarcity of game and all the problems imposed on the Hadza by encroaching neighbors and the interference of missionaries, charities, and adverse government policies, one cannot but wonder if an unusually significant decline in the 2007 baobab harvest period during the September–October height of the dry time might also have contributed to the following tragic scenario reported by Valentine Marc Nkwame in an October 26, 2007, newspaper article in *Arusha Times* of Tanzania titled "Tanzania: Bushmen lives threatened as famine hits":

> Hadzabe Bushmen ... are facing an acute shortage of food. Two of them died recently from what is believed to be famine related.
>
> At Sanola area, deep in the Yaeda Jungle a group of Bushmen mostly women could be seen gathered under [a] huge baobab tree, nursing about five sick children with what seemed to be high fever. The children were lying in the shade, covered with animal hides.
>
> An old Hadza lady, Tale Mudendee said they were also grieving because two old men had died within the last few weeks; she named them as Mbogosh and Endeku both being brothers from the Washema [group]...
>
> "We are starving, all the animals have disappeared and we Hadzabe only feed on meat," Kriti Gimbi an old man who was making arrows from sticks explained through an interpreter. His wife, Ntale Nzale was keeping a tin of maize flour in their grass hut. It is for the children, we mix it with water and they drink it. At least we force them to drink it because they also do not like the Swahili food. The adults here thrive on wild berries and roots to stay alive but now a strange sickness is sweeping the valley, we all going to die! she said.

The Hadza, Roots, and Seasonal Hunger

The desperate scenario presented by Nkwame is perplexing. Granted, the Hadza prize meat, but we are left to wonder why they did not gather roots, whose exceptional value is their year-round availability, or, for that matter, honey, meat, baobab fruit, or berries, which are also available year-round. Some researchers have argued that the amount of roots the Hadza consume through the year fluctuates in relation to the seasonal availability and procurement costs of other preferred foods. For example, Laden and Wrangham (2005:489), whose view of Kung and Hadza food seasonality was framed by the context of the seasonal polarity perspective, noted that in "the Kalahari, roots indeed conform to the concept of fallback foods because, although they are not preferred compared to meat, fruits, honey, or fatty seeds, for example, they become the dominant dietary item during the periods of food scarcity." They report that for the G/wi, "The worst time of year was early summer, when people complained of hunger and thirst, body weight was low, illness was more common, and there were even occasional episodes of starvation… All foods came from four species of tuber." Laden and Wrangham claimed a "similar kind of reliance on [roots]…was observed for the !Kung at Nyae Nyae" and cited Marshall (1976:108), Lee (1979), and Vincent (1985a). The worst time of the year for the G/wi of the central Kalahari with its unimodal rains occurs in the early summer, but this is not true for the Kung at Nyae Nyae, where the hunger season is more centered on spring. And Marshall (1976), Lee (1979), and Vincent (1985a) mentioned the importance of roots especially in the winter dry season, not the summer. For Vincent, most roots are taken by the Hadza in autumn rather than in the dry time generally.

Roots have long been theorized to represent a crucial food in human evolution, especially in association with the African savanna (Coursey 1973), and there are those like Landen and Wrangham (2005) and Marlowe (2010) who theorize that the ultimate evolutionary value of roots is that they are a "fallback food" when the preferred berries, baobab fruit, honey, and meat were less available. This view would suggest that the consumption of roots would be clear evidence of a hunger time, and one would indeed expect that more roots would be gathered in the hunger season than at other times. However, with the exception of Woodburn (1968a) and Vincent (1985a), published accounts of the seasonal harvesting of roots do not support the view that most roots are gathered in the spring hunger season. Within the framework of the seasonal polarity perspective, Woodburn (1968a) and Murray et al. (2001:3–4) reported that roots were mostly gathered in the November–March period of the annual rains. If roots are indeed an important hunger-season food, this would explain why they are gathered in significant amounts in November and December but not why they would be gathered in significant amounts in autumn, including the May–June dry half of autumn which is the start of the dry time.

Vincent (1985a) reported that the Hadza gather roots in significant quantities in the September–October height of the dry time and also during the period of the peak annual rainfall from the end of summer in March through the early autumn in April. If Vincent is correct, we can assume that in the late dry time, roots are eaten as a "fallback food" because other crops are scarce, and the extremely dry conditions mean high procurement cost. But in the period of the autumn rains, it is possible that roots are more easily dug, significantly lowering their procurement cost. However, Laden and Wrangham have made the fundamental point that roots are not a "fallback food" eaten only in times of desperation. They note that both Lee (1979) and Vincent (1985a) indicated in their ethnographic accounts "that the degree of difficulty in harvesting roots is a major factor affecting the degree to which they are [selected through the year]… Accordingly, the fact that [roots]… are eaten more when fruits are not available may have more to do with their being hard to extract from the ground than with their merit as a good tasting or highly valued food item" (2005:489). If this were indeed the case, we would expect large quantities of roots to be eaten during the autumn rains, but not at the September–October height of the dry time.

Unfortunately, Vincent's (1985a) overall perspective gave a contradictory view of the seasonal harvesting of roots. In her view, which is also the view presented in this work, May–October is the dry time and November–April is the wet time, and she says the "main rains" fall from February through April, which means minor rains or spring rains (which she does not specify) would fall from November through January. The contradiction is that in her Table 5 (1985:141) Vincent identified six kinds of roots and their "seasonal" availability in the following way: *//ekwa gadabi* and *matukwaiko* are gathered in the "Rains"; *shumuko* is gathered in the rains and through the "Early dry"; and *do'aiko*, *//ekwa hasa*, and *makalidako* are gathered during the "Early dry." There is no mention of these six important roots being gathered during the September–October height of the dry time, and yet she wrote: "Roots are available and edible throughout the year, although they are mainly eaten during the main rains and again during the late dry," which would be September and October. Few other researchers agree with this perspective that significant quantities of roots are being taken at the height of the dry time, the early part of the spring hunger season.

The distinction between roots as a regular food and a fallback food is ambiguous, since it does not clearly differentiate between hunger-time and hunger-only foods. If by a fallback food we mean something is only eaten in the face of starvation as a last resort, this would not be an adequate account of roots in the lives of the Hadza or the importance of roots in human evolution. There is no doubt that roots have been significant in hominin evolution, especially the evolved capacity for handiness in relation to the digging stick linked to their acquisition, the container to their transport, and cooking fire to their detoxification and palatability, and this is also clearly evident today in the many wild and domesticated species and varieties of roots that are important in Hadza life and worldwide. Roots (in the sense of ground crops in general) include such familiar and historically influential crops as the potato, sweet potato, cassava, yam, and Araceae species, of which the most important are taro and dasheen. Roots are not a desperation-only fallback food for the Hadza. Roots are simply the lowest ranked of their preferred staples and are best described as a dependable staple. When the other four preferred foods are not available or in short supply, there are always roots to be had. This is why most roots are gathered in autumn and, according to Vincent (1985a), in the September–October height of the dry time. According to Mabulla (2003:39), roots are taken from the end of summer in March to the start of spring in September but were not gathered in the rest of spring or in summer.

The views presented above are not adequate, however. Based on the data Marlowe and Berbesque (2009:754) present, roots are available year-round, with most gathered in summer and autumn and less in winter and spring, with spring the season when they are least gathered. This means that for the Hadza, roots are a staple of their diet whose dependability makes it a hunger-time food. This agrees with Vincent that most roots are gathered in autumn and it partly agrees with Mabulla (2003:89), Woodburn (1968a), and Murray et al. (2001:3–4), but the data do not show roots being significant in the spring hunger season, especially during the September–October height of the dry time, and only Vincent holds this view. The significant gathering of roots in the dry conditions of early spring also contradicts Woodburn (1968a) and Murray et al. (2001:3–4), since for them the second highest take of roots after the March–April autumn rains is in the November–April spring rains.

We can see that accounts of the harvest seasonality of roots vary, with explanations based on the annual rains, fallback foods, and seasonal procurement costs, but there is an additional factor to consider: the relative seasonality of Hadza preferred foods in terms of resource concentration and resource dispersion, particularly with regard to Hadza foods in general and water. If with respect to procurement costs, for example, root-harvesting was a significant phenomenon of the rains, with roots being more easily dug at that time, then more roots should be eaten in the November–December spring rains than is currently reported, and peak harvesting of roots should be November–December and March–April. Consequently, since most roots are taken in increasing amount from summer through autumn and in declining amounts in winter and spring, with the least amount taken for the year in spring, the harvest seasonality of roots cannot be simply a matter of lower procurement costs in the rains.

It is not just seasonal procurement costs that determine the amount of roots harvested throughout the year. The relative seasonal availability and distribution of preferred foods and water in winter, spring, and summer in contrast to autumn is a contributing factor. In autumn the seasonally available range of key foods is at its narrowest for the year, especially in comparison to the spring rains. From the vantage point of the seasonality of Hadza foods, it could be argued that gathering roots declines in winter because it is the peak harvest of baobab and the ambush-hunting of large game. It declines from spring through mid-summer because of the harvest of berries. *Hyphaene* and *Salvadora* berries are particularly important because they are available in September–October, when the harvest of *Grewia* berries ends and the *Cordia* berries of spring and summer are not yet in. The harvesting of roots might also decline from mid-spring through summer because tamarind ripens at this time and baobab fruits resulting from differential fruit fall can still be had, along with baobab seeds extracted from baboon dung.

The real question is this: Are the seeds of the baobab collected from baboon dung only in times of hunger, especially during the spring hunger season, or are they collected from baboon dung throughout the year? Baboon-dung seeds would be a hunger-time food if they were gathered year-round, but they would be a hunger-only food source if they were exclusively gathered in spring. The work of Peterson et al. (2013:1–3) suggests baboon-dung seeds are collected largely in spring only because the procurement costs are worth it during the hunger season. According to these authors, the Hadza recognize that they compete with baboons for baobab fruits and that this competition impacts the availability of fruits at the September–October height of the dry time:

> Our greatest competitors for food are baboons. We sing an old song that describes a year when very little rain fell, and when the tubers were thin, dry and shrunken like the people were and the baboons finished the baobab fruits. The song relates how we survived this difficult period by climbing up to the baboon roosts on the rocks to harvest baobab seeds from baboon feces. The seeds were second-hand, but intact and edible, and kept us alive.

In addition to the availability of berries, tamarind, and baobab fruit pulp and seeds, the gathering of roots might also decline in spring because of the taking of nesting birds and the start of the major honey harvest.

When the resource concentration of winter, spring, and summer ends, the Hadza have honey and the dependable roots which are taken in the greatest amount in autumn. At this time of year, the Hadza live in smaller, more dispersed camps that engage in more frequent residential moves, not only because surface water is more readily available and food resources are dispersed, but also because at the peak honey harvest, families try to keep more honey for themselves. Honey, which is dispersed and expensive to acquire, is the Hadza's most prized food and has commercial value. Honey is especially important to hunters' wives, both as a weaning food and trade good. This means that for the Hadza, roots are a staple of their diet whose year-round dependability also makes it a hunger-season food, though the hunger season is the time they gather it the least.

The baobab is the most prominent tree of the Hadza's mosaic savanna and one of their most valuable food trees. For Africans throughout its range, the baobab is an important seasonal resource that tides them over during the food scarcity of spring. It must have been the same for ancestral hominins. The baobab is generally an important tree for coping with seasonal scarcity, but in some cases, it might even be accurate to describe it as the principal food source for coping with seasonal hunger. The availability of its root, bark, leaves, flower, fruit pulp, seeds, honey, and other foods during the spring hunger season and in times of famine makes it worthy of being identified as among the top trees that provide food security in Africa. This is yet another important fact that contributes to the baobab's designation as the Hadza's and humanity's exceptional multipurpose tree of life.

Chapter 16
The Baobab and Health

Traditional medicinal plants worldwide have received a great deal of attention from ethnobotanists and other scholars, from commercial interest in search of new pharmaceutical products, and from individuals involved in a growing number of health-conscious movements. With these developments, no area of baobab use has received as much attention in the literature as its association with wellness, which includes the value of the tree as a whole, and the various uses of its root, bark, leaves, flower, fruit pulp, seeds, and fruit pod. This short chapter discusses Hadza health-related uses of the baobab.

The health benefits of the baobab, broadly conceived, include medicines, nutrition, soap-making, insect repellent, hair wash made from the fruit pulp as documented for East Africa (Hobley 1922), and the widely reported use of the tree as a shaded resting place. In their extensive literature review, Wickens and Lowe (2008: 81) noted that although poorly documented the "bark, roots, leaves, fruits and seeds are widely used by indigenous peoples" for treating both humans and animals, and they presented twenty-five subject headings under which the numerous health uses of the baobab were discussed. Despite the growing popularity of its fruit as highly nutritious and a source of medicines and other healthcare products, scientific studies have only recently begun to document the health benefits of the baobab.

The baobab can appropriately be identified as a healing tree given its historic reputation. It is noteworthy that the medicinal use of the baobab's fruit pulp has played an important part in the tree's dispersal in India. It was the presence of the fruit in Cairo markets that initially brought the baobab to the attention of Europeans. The link between "tree of life" and wellness is evident in the Biblical conception of the immortalizing fruit tree featured at the center of the Garden of Eden and the New Jerusalem tree with its healing leaves. But wellness, in general, and healthcare, in particular, are seldom given as reasons for identifying the baobab as the tree of life. A notable exception is an article by Simon Jackson and Anabel Maldonado (2015) titled "Baobab – The Tree of Life: An Ethnopharmacological Review."

J. Rashford, *Baobab*, https://doi.org/10.1007/978-3-031-26470-2_16

The Baobab's Health Benefits for the Hadza

Although systematic studies have not been done, there is no reason to doubt that the use of plants to promote health is as important to the Hadza as it is to all people. Directing his reader to Woodburn (1959) "for a fuller description" of Hadza medicinal plant use, Marlowe (2010: 140–141) wrote:

> There is a plant that is boiled and the liquid drunk to relieve the symptoms of malaria. Several plants are used to cause a person to vomit after being bitten by a poisonous snake. With the most poisonous snakes, there is little hope, since they can kill a person in minutes, but these plants can apparently work with less poisonous but still deadly snakes, and several people say they have been saved this way. Bark from a certain tree is boiled and made into a tea to treat syphilis and gonorrhea. One plant is given to someone who falls down with seizure, presumably epilepsy, which must be rare, given that there are only 1,000 Hadza. There is also a plant that is supposed to help men overcome impotence. Some say there are plants that can induce miscarriage in the early stages of pregnancy, though it seems that they are rarely used or that they are ineffective... There are no medical specialists or specialists of any kind among the Hadza. Every adult Hadza hunter knows about the various medicinal plants and practices. It is usually men who keep the snake medicine with them. Any adult present may treat someone with an ailment. (Marlowe 2010: 140–141)

Surprisingly, there are few reports of Hadza health-related uses of the baobab. In her discussion of the menstrual cycle of Hadza women and the use of menstrual soap, Fitzpatrick (2018: 128) wrote:

> At one of the nine camps, the interviews were conducted under a large baobab tree and some baobab seeds were scattered around us. The women described "soap made from this tree", "with this", "like these" and gesticulated the pounding of the baobab pods. As they were describing menstrual cleaning, women described "just water and baobab", "you grind (and) foam appears" and "we take the soap of the fruits; you put the fruits into water and produce foam. When a little foam is produced then you take your cloths and wash".

The baobab is also an indirect source of one recorded medicinal treatment. Marlowe (2010: 140) wrote, "When one is badly cut, a tourniquet is applied after boiled animal fat or honey is applied to the wound." According to Marlowe, the "Hadza do not like to wear bandages and believe it is better to let wounds have fresh air."

The Lack of Knowledge of Hadza Health Uses of the Baobab

This chapter, like the book as a whole, presents a view of the baobab as the Hadza's and humanity's tree of life that is at odds with the view of Watson (2007); he assumed there was a significant difference between West Africans and the Hadza in the baobab's nutritional, medicinal, and spiritual importance. But it is clear he was not aware of the literature documenting the baobab's importance in the Hadza hunter-gatherer way of life. What Watson regarded as an absence of baobab use for medicines and other health benefits by the Hadza I suspect is more a lack of documentation, and his view that foragers are by definition living an impoverished life has long ago been abandoned by anthropologists. Watson (2007: 159) wrote:

> [M]any Hadzabe hunter-gatherers will say that they have no medicinal use for any part of the baobab, other than the pith, which they eat more as food than medicine. Not even their gods are connected with the trees, but rather the sun and the moon. The Hadzabe live impoverished lives, surviving exclusively on what they can gather from their environment, and their minimal use of the baobab's resources is astonishing. Elsewhere, particularly in West Africa, the position is dramatically different, and the lush chemical factor of the tree in full leaf is an irresistible source of nutritional, medicinal or spiritual benefit.

Summary

The baobab is known more as a cause of death among the Hadza – with men falling from trees they climb to collect honey, fruits, and other resources – than as a source of medicines and other health benefits. Yet, given the diverse uses of the tree and all its parts, it is inconceivable that the baobab would not have found a place in the Hadza conception of wellness, especially as a source of medicines. The significance of the baobab in Hadza health traditions merits serious attention.

Chapter 17
The Baobab and Exchange

Human evolution is as much about the growth of the hominin phylogenetic tree with modern humans as its only extant branch, as it is about the global way of life that is the outcome of this evolution. These fundamental aspects of the human story are reflected in the lives of the Hadza. Contemporary researchers, especially cultural and evolutionary ecologists, have focused on Hadza traditional foraging. They have done so as a way of developing an adequate theoretical understanding of our long prehistory as foragers, but not as a way of understanding our now global way of life. Hence, the concern with the traditional hunting-and-gathering Hadza rather than with the nontraditional Hadza, even though the latter make up as the majority of the Hadza population; these nontraditionalists have been variously described as market integrated, assimilated, acculturated, modernized, or progressive (Mabulla 2003:35; Marlowe 2010:19). "Ethnographic studies," wrote Mabulla (2007:28), "show that at the end of the last century, approximately one-fourth of the Hadzabe population continued to practice a full-time traditional hunting and [gathering] … lifestyle (Bunn et al. 1988; Marlowe 2002; O'Connell et al. 1988b). The other three-quarters of the population practice a combination of part-time traditional hunting and [gathering] … and other production activities." Because the lives of the Hadza part-time foragers have been profoundly influenced by the local, national, and global economy, the research focus of ecologists and evolutionary anthropologists has made these part-time foragers largely ethnographically, historically and theoretically invisible.

Among the many overlooked Hadza uses of the baobab is its important contribution to their exchange relations, not only with their traditional farming and herding neighbors but increasingly with the diverse representatives of the monetized economy that has developed in their homeland. Transfer, as earlier noted, is the most general term for describing resource distribution within and between societies and it fundamentally involves exchange, borrowing, begging, and theft; and borrowing, begging, and theft as forms of resource transfer are best explained in the context of hominin-evolved capacity for exchange. Based on the history of anthropological thought, exchange in the form of pooling, delayed reciprocity, barter, redistribution,

J. Rashford, *Baobab*, https://doi.org/10.1007/978-3-031-26470-2_17

and monetary exchange are the most important aspects of resource transfer in hominin evolution. This has been so from the earliest hunter-gatherers right down to modern humans, and it included the development of the world market as the foundation of our increasingly global way of life (Malinowski 1922; Mauss 1925; Polanyi 1957, 1959; Sahlins 1972; Wolf 1982). A view of hominin evolution that incorporates ancestral environment, the fossil record, primate studies, archaeological documentation, and ethnographic insight suggests central-place living associated with the gendered division of labor included pooling, delayed exchange, and bartering as forms of internal and intersocietal exchange. And while the ideal form of delayed exchange is everywhere identified as gift-giving as an expression of selfless generosity, delayed exchange also includes the form in which there is a strong sense of entitlement – a right to a fair share of resources – as has been reported for some foragers.

The baobab has long had an exchange value throughout its range where people acquire desired resources by bartering and selling its various products. Baobab fruit pulp, as earlier noted, came to the attention of Europeans in the sixteenth century because of its sale in Cairo markets, and today it is exported from Africa as an ingredient in a rapidly growing variety of commodities. In addition to the Hadza trade in honey, beeswax, fruit, and handicrafts, the baobab's exchange value in other parts of Africa has also included whole leaves, mostly dried rather than fresh, and the cleaned and dried leaves crushed, powdered, or as a prepared sauce; whole fruits, chunky fruit pulp with embedded seeds, powdered fruit pulp, and seeds; prepared foods such as in candy and drinks; bark-fiber products, especially rope; and various craft items for tourists like baskets and mats, and small woven baobabs made of banana leaves or baobab fiber (Gustad et al. 2004; Kamatou et al. 2011). This chapter discusses the central importance of exchange in human evolution and in contemporary life as a necessary context for appreciating the current status of the baobab's exchange value in both traditional and contemporary Hadza society. The exchange value of the baobab links the Hadza to others near and far through both commercial products and tourism.

Hadza Exchange Relations

Whether local or worldwide, trade can simply be defined as the intersocietal transfer of resources. However, the nature of trade is always ambiguously discussed in the current literature. This is because important distinctions are often not clarified – for instance the differences between trade as delayed exchange in the sense of goods now for goods later, barter as immediate goods for goods, and trade as monetary payments involving goods and services (Adams 1974). The most important traditional means of Hadza resource transfer have been pooling, delayed exchange, and barter. Pooling and delayed exchange are especially associated with reproductive households, delayed exchange with camp-wide and inter-camp resource transfer, and barter with Hadza/non-Hadza resource transfer. Today, of course, monetary exchange is also important. As earlier noted, it is significant that men bring back almost as much baobab to camp as women do. When women and men return to camp, the baobab harvest is distributed within households by pooling, and within households and among co-resident kin and friends by delayed reciprocity.

Barter proper has generally been associated with trade as external resource transfer. This is especially so in its symbiotic form as an expression of interethnic specialization that becomes particularly important with the worldwide development of horticulture, pastoralism, and fishing. The end of the global expansion of modern humans in association with megafaunal extinction, followed by the "broad spectrum revolution" evident in the increasingly intensive use of aquatic resources, and later by the worldwide adoption of agriculture some 12,000 years ago, means the majority of foragers studied by anthropologists, with a few notable exceptions such as arctic and subarctic groups and native Australians, have traditionally lived in association with cultivators and herders of tropical and temperate latitudes (Kent 1992). In exchange for cultivcated carbohydrates such as cereals, roots, and bananas, foragers have offered meat, hides, honey, and various plant resources, including fruits, medicinal herbs, and fibers (Spielmann and Eder (1994). In the past, barter donimated Hadza trade, and early reports showed that they had little use for money.

The tendency to minimize the Hadza connections with the world in the early part of the twentieth century was already evident in the account of Bagshawe (1925:119–121). He was at the time a district officer of the British colonial government "who made several trips to Hadza country soon after the Germans were defeated in the First World War" (Marlowe 2002:9). Bagshawe recognized that members of neighboring groups had contact with the Hadza. He also recognized that the Hadza were informed about the English/German colonial conflict, but "had no intention of becoming involved in hostilities." This meant Bagshawe's "messengers obtained no results" regarding a small party of Germans "at large to the north of Mkalama about whom [he] .... badly wanted information." And Bagshawe recognized that the Hadza had good reasons for avoiding contact with European colonists; he wrote the Hadza "dislike the presence of strangers and are most unwilling guides, but no one else has a knowledge of more than a fringe of their country, so they must be used." Despite these observations, however, Bagshawe (1925:120–121) characterized the Hadza as isolated and exchange relations with others as minimal:

> The mentality and disposition of the Kangeju are quite without parallel in Africa. Other tribes, including even the Bushmen of South Africa, have adopted at all events some of the conveniences of life which they have seen in use among their neighbours. Excepting that he makes his arrow-heads by grinding down old spears obtained from other tribes, and that he likes to adorn himself and his women with beads, the [Hadza] ... is, and is content to be, exactly as he was a thousand years ago. He is a wild man, a creature of the bush, and as far as I can see he is incapable of becoming anything else. Certainly he does not desire to become anything else, for nothing will tempt him to leave his wilderness or to abandon his mode of living. He asks nothing from the rest of us but to be left alone. He interferes with no one and does his best to ensure that no one shall interfere with him. He needs but little from his neighbours; tobacco, old iron for his arrow-heads and a few beads are easily obtained in exchange for skins, honey and occasionally meat, from the surrounding tribes,

> and he wants nothing more. He has no use for money, his currency being arrow-heads. He is a hunter, and most of his time and energy is devoted to the search for game. If game fails him – as it often does – he falls back upon wild fruits, roots and even insects and reptiles. He is often hungry and sometimes starves, but he is happy and envies no man.

Instead of foragers like the San and Hadza being "isolated" and "pristine," as Bagshawe seems to suggest, and as sometimes represented "in the cultural ecology and behavioral ecological sense," Riches (1996:288) wrote: "[I]t is now clear that, perhaps for millennia, most hunter-gatherers have been engaged in activities other than hunting and gathering, such as trading with agricultural and pastoral neighbours or even practicing a small amount of cultivation or animal rearing themselves."

Written accounts of the Hadza indicate they have had exchange relationships with their traditional neighbors for some time. According to Mabulla (2003), the archaeological evidence suggests a forager/food producer coexistence dating back some 3000 years, and delayed exchange and barter would likely have been the essential forms of exchange. With respect to the Hadza homeland, these neighbors, belonging to distinct linguistic lineages, include Cushite Iraqw farmers to the southeast, the Nilotic Maasai pastoralists to the north, Datoga pastoralists to the east, and the agropastoralist Bantu groups including Isanzu to the south, Iramba to the southwest, and the Sukuma to the west. Little is known, however, about the range and historical depth of the Hadza trade relationships with their traditional food-producing neighbors. Early reports indicate the Hadza had no need for money. For example, Fosbrooke (1956:3) noted that the Hadza he visited in the company of the director of what was then the King George V Museum in Dar-es-salaam "would not accept money in payment for articles of their culture which we wished to purchase, but would only part on a basis of barter. Foreseeing this, we had armed ourselves with a bolt of cloth and a bag of beads in true 'Darkest Africa' style, and the Museum did not go short of specimens."

Delayed echange and bartering in time gave way to the development of monetary exchange as the essential component of an increasingly complex societal and inter-societal food-producing division of labor. This division of labor corresponds to the development of intensive agriculture worldwide, which required irrigation and other large-scale landscape modifications, and the emergence of social classes and the state. In time, merchants as circulators and distributors of commodities became capitalists by organizing labor and resources to produce commodities for worldwide sale.

For the Hadza, the development of a colonial monetized economy followed by national independence and the arrival of a great variety of outsiders to their homeland has had a significant impact on their way of life. These changes have brought with them government officials; commercial farmers and merchants; builders and suppliers of raw materials; salt producers and gemstone miners; coal makers and firewood collectors; missionaries and various charities; health, education, and economic services; a growing number of researchers and filmmakers; and, especially, tourists (Mabulla 1996, 2003; Marlowe 2010). The Hadza are increasingly involved in a cash economy that links them to national and international markets, and this development has also included Hadza services offered in exchange for goods and the sale of their labor. But nothing has threatened to undermine traditional foraging as the basis for the independent Hadza way of life more than their shrinking resource base. This change can be traced to the alienation of land and the decline of game. In his early accounts of the Hadza, Woodburn argued that the decline in game, and the desire for market goods not obtainable in exchange for the products of hunting and gathering alone, would naturally lead to the settlement of the Hadza and the end of their foraging way of life as they transitioned to farming (Woodburn 1962). Many Hadza have not given up foraging; they have opted instead for a more multifaceted way of life in which they can take advantage of all available opportunities. The Hadza who engage in "occupational multiplicity" (Comitas 1964) are semi-settled, part-time foragers who also engage in exchange with non-Hadza involving barter, services, and commerce. They may also practice farming and take advantage of charities and government assistance programs.

Given that researchers have focused on the Hadza who are still able to live as traditional foragers, and not on those whose lives have been profoundly influenced by local, national, and global economic forces, and that researchers generally write about the remaining traditional foragers in ways that suggest that they are describing the Hadza as a whole, the tendency has been to minimize the importance of trade in Hadzabeland. This is evident in simple statements such as this from Kaare and Woodburn (1999:201): "Few Hadza work for outsiders or are engaged in trade." In the face of traditional exchange with food-producing neighbors, the increase in market outlets within their traditional environment, and the efforts of missionaries, charities, and the state to turn them from "living like wild animals" into settled farmers, agro-pastoralists, or honey producers, it is the stubborn independence of the Hadza that is most emphasized in the literature. In an earlier publication, Woodburn (1968a:50) reported, "Although [the Hadza] … rely on trade and begging to obtain tobacco, cloth, beads, iron, and other goods, they have not entered into an elaborate dependence on, or interdependence with their neighbors. In this respect they stand in marked contrast to the Mbuti Pygmies of the Congo (Turnbull 1965, 1968) and the Kenya Dorobo described by Huntingford (1954)."

Peterson (1978:337) described the Mbuti as participants in a symbiotic forager/farmer exchange in which trading meat for carbohydrates benefits them and their agricultural neighbors, and he expressed a theoretical view that would become one of the main criticisms of the cultural ecology of the 1960s and of anthropology in general: describing peoples' way of life without adequate attention to historical and intersocietal circumstances. Marlowe (2002:247) concluded that much of this Kalahari debate (as this revisionist critique has come to be called) "turns on the issue of contact between the Ju/'hoansi and non-foragers (Wilmsen 1989, 1993; Wilmsen and Denbow 1990; Lee and Guenther 1995)":

> Most authors (cf. Lee 1972a; Woodburn 1968a; Maceda 1964) either do not explore or else directly minimize the extent of interdependence between two trading populations. Turnbull (1965) is most explicit in this. While he acknowledges the importance of the Mbuti in Bantu economy, he emphatically denies any absolute dependence of the Mbuti on their neighbors.

The case of the Mbuti is often described as different from that of the Hadza. According to Campbell (1983:57), who offered a view similar to Woodburn (1968a) and Tomita (1966:165), the Hadza "maintain only minimal trade relations with neighbouring groups and constitute an essentially independent socioeconomic system; in this they are different from the Mbuti pygmies." The Hadza trade with their neighbors for important resources, including cereals and sweet potatoes, but this trade has been so limited as not to fundamentally challenge their self-sufficiency fruits-and-roots forager way of life. According to Marlowe (2010:46):

> There is simply no routine division of labor except that based on sex. Some trade occurs, but there are no specialists who make [or acquire] certain products [just] for trade either with neighboring tribes or with other Hadza. The only thing that even comes close are the few men who are expert arrow makers, who spend a bit more time at it, and who sometimes give other men arrows in exchange for some other object, but even this is very minor because all men make almost all of their own arrows.

Although historic accounts of Hadza trading suggest they have largely been self-sufficient foragers, it is important not to underestimate the significance of trade in Hadza life, particularly when the concern is with the use of ethnographic insight in theorizing hominin evolution. Intersocietal exchange can mean improved conditions for reproductive success for foragers like the Hadza compared with prehistoric foragers, and, as a consequence, Hadza camps and total population are likely larger as well. The significant difference between Hadza foragers and true prehistoric foragers is that the Hadza have exchange relations with farming and herding neighbors who have more intensive production systems with larger, more densely settled populations. The Hadza have long depended on iron scraps from old hoes and spears obtained through trade with their food-producing neighbors. This opportunity for exchange has provided the Hadza with metal for ornaments, and especially for their arrow points, all-purpose knives, and indispensable axes; all are necessary to take advantage of the baobab, especially as a honey tree. Fosbrooke (1956:3) noted that for "personal adornment beads and wire are bought or bartered." More recently, reseachers have given the Hadza nails to make arrowheads (Marlowe 2002:14). In the past they also acquired brass neck rings from the Isanzu with whom they had close contact at the time of early twentieth-century reports. Exchange has provided the Hadza with containers in the form of the culticated *Largenaria* siceraria gourd and ceramic and metal cooking pots, and exchange has also been a source of beads for making personal adornments and decorating objects, and for exchange through gift-giving and sales of handcraft to tourists (Fosbrooke (1956:3; Marlowe 2002:14). With the developmentof a cash economy, the Hadza have increasing opportunities to earn money and spend it on various commodities.

The Hadza have also traded for cultivated food crops, including the sweet potato and especially corn and millet; tobacco and marijuana for use in the stone pipes they make.

Over the years, Hadza trade in animal products has included meat; the horns of rhinoceros, eland, and dik dik; elephant tusks; lion and zebra skins; zebra fat; and honey and beeswax (Bleek 1931:278). In the past, the Hadza traded lion and leopard skins for brass neck rings and clay pots (Bleek 1931:278), but this seems no longer to be the case (Marlowe 2002:265–266). Cooper (1949:14–15), who, as earlier indicated, was a colonial game ranger when he visited the Hadza, worried that "The only dangers with regard to [the Hadza's] … future relations with game lie in their contact with civilization, both at Isanzu and Mangora (Mbulu) limits of their range." He reported that at Isanzu he "found evidence of the existence of rather indefinite Kindiga-Mnyisanzu types living in more or less fixed abodes, yet retaining the hunting methods and powerful bows and poisoned arrows of their more typical brethren." Among these Hadza, Cooper reported, "some barter of game skins, for millet and other articles, was temporarily stopped but may begin again," and that "no action was taken, in any case, against genuine Kindiga who habitually use poisoned arrows for hunting although this is an offence against the Game Ordinance in force in the Territory." Rhinoceros horns, elephant tusks, and lion and leopard skins seem now to be trade goods of the past, for, as Cooper writes, the Hadza "were simply told that Giraffe, Kudus, Rhinos and such like are protected and asked not to hunt them. For the time being there can be no reason for enforcing the Game Laws in their area, except to discourage the killing of Rhinos in order to sell horn to traders." According to Tomita (1966:164), only a small part of the game the Hadza hunt, chiefly meat and skins, are bartered with their food-producing neighbors today.

The Baobab's Exchange Value

The baobab has been a factor in Hadza exchange relations through its salable products and its landscape and cultural significance in Hadza tourism. Hadza handicraft today includes necklaces and baobab fruit-pod containers with handles and decorations of strung beads and other objects. The exchange value of the baobab is clearly on the rise and this use of the tree contributes to its unrivaled status as the Hadza's multipurpose tree of life. For the Hadza, the baobab's chief trade good has long been honey (Woodburn 1968a:50). The Hadza are avid honey collectors, as earlier noted, and they barter honey for agricultural products and raw materials from their traditional food-producing neighbors; they also sell honey to merchants associated with the rapidly developing commercial economy within the boundary of what was, until relatively recently, part of their traditional territory. The neighbors they supply with honey "are eager to obtain it in trade" (Woodburn 1968a:50), primarily using it to make a honey beer called pombe. The Hadza also trade honey for pombe. Madsen (2000:41–42) identified "Datoga pastoralists, particularly the Barabaig," as representing "the front line of Hadzabe land alienation," noting that the "main source of conflict … is over water resources." Yet, we also learn: "Wild honey used to brew gisoda, a honey beer, is an important item of trade between Barabaig and Hadza, and there are signs the Barabaig recognize wild honey as being an exclusive

Hadzabe resource." Of the exchange of honey in Hadza camps, Marlowe (2010:234) wrote:

> Honey is shared widely when many people see it, but its distribution can be targeted to particular individuals more easily than meat because it is easier to hide and does not have to be cooked. I have seen a man sneak honey into his house, where he shared it with his wife and child plus two young single women he discreetly signaled. Still, it can be difficult to hide. Once a man slipped into camp and put his honey under my Land Rover to wait for an opportune time to fetch it and give it to his children. But when he finally retrieved it, others saw it, and he had to share it with everyone present.

Bleek (1931:278) reported as early as the 1930s that the Hadza "sometimes take [beeswax] to the Indian stores and sell it for money." In his article on "beekeeping in Africa," Hussein (2001:7) indicated that beeswax "is a very important by product of traditional beekeeping," noting that "Tanzania has been one of the largest exporters of wax in the world." While honey links the Hadza to their neighbors, beeswax links them to the world economy. Honey can be gathered year-round but is least available at the height of the dry time from the end of winter in September until the early spring in October. But from the start of the dry time at the end of autumn in June until the honey-harvesting ends in mid-winter in August, the Hadza live among baobab trees to gather fruit, honey, and beeswax.

Little has been reported on the precise ways in which the Hadza trade honey and beeswax. This is also true of baobab fruit for which there is only one report provided by Tomita (1966), who linked the baobab fruit trade to the acquisition of cereals. Most accounts indicate that cereals were obtained from missionaries or traded with neighboring groups for labor, meat, skins, and honey (Marlowe 2002:14). Mabulla (2003:48) identified the end of autumn in May until mid-winter in July as the time of the year the Hadza "depend more on baobab … fruits," and he pointed out that this period coincides with harvesting among inland agriculturalists and agro-pastoralists. He noted that some Hadzabe families and individuals "move close to the agriculturalists and assist in harvesting in return for food and other commodities. They construct simple shelters near corn farms and wet streams where water wells are easily dug." It is therefore noteworthy when Tomita (1966:169) reported: "When a new crop of … [baobab] fruit ripens, it is bartered [by the Mangola Hadza], in the same way as the *mhibe* fruit [*Tamarindus indica*], for corn with the agricultural people." Citing McDowell (1981a), Mabulla describes that tamarind is an "intensively utilized food" from the end of winter in September through the mid-spring of November. And from the end of spring in December until the end of summer tamarind is identified as a "rarely utilized food, but available." Like honey, baobab and tamarind fruit have probably long been traded by the Hadza. The exchange value of these fruits is also linked to the growing settlement of Mangola dating back to the 1930s and Yaeda Chini to settlement schemes of the Tanzanian government beginning in the 1980s. Hadza neighbors are, no doubt, able to obtain some baobab and tamarind fruit themselves. However, the fruit of the baobab and tamarind are hunger-season foods important to people throughout the drier parts of

Africa, and one cannot help but wonder if these fruits are traded by the Hadza because of their value for their neighbors during this critical time of the year.

It is not clear from current studies if the Hadza are merely collectors of baobab and tamarind fruit for local consumers. The Hadza may also be supplying baobab fruit to traders who sell to wholesalers for domestic use and for export. There has been significant commodification of the baobab since Sidibé and Williams (2002:44) wrote their book. Yet, even when their book was published, they noted that although harvesting and marketing of baobab products is not the primary activity for most people, these "activities reach their peak in the dry season when other field crop production is low. Marketing the baobab products is a secondary means of income generation for most people and can provide a much needed buffer in times of drought and famine."

At present, there appears to be a limited demand for baobab fruit in Tanzania. With regard to the baobab trade in East Africa, a "consortium of baobab experts" (North et al. 2014:3) consisting of "partners from research institutions, NGOs and the private sector in Kenya, Sudan, Malawi, UK and Germany," whose aim was "the exploration of sustainable use and commercialization of products based on baobab to improve food and nutrition security and to combat rural poverty in East Africa," wrote:

> [I]n Eastern Africa, the species is regarded as underutilized as its potential for improving local diets and livelihoods is not yet fully recognized. Also, value chains and marketing pathways are poorly developed and the species is largely neglected by research, development and extension institutions. Consequently, little information is available regarding the abundance and productivity of baobab trees in Eastern Africa including the present contribution of baobab to local food security and livelihoods.

As earlier noted, however, the potential impact of the growing international trade in baobab fruit and other products should not be overlooked (Blurton Jones 2016:21). We have previously noted the rising commercial interest in baobab products, especially the fruit pulp and seed oil, but the extent to which the international baobab market has touched the Hadza is unknown.

Besides, baobab, tamarind, and berries, the only kinds of fruit trees catalogued by Marlowe (2010:110) are the desert date palm and doum palm, and the fig, marula, and *wika*. None of these have been reported as traded by the Hadza, however. This means the baobab and tamarind are distinguished as important sources of tradable goods in the Hadza environment, and the baobab is clearly the more important of the two. The paucity of reports of baobab fruit as a Hadza trade item is surprising, given that the largest category of food the Hadza gather is baobab-associated products, especially fruit and honey. If to this we add roots, berries, and trade-acquired cereals, plant-sourced foods constitute more than three-quarters of the Hadza in-camp diet, with meat making up the rest. The vast majority of the Hadza plant-sourced foods – everything except roots, berries, and small amounts of cereals – are gathered from the baobab tree (Marlowe 2006:366).

Baobab and Tourism

Today, encroaching food-producing neighbors, the growth of largely immigrant settlements in their homeland, and the tourist-driven commercialization of their landscape and way of life, especially their material culture and inspirational activities, imperil Hadza traditional foraging. In his recent ethnography, Marlowe (2010:287) singled out tourism as a principal threat to the sustainability of the Hadza way of life. Speculating on the ambiguities in the influence of the work of researchers on the development of Hadza tourism, Marlowe wrote (2002:268):

> It is possible that the interest of researchers beginning in the late 1950's and growing up to the present may have prolonged Hadza foraging (for reviews see Ndagala and Waane 1982; McDowell 1982). This might have occurred from the many gifts that researchers are expected to give. This may have made it less likely that some other tribe would subjugate the Hadza. Others also want outsiders to come and provide goods, and they recognize that the outsiders are coming to see or study the Hadza. For example, recently several villages have instituted fees for researchers and tourists just for passing through on their way to see the Hadza. It would hurt these neighbors if outsiders stopped coming because the Hadza were no longer foraging. On the other hand, it may well be research publications that have contributed to the awareness responsible for attracting tourists and hastening the end of foraging.

With respect to the impact of researchers, and especially following the successful 2001 PBS and BBC Hadza documentaries, the commercialization of Hadza life, as a consequence of what is known today cultural tourism, ecological tourism, and historical tourism (Mabulla 2000), led many Hadza to work as ethnic performers, tour guides, and handcraft producers. In the process, they are becoming increasingly dependent on a monetized economy rather than on foraging. Mangola and Yaeda Chini, in particular, where the Hadza have lost significant territory to immigrants and commercial development, have become significant tourist destinations. With this development, the baobab is now part of Hadza exchange relations involving tourism. The Hadza are the people of the baobab, and it is not at all surprising that their intimate connection to the baobab is frequently mentioned in accounts of tourists. What visitor to the Hadza homeland could miss the importance of baobabs as landmarks of the environment? Who could forget a trek with the Hadza to a baobab and the chance to witness a hunt or collect and taste the baobab's fruit? What could be more striking than a Hadza using a peg ladder to harvest honey or collect water from a towering baobab, or more curious than tourists attempting the same? Learning about the baobab's importance for the Hadza and exploring the interior of hollow trees identified in travel blogs as Hadza "birthing tree" and "temples" are familiar aspects of Hadza tourism today.

Summary

The hominin's evolved capacity for handiness and bipedality in support of gender-specialized foraging (including central-place living) is theorized to have been responsible for humans' extraordinary reproductive success. This success has necessitated global migration; the greater use of aquatic resources; the intensification of food production brought about by the near-simultaneous worldwide development of agriculture; the development of social classes and the state; and the production efficiency that came with the success of merchants whose long-distance trade gave rise to capitalist producers and our present industrial world. Hadza foraging provides insight into the organization of human life before agricultural intensification and markets, while Hadza exchange offers insight into the influence hominin reproductive success and the development of agriculture and markets have had on the sustainability of ancestral foraging over the past 12,000 years. Because the baobab is unevenly distributed in the environment of the mosaic savanna, it is possible that with the rise of food producers, foragers trafficked in baobab fruit with neighboring cultivators and herders who lived in more settled communities, as the Hadza do today. The next five chapters of Part Six examine the baobab's inspirational value for the Hadza in the context of similalr uses of the tree elsewhere in Africa.

Part VI
The Inspirational Value of the Baobab

Chapter 18
The Baobab in Hadza Inspirational Life

Central-place provisioning was earlier identified as a system of social production in which resources acquired by male hunters and female gatherers are brought back to camps and made available to offspring and caregivers by delayed exchange and by pooling and sharing. However, the idea of central-place camping and its associated gender-specialized food production (also identified as central-place foraging and central-place provisioning (Marlowe 2006)), should be expanded to include the broader conception of central-place living. This broader view is necessary because Hadza camps, like those of other foragers, involve more than the practical requirements of a secure, sheltered, residential environment that is centrally located for kin-based and camp-wide exchange; more than a central place for processing, cooking, and eating food; and more than an environment for the care and provisioning of offspring, especially dependent children, and their caretakers, as well as the old, sick, and injured.

A Hadza camp is also the space for various inspirational activities such as storytelling and recreation. Humans are drawn to the baobab by its aesthetic appeal, both as a curiosity and an ornamental. Its striking size and shape have earned the baobab a featured place in books like *Fantastic Trees* (Menninger 1967), *Bizarre Plants* (Emboden 1974), and *The Remarkable Baobab* (Pakenham 2004). Trees, like other things, are identified with the spiritual realm because they are unusual in some way. As discussed earlier, many trees around the world have been identified as trees of life because of the exceptional vitality perceived in their size, shape, age, leaves, flowers, fragrance, or fruitfulness; as such, these trees are associated with manifesting, imparting, and representing life. In Africa and other parts of the world, the baobab's conspicuous appearance has made it a tree of inspirational importance with significant ties to aesthetics, religion, recreation, and oral traditions, the latter including mythology, ethnosystematics, proverbs, riddles, word games, humor, folk tales, common beliefs, and popular expressions. Other inspirational uses include the baobab as a performance venue for storytelling, music, dance, and theater, and the baobab as the subject of various forms of aesthetic expression, most notably the graphic arts, especially drawing, painting, and personal ornamentation. Other inspirational uses include recreation, gathering in its shade for relaxation, socializing and

J. Rashford, *Baobab*, https://doi.org/10.1007/978-3-031-26470-2_18

play, and providing a sheltered learning environment where the transmission of culture takes place. The historical significance of the baobab, given its longevity and its rich symbolic value as an icon, commemorator, and cultural marker, also contributes to its status as an inspirational tree. This chapter discusses the baobab's inspirational significance for the Hadza along with its inspirational significance elsewhere in Africa and in other parts of the world. The discussion includes a summation of the kinds of spirits commonly identified in the anthropological literature, a review of the concept of a cosmic tree based on this summation, a critique of the concept of "tree worship" applied to the baobab, and the use of the baobab as a cosmic tree that figures in various creation narratives, including those of the Hadza.

Inspirational life provides motivation for practical activity, just as practical activity provides the material basis for inspirational life. These complementary aspects of human adaptation, which find expression in the baobab's manifest vigor and status as the all-purpose tree of life, have made it one of the most widely distributed trees in the human environments of the African savanna. In addition, the baobab's inspirational value has played an important part in its dispersal around the world. The baobab is one of many trees intimately associated with human inspiration, and Layser (2001:1) correctly states that the "traditions surrounding the baobab are not unique but parallel other cultures' ancient relationships with trees since time immemorial." At the same time, it is likely that baobab traditions preceded all others since it is the primeval tree of life in the mosaic savanna of hominin evolution.

Difficulties in Understanding Hadza Inspirational Life

Although not reported in early accounts or recent academic studies, an intriguing expression of the baobab's significance for the Hadza today are old baobab trees with bows and arrows and skulls of baboons and other game fixed to their trunks. The great girth of the baobab, which provides a broad relatively flat surface, and its thin bark and soft water-storing wood make its trunk an ideal place for drawing and writing, and for attaching objects. Around the world, old baobab trees in heavily trafficked areas invariably become message boards. Visitors to Tanzania who spent time with the Hadza have posted photographs of these decorated baobabs on blogs, but offer no explanation. Seale (2012) likened this baobab/animal scene to "artwork in a museum," but it was clearly an exhibit she did not enjoy. "As I neared enough to realize what they were," she wrote, "I was at once shocked, intrigued and disgusted. Skulls, bleached white from the sun and cooking, dotted the baobab's trunk. Small and large, with teeth hanging from the jaws haphazardly and empty eye sockets staring woefully." As interesting as Seale's reaction may be, it would have been far more meaningful to learn about the display's significance for the Hadza. Is it purely aesthetic; a work of art in the form of a trophy exhibit? Is it religious in nature? Does it mark a boundary? Is it a memorial commemorating something of social importance?

We know little about Hadza inspirational life for three important reasons. First is the lack of serious attention to this aspect of Hadza culture, starting with the dismissive claim in early accounts that the Hadza had no religion. For example, Bagshawe (1925:126) wrote:

> The Kangeju [Hadza], I think because he never lives in anything like community with his fellow men is a law unto himself, and refuses to be bound by customs such as regulate the lives of most African tribes. He does not appear to be very superstitious, and I have never succeeded in discovering any trace of a tribal religion or worship. He makes no sacrifices and apparently acknowledges no god. I have not come across anything approaching a tribal magician or witch doctor.

If such claims had merit, we should not expect the baobab to have deep spiritual significance for the Hadza, which it clearly does. We have learned enough about Hadza culture to state that they are a religious people. This is especially so if religion is defined, as it is in this work, as a symbolically dualistic framework within which members of a community believe there is a spiritual dimension to ordinary life, a dimension with which they can establish contact to ensure the positive outcome of their most important concerns.

The second reason we know little of Hadza inspirational life is the materialist orientation that has dominated the study of foragers, both in the older cultural ecology approach and in current studies by behavioral and evolutionary ecologists. Although these materialist studies have largely focused on adaptation apart from inspirational life, they offer a more generous view of Hadza spirituality than Bagshawe presented. For example, Marlowe (2004:689) has reported that the Hadza do not have an "organized religion," and as such they can be adequately characterized as "minimalist." The work of Woodburn (1964), Marlowe (2010), and Power et al. (1997) make it clear that the Hadza are religious, and the Epeme dance and Maitoko female initiation are their most important ritual occasions, constituting the essence of their spiritual tradition. According to Marlowe (2010:60–63), much of what we know about Hadza religion can be attributed to Woodburn (1964). But religion was not Woodburn's primary interest. His focus was on Hadza forager adaptation and its links to an egalitarian social organization based on a fluid residential pattern characteristic of what he identified as immediate-return foragers. Nevertheless, Woodburn's (1970) research documented the centrality of the baobab in Hadza practical life, and he provided important information on the tree's significance in Hadza inspirational life.

Finally, we know little about Hadza inspirational life because they are reluctant to discuss it with outsiders (Ndagala and Waane 1982). This is understandable, given the history of prejudice shown by their traditional neighbors, European colonists, government officials, missionaries, and charitable organizations. Similarly, the Hadza are leery of researchers curious about their inner life. According to McDowell (1982:104), the "Hadza are reluctant to cooperate with researchers" because they fear they will provide evidence that could lead to accusations of poaching, and because of "their conviction that the arrangement [with researchers] is economically unfair." McDowell (1982:104) wrote:

> The … reason Hadza are reluctant to cooperate with researchers … is fear of disrespect towards cultural "secrets". Information about myths, religion, beliefs and practices is only grudgingly imparted to researchers or to people of neighbouring ethnic groups. Many Hadza men reacted to my inquiries into traditional knowledge by claiming they were ignorant, citing their youthfulness or association with Bantu neighbours as evidence that they had not learned Hadza folklore. In reality they preferred not to be responsible for dispensing information of inherent value to their society.

Despite all these limitations, and all that remains to be documented, it is clear that the baobab is inspiring to the Hadza, and can genuinely be identified as their inspirational tree of life.

The Kinds of Spirits Associated with the Baobab

That Africans associate the baobab with spirits is well known, but the nature of these spirits and their relationship to the tree is rarely made explicit. Twelve categories of spirits commonly discussed in the anthropological literature are presented here to facilitate the discussion of the baobab's religious significance and to clarify the meaning of "worship" as it has been used in relation to the baobab.

An important conception of "spirit" in the human imagination is (1) God, the Supreme Being, whose representation in cultures worldwide has been the subject of significant debate in anthropology and other fields. Next is (2) the animating inner spirit or soul of every living human being. The concept of an animating spirit is integral to two related ideas of spirits: (3) a distanced spirit, which is an animating spirit that has been lost, stolen, or otherwise separated, in part or wholly, from its body, and (4) the spirit of one who has died, especially an ancestor, esteemed political leader, exemplary warrior, or some other culturally recognized hero or heroine. A distanced spirit is associated with dying, dreaming, or visions. The spirit of a person who has died leaves behind a corpse.

What is variously referred to as (5) lower, lesser, minor, or secondary gods or goddesses (or as lower, lesser, minor, or secondary deities, divinities, superhuman beings, or supernatural figures) is one of the most important categories of spirits associated with the baobab. These are the celebrated spirits of a people identified in this work as contact spirits. They are the ones people ritually engage on a regular basis. Contact spirits are typically conceived as intermediaries between God and humans and are variously identified as God's family, assistants, messengers, delegates, agents, representatives, or ministers. The social significance of contact spirits is evidenced in the fact that they are usually well-defined, with individual names, familiar images, representative colors, ritual dress, identifying implements, religious narratives, specific altar sites, ceremonial centers, organized devotees, and festive occasions. Important subcategories of contact spirits include nature spirits, place spirits, and desire-fulfilling spirits. Nature spirits are particularly associated with astronomical bodies, climate, terrestrial processes, and living things; place spirits, with locations such as mountains, caves, crossroads, and entrances; and desire-fulfilling spirits, with the wish for such things as protection (tutelary spirits and personal guardian spirits), occupational gain (patron spirits), reproductive success (fertility spirits), and insight into the past, present, and future (divination spirits). Desire-fulfilling spirits are also associated with many things such as luck, assurance, love, health, prosperity, longevity, justice, and power.

Other important categories of spirits are (6) harmful spirits, variously identified as spirits that are mischievous, malevolent, malicious, evil, vengeful, dangerous, demonic, or devilish; (7) embodied spirits, also described as indwelling spirits; (8) personified spirits, which are non-human entities regarded as human beings; (9) residing spirits, which are thought of as dwelling in particular places; (10) trickster spirits, which may be religious or folkloric and are found in cultures worldwide; (11) folkloric spirits, usually of a frightening nature and most often presented through oral traditions rather than religion; and (12) non-specific spirits, typically, as the label implies, vague and marginal. These 12 categories of spirits as discussed here are distinct, but not necessarily mutually exclusive. African religious traditions especially show great flexibility in the way contact spirits are viewed. In Candomble, for example, the orisha Iroko, who is the guardian of the cosmic crossroads and promoter of peace and reconciliation, is often – but not always – thought of as a residing spirit. In their book on Iroko, Martins and Marinho (2002) identified him as "the orisha of the tree" (suggesting an embodied or residing spirit) and "the tree orisha" (suggesting either the tree as a being with its own inner spirit, or the tree as a personified spirit). The baobab has acquired, through its association with all of these kinds of spirits, a religious and folkloric significance in Africa and around the world.

Trees and the Spiritual Dimension

Understanding the baobab's inspirational value for the Hadza and other Africans is severely hampered by the uncritical use of concepts from now-discredited nineteenth-century socio-cultural evolutionary theories. This limited perspective includes the work of leading scholars of their time like Edward B. Tylor (1871), Robertson Smith (1894), and Sir James Frazer (1976), for whom the religious importance of trees and other natural phenomena represented the "animistic" or "magical" origin of religion. It was this "origin" that laid the foundations of polytheism and monotheism in the progressive development of human religiosity (Haberman 2010). Frazer characterized this progression as going from magic to religion, and eventually to science.

Tylor argued that all religions developed from a belief in spirits based on the universal human experience of death, dreams, visions, and psychic phenomena. The idea of human beings having an animating inner spirit expanded to include a similar concept for other animals, other living things in general, and ultimately a wide range of natural phenomena, including the cosmic whole. The earliest systematized form of this belief Tylor named animism, a belief that attributed "spirit" to nature.

From animism developed the more elaborate form of polytheism which recognized "many gods." The final form of this evolutionary development was said to be monotheism, defined as a belief in the Supreme Being. Although Tylor's elaborate theory of "a belief in spirits" has been largely rejected or ignored in modern anthropology, the discussion regarding the nature of spirits continues because they are important in the cultural traditions of the vast majority of people around the world that have been studied by anthropologists.

The widely recognized inspirational significance of the baobab for Africans has led many commentators to frame their discussion in the ambiguous and pejorative terms of "tree worship" (Porteous 1968:235; Emboden 1974:21; Owen 1974:27). Simoons (1998:272), for example, described baobabs as "objects of sacrifice and worship." On their website "Coppermine Photo Gallery," Robert and Riza presented pictures of the Hadza and their environment, including one of a magnificent baobab identified as the tree "worshipped" by the Hadza. Similarly, Wickens and Lowe (2008:49) identified the baobab as a "fetish tree." Fetishism is generally understood to be characteristic of the idolatry associated with religions categorized as "animistic" and "animist," and as "polytheistic." The word, which means a charm, was adopted from the Portuguese *feitico*; in the fifteenth century, it was applied to relics, rosaries, and images, which were thought to have magical properties. Portuguese explorers in West Africa used it to describe stone figures and objects "worshipped by the natives," and it was adopted by Comte in his theory of the origin of religion. Fetishism for Comte was essentially nature worship based on the idea that all of nature was animated in a way analogous to human life. From this perspective, the baobab is a fetish tree with an animating inner spirit and, as such, is a personified being worshiped by people.

Human beings worldwide seek interaction with spirits associated with trees, but not all such interaction is explainable as "worship." Lexically defined, "worship" is religiously inspired reverential devotion. Applied to trees, the term would imply the tree itself was the object of prayer, offerings, ceremonies, and other ritual activity. There is no evidence that the Hadza worship the baobab in this sense. The same is true of "ancestor worship," a term used in 1885 by Herbert Spencer in *Principles of Sociology* to designate a kind of religion. According to Mbiti (1970:11), "Many books speak of 'ancestor worship' to describe African religions." Mbiti wrote:

> Certainly, it cannot be denied that the departed occupy an important place in African religiosity; but it is wrong to interpret traditional religions simply in terms of "worshipping the ancestors." … [T]he departed, whether parents, brothers, sisters or children, form part of the family, and must therefore be kept in touch with their surviving relatives. Libation and the giving of food to the departed are tokens of fellowship, hospitality and respect; the drink and food so given are symbols of family continuity and contact. "Worship" is the wrong word to apply in this situation; and Africans themselves know very well that they are not "worshipping" the departed members of their family.

In fact, the small number of published accounts that discuss the spiritual significance of the baobab and reflect a familiarity with African religious traditions suggests that in general, Africans do not "worship" the baobab. Rather, the tree is everywhere associated with various kinds of spirits that people seek to engage or avoid. Engaged spirits especially include ancestral spirits and contact spirits; both are thought to reside in or be embodied by the baobab and are linked to the promotion of kin ties, health, fertility, and political authority. It cannot be said without qualification that Africans worship the baobab without evidence that they conceive of the tree as a spiritual being with its own inner animating spirit or as a personified being.

In addition to tree worship, some authors have identified the baobab as the object of "superstitious veneration" and as "bewitched" with ties to the malevolent activities of "witchcraft" and "sorcery." Others uncritically identify it as a "sacred tree" which suggests that it is only associated with the positive spiritual dimension. From a broadly comparative approach, however, the baobab is associated with both the positive and negative spiritual dimensions. The fundamental point is that the baobab is a tree of great spiritual significance for Africans, not merely a fetish tree, bewitched tree, or sacred tree.

The Baobab as the Cosmic Tree of Life

In Africa, the tree-of-life baobab is widely associated with modeling the cosmos and with fertility. As an attractive resource environment for human beings and other species, the baobab is a powerful symbol of nurturing that has been incorporated into cultural narratives of the cosmos and its generative power, including the origin of human beings. The conception of a tree-like cosmos is an important value of trees, and some version of the cosmic tree appears in religious cosmologies worldwide (Reno 1977). The cosmic tree is often represented by a fig (Beech 1913; Maguire 1931; Evans-Pritchard 1940; Von Hagen 1943; Gautier 1996; Condit 1969; Simoons 1998; Prussin 1999; Kunwar and Bussmann 2006; Kislev et al. 2006; Karangi 2008; Rashford 2022). We have only to think of *Ficus natalensis* and other figs of Africa, especially *Ficus sycomorus* of the ancient Egyptians; *Ficus carica* of Western Asia and the Mediterranean; *Ficus benghalensis*, *Ficus religiosa* and *Ficus rumphii* of India and Southeast Asia; *Ficus prolixa*, *Ficus obliqua*, and *Ficus tinctoria* of Oceania; and *Ficus tecolutensis*, *Ficus cotinifolia,* and *Ficus petiolaris* of Mesoamerica, the amate figs of "Aztec and Maya papermakers" (Von Hagen 1943). Recent ethnobotanical accounts of Oceania document the practical uses of figs (Whistler 2000, 2009; Balick 1988), but generally have little to say about their inspirational value. Like figs, the baobab is widely represented as a cosmic tree in Africa because of its imposing presence – gigantic, oddly shaped, and long-lived. These attributes, together with its numerous uses, have ensured it a prominent place as the cosmic tree in the inspirational life of many Africans, and a likely candidate for being the cosmic tree in the inspirational life of early hominins.

Africans, as earlier noted, have been unwittingly and intentionally encouraging baobab trees throughout their settlement environments for millennia, especially as a part of their religious landscapes. The conspicuous presence of a baobab defines a spiritual space intended to invoke the structure and dynamism of the cosmos. It is also recognized as a place of spirits and a source of spiritual power. The baobab especially defines a spiritual place when it provides a sheltered environment for religious activities, when it is used as an altar site, and when it serves as an altar. The baobab is everywhere the basis for the development of shrines.

In Africa's diverse religious traditions, altars and their associated shrines are sometimes established for the Supreme Being, as with the Ashanti. And while these shrines and altars are especially linked to ancestral and celebrated spirits, they are also established to placate harmful spirits. Ritual objects at the base of trees and trunks stained with offerings are widely reported in the literature (Bolandier and Maquet 1974:144; Parrinder 1949:28, 1954:52; Wickens and Lowe 2008:52–53). Bolandier and Maquet (1974:146) noted, for example, that the baobab "stained with libations is a common sight in West Africa." According to Weiss (1979:40), "many tribes" in East Africa "would place pieces of cloth tied to sticks and offerings of food at their base." Parrinder (1970:28) reported on a visit to what he described as a "compound" or "convent" for mediums and priests near the village of Dassa Zoume in West Africa. There he saw a large baobab whose trunk was "blackened in parts with blood and oil." That a baobab serves as an altar site or altar does not in every case mean the tree is the object of spiritual concern. Only when the tree is considered to be a spiritual or personified being can it be said that Africans regard it as the object of their religious activity.

The classic view of the cosmic tree owes much to Mircea Eliade, one of the most influential historians of religion, who emphasized the tree's symbolic role as the center of the universe, connecting sky and earth. Summarizing his view of the cosmic tree and the religious importance of trees generally, Eliade (1991:44) wrote:

> The most widely distributed variant of the symbolism of the Center is the Cosmic Tree, situated in the middle of the Universe, and upholding the three worlds as upon one axis .... It may be said, in general, that the majority of the sacred and ritual trees that we meet within the history of religions are only replicas, imperfect copies of this exemplary archetype, the Cosmic Tree. Thus, all these sacred trees are thought of as situated in the Center of the World.

This conception has extensive ethnographic support and is widely presented in scholarly publications, especially encyclopedias and other reference works. It is also the way the cosmic tree is most often represented in popular accounts.

The "cosmic connector" perspective is limited, however. If the cosmic tree is truly universal, it must adequately represent the cosmic whole as center and periphery – the cosmic connector must include the sky and the earth that the tree connects. The cosmic tree as cosmic whole is an alternative perspective not as widely referenced as Eliade's (1958:273–274) conception. However, it has been discussed by researchers for more than a hundred years (Anonymous 1890:4; Simoons 1998:41–99). The cosmic tree as cosmic whole has also been identified as the cosmogonic tree (Philpot 1897:109), world tree (Philpot 1897:109; Jobes 1962:375; Altman 2000:23; Anderson 2003:16), cosmic world tree (Leeming 2001), universe tree (Philpot 1897:109; Collis 1954:80), tree of the knowledge of good and evil (Jobes 1962:375; Encyclopedia Britannica 2015), tree of life (Jobes 1962:375; Cook 1974:9; Encyclopedia Britannica 2015), and cosmic tree of life (Reno

1977:79). With its branches above and roots below, the tree has been widely chosen to represent the universe, probably because it fits well the human conception of the sky above and earth below. The cosmic whole is the grounding motif of the cosmic tree, and ideas of the cosmic tree as cosmic center or cosmic connector, or as cosmic center-connector, are only its varying expressions. For a more coherent understanding, the conception of a tree-like universe provides a more adequate framework than Eliade's emphasis on the cosmic tree as cosmic center.

Although primarily focused on architecture, Prussin's (1999:430) discussion of "African models" of spiritual sites reported that in the funerary customs of Mali's *Bamana komo* societies (described as dealing "with the acquisition of sacred human knowledge"), a branch of "the fig tree" was buried with the dead "in order to guarantee rebirth." Prussin continued: "This particular sacred attribute of the fig tree suggests an analogy with the Tree of Life, the Cosmic Tree, and axis mundi, familiar to the Western world in its various cosmologies but equally at the core of African religious thought, as evidenced by the towering singular trees that mark both sacred places and village centers alike.... Eric Ross (1995:43) has shown how the Tree of Life, or Tree of Paradise, appears to be firmly rooted in African religious thought as an eschatological and cosmological symbol."

The Cosmic Tree of African-Brazilians

Voeks (1997:184–185) rightly noted that the Iroko tree is at "the center of [Candomblé] ceremonies," and this is understandable. As a representation of the tree-like universe, it shelters all orishas, whose domains, residences, seats, altars, and offerings are associated with its various parts. The branching trunk, which forms the sky, is the realm of Olorun the Creator. The branched sky is also associated with the orishas Oxalá, linked to high places and the creation of humanity; Xangô, linked to lightning and thunder; Iansã, to wind and storms; and Oxumarê, to rainbows and snakes. The horizontal axis where the sky and earth meet to form the earth's surface is associated with Nanã, linked to swamps and ruins; Oxum, to rivers and lakes; Obá, to waterfalls; Iemanjá, to the ocean; Ossaim, to leaves indispensable for healing and the conduct of rites; Oxóssi, to forests and hunting; Ogun, to metals and opening the way; and Exu, to roads and entrances. The branching root, which represents the earth, is the domain of Omulu, the guardian of cemeteries and guide to spirits of the dead.

The baobab is present in the shrines of Brazil's Candomblé religious traditions; a few old trees have been identified in the northeast states of Bahia and Pernambuco, but many young baobabs have been planted in and around Salvador, Bahia, over the past 20 years, and these are now flowering. In the historic tradition of Candomble, it is not the baobab, but several species of Brazil's native fig trees that have long been important. In Candomblé, Iroko is the orisha of the tree-like universe which is his domain. The intersection of the base of the tree and the earth's surface forms a vertical/horizontal axis that is the center of the universe, the cosmic crossroad where his seat and altar are ritually established. Here spirits gather, especially the orishas and ancestors, and cosmic spiritual energy, called Axé, is most concentrated. Walker (1990:123–124) argues that "to increase human participation in the *ashe*, the spiritual force of the universe" is the "raison d'être of … Candomblé." According to Abiodun (1994:71), "The concept of *ase* has intrigued many scholars of Yoruba culture both in Africa and the African diaspora. Still keeping more or less its original Yoruba meaning … *ase* remains foundational for religio-aesthetic discourse in Brazil, the Caribbean islands, and the United States … [and] is fast becoming a Pan-Africanist term."

The Baobab as the Cosmic Wish-Fulfilling Tree of Hinduism

Wickens and Lowe (2008:268) noted that the baobab "is often associated with temples, shrines and religious ceremonies" in India, and this is particularly true of local communities that identify the baobab as the mythic *kalpa* tree (Sastry 2000:315–318). Of the 13 Sanskrit names for the baobab catalog by Wickens and Lowe, the most important is *kalpa* tree (*kalpa vriksha*), identified in the literature as *kalpavriksha*, *kalbriskh*, *kalpbriskh*, *kalapbirchh*, *kalpataru*, *karpaga viruksham*, *kalpadruma*, and *kalpapadapa* (Gupta 1980). The *kalpa* tree, variously described as heavenly, holy, divine, and sacred, is the long-lived wishing tree of ancient Hindu religious texts. It is also featured in the tradition of Jains and Buddhists (Vaid 1964:789, 1978a, b; Hema and Ghate 2000:283–296; Kharia 2007).

In the chronology of Hindu mythology, *kalpa* means lasting for a single day of Brahma, a creation-to-destruction cycle of 4,320,000,000 years identified as one Brahmanic eon. In reality, this mythic conception identifies the baobab as a long-lived tree – the longest-living flowering tree in the world. In mythology, *kalpa* means indestructible, and this identifies the baobab as tenacious in its grip on life. *Kalpa* also means wishing-fulfilling, and this is a tribute to what can be described in ideal terms as the baobab's all-purpose use. And in a general sense, *kalpa* also stands for extraordinary things of a mythical or imaginative nature, which alludes to the inspirational value of the baobab's massive size and bizarre shape, which make it not only a curiosity and ornamental but also a presence that seems otherworldly. What is truly significant about the baobab's *kalpa* name in India is that even when the baobab is not called the *kalpa* tree, it is still widely regarded as the wishing tree, especially valued for its health benefits and religious significance. For example, Reddy et al. (2002:346) report that in the Jhalawar district of southeastern Rajasthan, people call the baobab *maansapooran*, meaning the "one which fulfills the aspirations." In the Ajmer district of northern Rajasthan, Reddy et al. (2002:346) noted that they "tie a small stone with red colored sacred tag (*lachha*) to the branches after praying for their wishes to be fulfilled."

Vaid (1978a) offered a reasonable alternative perspective to the idea that Arabs, Muslims, Mughals, or Europeans were responsible for the early introduction of the baobab to India, where it is now widely distributed. Vaid (1978a:35) was concerned with identifying the botanical species that was the basis for the *kalpa vriksha* name. In his influential article, he wrote, this "question that has been haunting the mind of the scholar as well as the imagination of the layman for millennia – where is [India's] … mythical wishing tree, the *kalpa-vriksha* mention in all our *puranic* literature?" Vaid argued that the baobab was India's mythological "wish-fulfilling tree" called *kalpa vriksha*, the tree associated with India's ancient Hindu religious sites, sacred literature, temple sculpture, folklore, and contemporary ritual practices. According to Vaid, "all ancient literatures describe it as an immortal tree of colossal dimensions, laden with fruits, its roots penetrating into unreachable depths, … its flowers are of unusual shape and they have a strange aroma, and … the tree fulfilled every wish made under its bough" (1978a:35).

While Vaid accepted the conventional view that the baobab is an exotic tree in India, he argued (based on what he described as an "interdisciplinary approach" involving botany, archaeology, mythology, and phytogeography) that the tree has been in India for over 4000 years and was probably introduced by "Indian seamen" who made "circuitous voyages all around the Arabian Sea and adjacent parts of the Indian Ocean" (1978a:37). Vaid's view incorporated the famous Hindu story of the *amrit manthan* or *sagar manthan* (churning of the ocean), which, he thought, identified the baobab as one of the 14 rare treasures from the sea, described in the Bhagavat Gita as "eternal, the heavenly tree which has its roots upward; … its branches extend both downward and upward; … its size so massive, none can understand where it begins, where it ends or where its roots are; …the real form of this tree cannot be perceived in this world." If we conceive of the upside-down baobab as a representation of the cosmic whole, then its upturned roots as branches form the sky and its downturned branches as roots spread out to form the earth. The sky, as the source of sunlight and rain, is also the source of energy that allows the tree to support the earth and all living things. In this sense, the baobab is the mother of all things – the long-lived wish-fulfilling tree of Hindu thought.

The Baobab and Creation Narratives

The Yoruba Creation Narrative

A number of cultures, including the Hadza, identify human origins with the baobab. This idea has been popularized by Peter Matthiessen in his well-known book, *The Tree Where Man Was Born* (1972). "According to the Nuer," wrote Matthiessen, "the tree where man was born … still stood within man's memory in the … South Sudan…. A great baobab thrust up like an old root of life in those wild grasses that blow forever to the horizons" (1972:214). Blench (2007:7) reported that the "Dompo people of western Ghana claim to have arisen from a horseman who emerged from a split baobab that is still alive a short walk from their settlement." The Yoruba also associate their origin with the baobab. In a condensed version of the Yoruba creation narrative presented by Anderson (1990:23) from which all quotations below are taken, we can see the baobab's link to different kinds of spirits, including the Supreme Being, contact spirits called orishas, and the animating spirit of all human beings. The first human beings were created by the orisha, but they received their animating spirit from the Supreme Being.

The sky/earth opposition is central to African religious cosmologies, as it is to religious cosmologies worldwide. According to the Yoruba creation narrative, before there were human beings all life was in the sky. The Supreme Being and the orishas "lived around a young baobab" where they "found everything they need for their lives," including beautiful clothes and gold jewelry. Free to explore the vast sky, all the orishas except Obatala "were content to stay near the baobab tree." Not satisfied to live blissfully by the all-purpose baobab, Obatala realized that there was a vast ocean beneath the mist and "asked Olorun to let him make something solid in the waters below. That way there could be beings that Obatala and the orishas could help with their powers." Olorun agreed and Obatala went to the orisha Orunmila to learn how to prepare himself for the task at hand.

> Orunmila brought out a sacred tray and sprinkled the powder of baobab roots on it. He tossed sixteen palm kernels onto the tray and studied the marks and tracks they made on the powder. He did this eight times, each time carefully observing the patterns. Finally, he told Obatala to prepare a chain of gold, and to gather sand, palm nuts, and maize. He also told Obatala to get the sacred egg to carry the personalities of all the orishas.

Following Orunmila's instructions, which included "a little baobab powder" along with "palm nuts, maize, and other seeds that he found around the baobab tree," Obatala used the chain to reach the water where he sprinkled the sand to make the land. His heart pounded so much that it broke the egg that he had wrapped in his shirt "close to his chest so that it would be warm during his journey." From the egg "flew Sankofa, the bird bearing the spirits of all the orishas," and the orishas gave the land character. "Finally, Obatala let go of the chain and dropped to this new land, which he called 'Ife,' the place that divides the waters. Soon he began to explore this land, and as he did so he scattered the seeds from his pack, and as he walked the seeds began to grow behind him, so that the land turned green in his wake." Quenching his thirst in a pond, Obatala saw his image and decided to create beings of clay in his image, but the effort made him even thirstier. He drank the juice from the newly growing palm trees; it had fermented and made him drunk, and he did not realize that the beings he was creating in his intoxication were malformed. Olorun dispatched Chameleon to report on Obatala's progress and learned what had happened. "Gathering grasses from the space beyond the sky, Olorun sparked the grasses into an explosion that he shaped into a fireball. He sent that fireball to Ife, where it dried the lands that were still wet and began to bake the clay figures that Obatala had made. The fireball even set the earth to spinning, as it still does today. Olorun then blew his breath across Ife, and Obatala's figures slowly came to life as the first people of Ife."

Sandawe Creation Narrative

Similarly, the baobab's association with different categories of spirits is evident in the creation story of the Sandawe of Tanzania for whom all life, including the first people, came from the baobab. The Sandawe identify the baobab with the Supreme Being and with the moon and sun as personified ancestral spirits. In the beginning, according to Sandawe tradition, when the scorching sun was not yet up in the sky, there was only the moon identified as female and an emblem of coolness and life-giving rain. When the sun saw the beauty of the moon, he rose up in the sky in pursuit and married her. Raa (1971:138:138) wrote:

> The moon, through her womb, the baobab ... began to bear fruit. But she did not do this on her own: it took the Creator [Mathunda] to cause her fruit to assume life, for it was he who created. Mathunda stood by the baobab tree and created all living beings. He opened the baobab, and out of it emerged all the animals: the elephant, the giraffe, the zebra, the hyena, and all the other wild animals. He let them go and they ran into the bush which Mathunda had presumably created before the animals. Then came the domestic animals, and those he held fast and he kept them. And out of the baobab came also a woman. Mathunda asked her: "Where is your husband?" to which she replied: In the tree". And then he, too, came out of the tree. He was the first man to be created on earth, and his name was Wangu. Wangu and his wife then began to procreate and begat many children, and Mathunda himself married Wangu's sister, and they too began to procreate. Mathunda's and Wangu's children intermarried and multiplied, and some of them went to the north and became the Barabaiga, others went to the east and became the Burunge, some went to the south and became the Wagogo, and yet others went to the west and became the Wanyaturu. And those who remained on the spot became the Sandawe. And so creation was completed and the world became populated.

Hadza Creation Narrative

A baobab-modeled cosmos is central to the Hadza conception of themselves as a people. It is within this framework that they explain their origin as a people. In the Hadza tree-of-life view of the universe, the baobab represents the cosmic connector and the residence of their ancestors, for whom the baobab was the place where women gave birth. Human beings, the Hadza say, descended from the sky above to the earth below via the neck of a giraffe, "but more often," according to Matthiessen 1972:214), they say that human beings "climbed down from a baobab." Of another version of the Hadza origin narrative Matthiessen (1972:214) wrote:

> The Hadza themselves came into being in this way: a giant ancestor named Hohole lived at Dungiko with his wife Tsikaio, in a great hall under the rocks where Haine, who is God, the Sun, was not able to follow. Hohole was a hunter of elephants, which were killed with one blow of his stick and stuck into his belt. Sometimes he walked one hundred miles and returned to the cave by evening with six elephants. One day while hunting, Hohole was bitten by a cobra in his little toe. The mighty Hohole died. Tsikaio, finding him, stayed there five days feeding on his leg, until she felt strong enough to carry the body to Masako. There

> she left it to be devoured by birds. Soon Tsikaio left the cave and went to live in a great baobab. After six days in the baobab, she gave birth to Konzere, and the children of Tsikaio and Konzere are the Hadza. The Hadza, as the people say, is us.

With this backdrop in mind, we can understand why James O'Connell, a University of Utah anthropology professor, photographer, and leading Hadza researcher, chose "Children of the Baobab: Growing up Hadza" as the title of a photographic exhibition produced by the students of Hillcrest Elementary South as a joint effort of the University of Utah and the University of California Los Angeles.

Chapter 19
The Baobab as a Fertility Tree

Female Resemblances

In Africa, the baobab is widely regarded as a fertility tree. Although researchers have documented its use as a male symbol for what is "big and strong," many African cultures place the baobab on the female side of their oppositional symbolic system, associating it with women as mothers who conceive, carry, give birth to, and nurse children. Given the profound significance of fertility in the practical and inspirational life of all people, it is understandable that highlighting, promoting, and celebrating it is a fundamental feature of human religiosity (Soni 2015; Behjati-Ardakani et al. 2016). Broadly conceived, fertility rites seek to ensure the timely arrival and sufficiency of the annual rains, and the productivity of the spring and summer growing seasons; to increase the number of useful plants and animals; and to promote human reproductive success, both individually and socially. It is against this background that Emboden (1974:21) wrote: "So important has [the baobab] … become that in the Sudan and West Africa it is worshiped as the fertility tree and assigned magical properties." Owen (1974:27–28) pointed out that in southern Nigeria it is "common to find villages named after some dominant tree in the immediate vicinity," and he offered the following account of a baobab tree in a Yoruba village that was called Mother of Children:

> People came from as far afield as Lagos (100 miles to the south) to sacrifice to the tree; many have attested to its power in relieving them of barrenness. With the arrival of Islam, worship of the tree was said to have been abandoned but, one of the informants, a woman of about 65 years stated that people desirous of having children continue to sacrifice goats and cocks at its base.

Several reasons have been given for the baobab's link to fertility throughout its range. The most obvious are its fruitfulness and prolific seed production. These, like the tree's extraordinary size and shape, are expressions of the baobab's vigor; they contribute to its characterization as a manifest tree of life. Some large baobab fruit can have as many as 300 seeds. But fruitfulness and prolific seed production are not explanations found in the literature of the baobab as a fertility tree.

J. Rashford, *Baobab*, https://doi.org/10.1007/978-3-031-26470-2_19

Parrinder pointed instead to the baobab's link to spirits. The tree is believed to be associated with a great variety of spirits, but also to have specific ties to ancestors, contact spirits, and the Supreme Being. According to Parrinder (1970:54), some groups believe spirits living in the baobab and other trees are "particularly concerned with fertility," since "the souls of those about to be born may be said to live in trees, or thickets of bamboos. Women often pray to spirits in Iroko, baobab, silk-cotton, acacia, mahogany and palm trees to send them children."

Fruitfulness, prolific seed production, and a link to spirits are still not sufficient to account for the baobab's widespread representation as a fertility tree. Especially important are the baobab's female imagery and female-associated symbols. Within the framework of oppositional symbolism, the baobab's female resemblances have made it a powerful figurative representation of fertility, women being everywhere the immediate expression of generative power. Female imagery, as earlier indicated, includes fatness, pregnant shape, soft bark and wood, vagina-like holes at the base of the tree, womb-like hollow trunk, breast-shaped fruits, milk-colored fruit drink, and nurturing environment. Citing Owen (1970) and John Watt (1972), Simoons (1998:272–273) pointed to the baobab's female resemblances for its status as a fertility tree.

> In the Sudan and West Africa, for example, the baobab is revered as a "fertility tree," likely because of the extraordinary girth of its trunk and the widespread belief that it can make barren women fertile. Such thinking has not been restricted to those two regions. In a cave in the Limpopo Valley of southeast Africa, rock drawings depict women as having "life-sized 'seedpods' of the baobab instead of breast [Simoons cites Owen (1970) and Watt (1972), and the cave to which he refers is the Cave of Spotted Giraffe at Mabola Tutwa in the Zoutpanberg area of northeastern South Africa].

Female-Associated Symbols

In addition to female imagery, the baobab is also a rich source of fertility beliefs and practices because of female-associated symbols, especially water. The tree's many connections to water include its water-storing wood, its use as a living cistern, its value as a rain indicator and planting cue, and its use in rain-making rituals, especially in times of drought and routinely in anticipation of the arrival of the spring rains that mark the start of the annual growing seasons. Jallow (1995:26) wrote:

> The appearance of some deciduous species of vegetation such as *netteh* (*Parkia biglobosa*) and the baobab (*Adansonia digitata*) [in local communities of the Fulladu West District of The Gambia] are closely observed towards the end of the dry season … [identified as early to mid-spring which is April to May] to see the rate of regrowth of leaves and fruits. Abundance of leaves and fruits in the trees giving a thick, lush and green appearance is taken by local people to indicate adequate rainfall during the approaching season for the range of crops normally cultivated. If, on the other hand, the trees are thought to have fewer leaves and fruits, it is an indication that rainfall during the approaching season will be inadequate for normal agricultural practice.

Writing of the Sudan where the tradition of using natural and artificially hollowed water-storing baobabs has a long history, Hollis noted (1963:117):

> Every Tebeldi in Western Kordofan which is in use has a name, invariably of the feminine gender. Each tree is listed under its name in the Merkaz registers. The great majority of the names are compounded with "Um" (= mother) and a substantive, e.g. Um Khiban (from "Khaban" to store provender for a time of want), Um Asal (= full of honey), Um Tiyur (= home of birds), etc. The names range from the poetic and picturesque, e.g., Um Fakha Khir (= Mother of glories) to the frankly coarse.

The symbolic significance of darkness, upside-down orientation, and nocturnal visitors to the flowers also associate the baobab with female symbolism. The baobab comes to life in the mysterious darkness of night. This is the time when the tree's large pendant glossy-white flowers bloom and are visited by diverse "creatures of the night," especially bushbabies and upside-down bats.

The Baobab as the Hadza Birthing Tree

In his semantic analysis of the gender classification of nouns in the Hadza language, Edenmyr (2004:11) reported that although the general term for "tree" is masculine, "trees and plant species" are generally feminine, and this is also true for most "tuber and edible roots." The Hadza name for the baobab, ɧ//oba-ko, marks it as feminine and *Kesati-ko*, the name for a water-storing baobab, is also marked as feminine. This is understandable when we consider that the Hadza raise their children on what they call "Hadzapi milk," which is made from the fruit of the baobab; they identify the baobab as the place where they were "born" as a people, and a place where some Hadza women give birth.

The birth of children as a critical moment in the overall reproductive process is clearly a practical activity. However, the subject is discussed within the framework of Hadza's inspirational life because of its religious and symbolic significance. Marlowe (2004:689) claimed in no uncertain terms that Hadza women "give birth at home, squatting." Nonetheless, not only were the originating ancestors of the Hadza as a people born inside a baobab – "the baobab being the place where man was born," according to Matthiessen (1972) – the baobab is still used as a birthing tree. In the film Bushcraft, the BBC video documentary on the Hadza produced by Ray Mears one Hadza woman recounted:

> I have nine children, eight girls and one boy, and I gave birth to all of them inside the baobab tree. I stay in the tree for a month each time until the baby's umbilical cord came away. It is part of our culture. I and my ancestors have given birth in the baobab tree. I think if the custom dies away you will see the Hadza people die away too.

The extent to which the baobab is used as a birthing tree has never been systematically studied. Nevertheless, the baobab as a birthing tree has taken on a life of its own on the Internet, where we learn that the Hadza give birth in a particular hollow baobab or in hollow baobabs in general.

According to Margaret Silf (2008) in a column in the *Jesuit Review*, the "importance of the baobab" first caught her attention when she "watched a television documentary filmed in Tanzania," and she presented what amounts to an inspirational view of the baobab as the birthing tree in hominin evolution:

> According to the Swiss painter Paul Klee, "the artist is like the trunk of a tree, drawing up through its roots in the unknown soil below what will bring life to the branches above: leaves, flowers and fruit, a life of which he or she knows nothing." This strikes me as a pretty apt description of the priestly calling, as well as the call of every believer, to be deeply in touch with the power of an invisible reality … My personal reminder of this call is a baobab tree, or at least a raffia representation of one that I brought back from South Africa … In a remote rural village in [in Tanzania] stood an ancient baobab, already completely hollowed out by the passing of the centuries. Its roots reached deep into the African earth. Its branches stretched out to the brilliant blue skies and the star-laden canopy that have captured the human imagination since *Homo sapiens* took his first bipedal steps here and told stories around the campfire in the equatorial night. This was a sacred tree, not only because of its ancient lineage, but because it was the community's birthing tree. Whenever a pregnant woman came to term, she would enter into the hollowed-out sanctuary of the baobab, give birth to her child and remain there with her young until the umbilical cord fell away. Every child in the village had first seen the light of day within the enfolding shelter of that tree. It had literally borne the fruit of the human family in that place, delighted them with its large white flowers and nourished them with its gourd-like fruit.

The Internet accounts of tourists who blog about their "adventure trekking with the Hadza" show that the baobab is among the important attractions of the Hadza landscape. They post images of the Hadza climbing for honey and other resources, collecting fruits from the ground, hunting in the vicinity of the tree, or resting in its shade. They also report being taken to see the Hadza birthing tree. For example, one of the tour opportunities the travel company Mountain Madness offers on its website is an "African trekking Safari." On day 14 of a 16-day itinerary, trekkers get to camp "near Lake Eyasi," where the "whole day is designed to be spent visiting and walking with the Wahadzabe. The Hadza will take us to their birthing tree and help us search for honey."

On a website where readers posted their response to the Ray Mears documentary, one viewer stated what has now become the general view in popular discourse: "Hadza babies are born inside an old baobab, what better place to come into the world!" Given the global reach of the Internet, it is not surprising that Olivia Southey (2014), self-described as a "West London working mama and qualified Natal Hypnotherapy Practitioner," should name her business Baobab Birth. She also referenced the documentary presenting one of the many permutations of the tree-of-life theme. The baobab, she noted, is "often referred to as the Tree of Life," but it "is actually the Tree of New Life" – the baobab as the Hadza's birthing tree offered "safety, protection, privacy and warmth." The aim of Baobab Birth, Southey (2014) reported, was to "offer London mothers-to-be a service that suits both their urban life and their changing needs during pregnancy and birth, a transformative and hopefully empowering time in a woman's life," we learn from her website:

> The name Baobab Birth is inspired by the way in which the Baobab tree supports life in so many different ways, and indeed Baobabs are called the Tree of Life in some areas. The African Hadza tribe gives birth within Baobab trees, whose huge trunks can be hollow, and offer the women a private and secure place to birth. This birth tradition, in the heart of the tree, is considered essential to their cultural survival. Natal Hypnotherapy workshops enable women giving birth in more modernized circumstances – whether in hospital, at a birth centre, or at home – to create a similar feeling of connection with their natural selves, safety, and privacy. The Baobab tree is a powerful symbol that women can connect with during their birth preparation, both in imagining their own strength and as a metaphorical birth environment.

Summary

The inspirational value of the baobab includes it identification as a fertility tree associated with female resemblances and symbols and its use as a birthing tree. Other inspirational uses of the baobab are discussed in the next chapter.

Chapter 20
Other Inspirational Uses of the Baobab

The Baobab and Hadza Aesthetics

As a fiber tree, the baobab enters into Hadza aesthetics by its contribution to the making of wellness objects, personal ornaments, decorations, and musical instruments. The Hadza are fond of jewelry, as earlier noted, especially necklaces and headbands made with beads and other natural and acquired materials. Beadwork is also used to decorate containers and other objects. According to Matthiessen (1972:234): "Sometimes the seeds are left inside the hull to make a baby's rattle." Woodburn (1970) had earlier mentioned this toy rattle and provided photographic illustrations. One wonders if this baobab rattle is ever used in the Epeme Dance, if only as a last resort when cultivated *Lagenaria* gourds are not available. Of course, baobab and wild *Lagenaria* rattles would also have been available prior to the development of agriculture some twelve thousand years ago.

The Baobab and Hadza Oral Traditions

The imposing presence and settlement amenities of the baobab, especially as a shade tree, underlie its inspirational value as a spiritual, aesthetic, and recreational place. As the preeminent multipurpose tree of Hadza central-place foraging, it provides the Hadza with a suitable environment for music, dance, and theater, and for diverse activities associated with oral traditions, especially storytelling. Marlowe (2002:250) pointed out that Hadza "men tell endless stories about how things came to be," and this storytelling often includes the baobab. The baobab is popularly known as a tree of legends and the place where legends are told; it enters into Hadza storytelling in part because they prefer to establish their camps close to baobab trees.

The Epeme Dance and Maitoko female initiation rite are the Hadza's two most important religious ceremonies. As earlier noted, the Epeme Dance especially

J. Rashford, *Baobab*, https://doi.org/10.1007/978-3-031-26470-2_20

occurs at the height of the dry time, which is September and October when baobab fruits are still plentiful; the female puberty rite is held at the annual peak harvest of berries in summer and autumn. The Epeme Dance, which involves special dress, leg bells, and dance rattles, is performed during the dark phase of the lunar cycle and is associated with the Hadza's most important practical and inspirational concerns, including promoting general well-being, especially social solidarity, and "above all, good health and successful hunting" (Woodburn 1982b:190; Power and Watts 1997). As Marlowe (2010:61) pointed out, all "the beliefs and rituals associated with the *epeme* dance and *epeme* meat eating are at the heart of Hadza religion."

Three important stories about the origin of the first Hadza that have been published all involve the baobab in one way or another. The first is the Hadza tale that refers to the use of a hollow baobab and explains how men came to monopolize the Epeme meat. It is a story about an old woman named Mambedako recorded by Woodburn (1964) and paraphrased in condensed form by Marlowe (2010:62):

> Long ago *epeme* meat belonged to the women. The owner of a pot in which *epeme* meat was cooked was an old woman called Mambedako. She dressed like a man, wearing a wildcat skin in front and leaving her buttocks naked. Under the cat skin she had tied a zebra penis. She was married to beautiful wives and had intercourse using the penis. She had a man's bow but hunted zebra only and cut off the penis to replace the one she was wearing when it got old. When men killed animals, they had to carry the *epeme* meat to her hut, and she ate it with the women. One day, mongooses and wildcats who were cooking epeme meat inside the hollow of a baobab tree called over a man who had gone hunting. He stopped, and they offered him some of the *epeme* meat, but he refused and told them that men did not eat that meat. They gave him some regular meat and he ate, explaining that *epeme* meat was eaten only by women, with Mambedako as their leader. They told him to kill an eland, and that they would come. After he did kill an eland, Mambedako sent some unmarried women to get water to cook the *epeme* meat. Once the women cooked the meat, big red flies stung the women around their genitalia; then the mongooses and wildcats tied the legs of Mambedako and the other women to the trees, filled their vaginas with sand, beat them with thorny branches, and broke the women's gourds and pots. The mongooses and wildcats gave the *epeme* meat to the men, and from that time on, the *epeme* meat has been the property of men.

The second story which was earlier discussed is that of the Hadza ancestral couple who descended from the sky via a baobab as the cosmic connector. The third story is another Hadza creation narrative recorded by Peterson et al. (2013:9–10). In this story, the Hadza say that they were baboons until the day God, whom they call Haine, sent some baboons to gather food, and others to the river to collect water. When the food gatherers returned to Haine loaded with fruits and roots, they were very thirsty and grew increasingly impatient waiting for those who had gone to the river for water. Haine, too, lost his patience, and they all went to the river to see what had happened. As "they neared the riverbank they could hear the screeching and splashing of baboons having a good time … frolicking in the water, both Haine and friends all but forgotten." Peterson et al. (2013:9) added:

> Haine was furious and called all the baboons to a meeting and separated the ones who had been sent to find food from the ones that had been sent to fetch water. To the first group Haine said, "From now on, you will be Hadzabe," to the others he said, "and you will continue to be baboons." … Once Haine had separated both groups, the order was given to the Hadzabe, "You will live in the bush and eat berries and tubers and the fruit of the baobab

> tree. But first and foremost, you will eat meat." Then Haine turned to the baboons that were to remain baboons and said, "And you will live in the bush and eat berries and fruits and tubers, even maize and sweet potatoes if you find them. When you see a Hadza you will bark and run and cry out in warning, for they eat meat."

In the above narrative, Haine did not tell the baboons that they, too, would eat the fruit of the baobab. But they do – the baobab is one of their most important foods, which explains why the Hadza collect baobab seeds from baboon dung.

In addition to the first story mentioned above, there is yet another interesting way in which the baobab's association with the Epeme Rite enters into the activities of a "very small minority of Hadza with close connections with the Isanzu." The fruit pulp mixed with honey is used as a spirit-placating ritual offering (Woodburn 1982b:194). The Isanzu are the Hadza's agropastoral neighbors to the south of Lake Eyasi with whom they have had a long-established contact through trade and intermarriage. Of this "very small minority" expressing a "comparatively recent importation," Woodburn (1982b:194) wrote:

> [They] believe themselves to be possessed, following a severe illness, by alungube (the Hadza form of the Isanzu word for spirit of the dead). Men possessed by alungube occasionally run naked into the night and are believed to be guided by these spirits to select particular pieces of plant or wood or stone as medicines. Sometimes they may believe themselves to be possessed by specific named ancestors who have to be placated by gifts of honey mixed with baobab pulp at an epeme dance.

The Baobab and Hadza Recreation

The recreational value of the baobab in Hadza camps is especially associated with its use as a shade tree (in the sense of a relaxing retreat, socializing hub, play area, and performance space), with ties to gambling. Woodburn documents the use of baobab bark to make a gambling disc referred to as *lukucuko*, which is also the name of the game (1970:33–34). Hadza men have been reported to spend a great deal of time gambling in the large camps of the winter dry season. Camps are larger at this time of the year because they are more likely to take bigger animals based on ambush hunting. And large camps are also characteristic of the peak berry season of spring and summer, and even when researchers have shot large game for a camp (Marlowe 2010:66). In the dry time of the year, Woodburn (1968a:53) noted, the Hadza spend "much more time gambling than hunting." Similarly, one informant told Peterson et al. (2013:130) that "*Lukuchuko* is a bad game. Men can become so addicted to playing that they have no time to spend looking for food. This is why it can only be played when there is plentiful food such as [during the time of] the undushibi or tafabe berries or if large animals have been killed [as most often occurs with water-hole ambush hunting in the dry time]."

Most of the objects Hadza men value and for which they gambled are not made from readily available materials. Matthiessen (1972:226) wrote in his discussion of the personal items of Hadza men:

> Each carries a hide pouch with shoulder strap containing scraps of skin and tendon, tobacco leaves and hemp, *a disc of baobab wood, lucocuko* [my emphasis], used in gambling, a hunk of vine tuber which, when chewed serves as a glue for binding arrow vanes, some rag-wrapped hornet larvae medicine or dawa, useful for chest pain, and snakebite dawa, of ingredients known only to a few, which is used in trade with the Mbulu and Mangati, spare arrowheads and scraps of metal, a chisel tool made from a nail, a pipe carved from a soft stone in the river.

According to Woodburn (1999:12), the following personal items were among the things for which Hadza men gambled.

> The best type of wood for making a *//'anako* arrow, the poison for the head of a *//'anako* arrow, the stone for making a stone pipe – each of these materials is not easily obtained in most areas. It is no coincidence that *//'anako* arrows, stone pipes and the valued articles obtained by trade or incorporating traded material [such as beads, gourds, pots, cloth, and iron blades for axes, knives and *kasama* arrows] are used as stakes in the gambling game of *lukucuko* … This is a game of chance – there is little opportunity for using skill – and it serves to distribute these scarce objects throughout Hadza country.

Peterson and coauthors (2013:129) noted that the winner of a game "takes all the goods gambled and the men must put in another set of goods for the game to continue. Historically, men would follow their lost possessions as they moved from camp to camp seeking to recover them again through *lukuchuko*."

To play the game, Woodburn (1970:12) said each contestant cuts himself a bark or wood disk called *male*, but the larger disk called *lukucuko*, "which is almost always made of baobab bark, is cut to serve as a master disc." Woodburn (1970:34) described the game in the following way:

> Two, three or four people may play together. A loser in the previous contest or a newcomer to the game takes a disc for each player (including himself) and the master disc and, holding the discs in a pile all one way up with the master disc either on the top or the bottom of the pile, throws them against a tree. The various discs drop to the ground and roll about. The winner of the contest is the player whose disc lands the same way up (either 'outside' up or 'inside' up) as the master disc provided that the other player's discs are all the other way up. If no player's disc or more than one player's disc lands the same way up as the master disc, all the discs are thrown again until a win is obtained. In general about three throws are needed to secure a win.

Unfortunately, no explanation is given for why "the master disc" is "almost always made of baobab bark."

Surprisingly, Peterson et al. (2013:129) did not mention either the player's disc or master disc being made from the bark of the baobab. What Peterson and colleagues did mention was that "the game is played with a large mother disc and smaller individual player's disc carved from the bark of a tree. In this game of chance, the participants take turns hurling all the discs against a baobab tree until only one individual's disc lands in the same position as the mother disc – either with outer or inner bark up and all the other discs different."

Marlowe (2002:15) reported that unlike in the 1960s and early 1970s, there "is less gambling by men nowadays. I have seen Hadza men play their gambling game, lukucuko, only at one camp in one season." Peterson et al. (2013:129) agreed:

These days, the game has waned in most areas possibly due to the far greater use of money from tourism and researchers. Ironically, many men (even those who play it) consider *lukuchuko* to be a bad game because its addiction can result in marital discord and the loss of harmony within the community. Others of us associate *lukuchuko* with more harmonious and plentiful times of the past.

The Baobab in Hadza Education

The Hadza regard the baobab as a source of knowledge-based power. This is evident in the comments of an old, blind Hadza hunter named Mzee Memelan, who lived at Endofridge, a small camp in an oasis on the northeast edge of Lake Eyasi. In describing a conversation with Mzee Memelan, Stephenson (2000:22) wrote:

> When Mzee Memelan could not find animals, he went to a sacred baobab tree and entered the tree through a small hole. The snakes in the tree bit him, giving him the sight of the tree. The tree knows everything, he said. It knows the times the animals will pass, when the bees come to make their nests. After the snakes bit Mzee Memelan, he sat and ate medicine so he would not die from the poison. And if he lived, he knows nothing could harm him with the knowledge of the tree and snake; no spirit or animal could kill him.

However spontaneous the process might seem for the Hadza and other foragers, education is a fundamental aspect of hominin adaptation. As a shade tree, the baobab is an important learning environment in which practical knowledge, together with religious customs, oral traditions, aesthetics values, and recreational pursuits is transmitted to the younger generation. While it might be relatively easy to acquire foraging skills, there can be no doubt that such acquisition follows upon a period of development. According to Sugiyama (2011:1), "Although human life history theory tends to characterize … [parental] investment in terms of food and care, ethnographic research on foraging skill transmission suggests that the flow of resources from old-to-young also includes knowledge. Given the adaptive value of information, parents may have been under selection pressure to invest knowledge – e.g., warnings, advice – in children; proactive provisioning of reliable information would have increased offspring survival rates and, hence, parental fitness." In terms of life-history theory, however, prolonged juvenility is more than simply parental investment in food and care, and in knowledge. Prolonged juvenility is the parenting time required for the sufficient physical development of children, which is the time it takes for them to become productively and reproductively successful. This especially involves the development of the brain and its associated knowledge acquisition and the development of handiness. Sugiyama noted that "One way that foragers acquire subsistence knowledge is through symbolic communication, including narrative. Tellingly, oral traditions are characterized by an old-to-young transmission pattern, which suggests that, in forager groups, storytelling might be an important means by which adults transfer knowledge to juveniles." From an anthropological perspective, education, broadly speaking, is enculturation, the process by which an individual acquires the customs of the culture into which he or she is born. The baobab, in one way or another, is part of the education of all Hadza as the "Children of the Baobab."

Summary

The documentation of Hadza baobab use suggests that it is indeed a central feature of their inspirational life as is the case with other African cultures. There is every reason to theorize that the baobab would also have been inspiring to early hominins of the African savanna.

Chapter 21
The Baobab and Danger

The baobab is widely recognized as a tree of life, but it is not without a variety of associations with danger and death. Monstrous, grotesque, frightening, and threatening are characterizations that identify the baobab as a disturbing tree. They highlight the baobab's ability to destroy built structures, its perilous challenge to climbers, its links to spirits in general and to harmful spirits in particular, and its deadly ecological associates. There are no reports of the Hadza fearing the baobab or being preoccupied with its real dangers. Nevertheless, the negative experiences associated with the baobab are part of the tree's overall significance in the human environment. This chapter discusses the fearful impressions and potential harm associated with the baobab in Africa and it serves as an introduction to the next chapter focused on the relation between the baobab and death, which is significant for the Hadza.

The Baobab's Destructive Vigor

In Antoine de Saint-Exupery's *The Little Prince* (1943), the baobab's size and vigor menaced the Little Prince's asteroid. Every morning he had to pull up small baobabs to protect his beloved rose and the all-important volcanoes of his asteroid, both the active one he used for cooking and the inactive one in which he lived. The inactive volcano he kept free of baobabs for the benefit of the good plants that sprang up with baobab seedlings and looked so much like baobab seedlings that it was hard to tell them apart. Nicholas and Scott Thompson (2000:11) adopted the Little Prince's view of the baobab's destructive vigor, which they applied to their dualistic symbolic contrast between bad politics in Ghana as the baobab and good politics in Thailand as the mango.

> In *The Little Prince*, Antoine de Saint-Exupéry tells of a prince, alone on a tiny planet, desperately trying to keep it free of baobab trees. … The state in Ghana, as in the rest of Africa, has been a baobab tree. Its role hasn't been to help people, but simply to swallow, to

J. Rashford, *Baobab*, https://doi.org/10.1007/978-3-031-26470-2_21

> consume, to obstruct, to stick its branches aggressively in the air and thrust its roots through the ground, sucking up every available resource… Thailand – the Mango tree – may have been corrupt and even careless at times, but at no point was the government consciously and intentionally acting against the interest of the people or of development.

Contrary to de Saint-Exupery's fantasy in the form of the Little Prince and to Nicholas and Scott Thompson's (2000:11) ethnocentrism, Leopold Senghor of Senegal is remembered as a "great baobab" (Vaillant 2002:17), and when Nelson Mandela died the African National Congress reported that "The large African baobab has fallen" (Mail and Guardian 2013). Nevertheless, the argument that the baobab's vigor is the reason for its menacing character has a basis in fact, but only after the nomadic lifestyle of hunter-gatherer camps began to give way to more permanent settlements that developed in association with cultivation, herding, and trade. The baobab will have its fair share of space in the human environment since it is indeed destructive when planted too close to human structures. People who buy baobab seeds, seedlings, and saplings are often warned against planting them in small gardens or too close to buildings, walls, and paved walkways. Although this competitive aspect of the baobab is not usually discussed in the literature, it has been frequently noted in fieldwork conducted in Florida, Jamaica Antigua, Barbados, Tobago, and Brazil. Many baobab trees have been cut down because they outgrew the space in which they were planted.

Climbers Falling

The hazards of the baobab stem not only from its destructive vigor but also from the danger of climbers falling as noted earlier. The baobab as a place of death is especially associated with climbing the tree for honey. In such a case, Woodburn (1982b) reported that the body would go undetected long enough to be scavenged by animals or rot in place.

The Danger of the Baobab's Link to Spirits

A common reason for the baobab's association with danger is its identification with spirits, especially death spirits and harmful spirits. As previously noted, the baobab is linked to death spirits of various kinds, especially when its hollow trunk is used as a tomb, when the vicinity of the tree is used as a burial or disposal site, when religious narratives and oral traditions link it to ancestors, or when it is regarded as the residence of diverse spirits. With respect to harmful spirits, Frazer (1976:34) noted that "On the Tanga Coast of East Africa mischievous spirits reside in great trees, especially in the fantastically shaped baobabs … as a rule they enter unseen into people's bodies, from which, after causing much sickness and misery, they have to be cast out by the sorcerer." Aitken (1951:458), reporting what he regarded as a widespread Africa belief, noted that the baobab "is inhabited by evil spirits." He claimed some Africans avoid staying under the tree at night, or even passing close to it. There is also the flower-picking taboo that is frequently mentioned in the literature. Spirits, it is said, inhabit the baobab's flower, and they are sometimes specifically identified as "ancestors" (Olasky 2005) or "evil spirits" (Layser 2001:152), and anyone who picks the flower will be eaten by a lion (Palmer and Pitman 1961:234, Hardy and LaFon 1982:52).

One of the most intriguing of the baobab's many common names that link it to both good and bad spirits is the upside-down tree, and according to Watson (2007:9), the baobab is "the ultimate upside-down tree." This name alludes to the baobab's massive trunk and stubby branches, especially when leafless in the dry time, appearing to be like roots growing into the sky. In their comprehensive listing of baobab vernacular names recorded in the literature, Wickens and Lowe (2008: 337–357) identified "upside-down tree" as an English common name, which they included among the "European names" for the species. They made it clear, however, that the name originated in African oral traditions where the baobab is widely identified as growing upside down.

Judging from African oral traditions, the seemingly inverted nature of the baobab is so obvious that it is not the subject of cultural elaboration. The principal goal of African oral traditions is to explain not that the baobab is upside down, but why. The most widely recurring explanation reported in the literature attributes the baobab's peculiar orientation to God's actions. Ivy Newton titled her book *Baobab, the Tree God Planted Upside Down: African Tree Tales* (2015). But the idea that God planted the baobab upside down is perplexing. As the story goes, when God created the world, he planted the baobab upright like all other normal trees, with its branches in the sky and its root in the earth. It is the behavior of the baobab that caused it to be inverted. In West Africa, a Burkina Faso folktale says the baobab was inverted by God to stop it from walking about; in East Africa, a similar folktale has been reported in Tanzania. Another traditional story from southern Africa reported by Wickens and Lowe (2008:50) "tell how God visited the Sabi Valley in Zimbabwe and became thirsty and sought water from the *muuyu* [baobab]. Angry at finding none, he uprooted the baobab and replanted it head down, with the roots in the air."

Other folktales indicate that God punished the baobab for its pride, jealousy, and incessant complaining by uprooting it and replanting it upside down. From Kilimanjaro comes the story of a vain, fat, and wrinkled baobab beside a pool. Angered by the more colorful reflections of neighboring trees, it continuously complained to God. Watson (2007:9) presented a version of this tale, which he identified as a "popular legend" of which he wrote "one of the first trees God created was the baobab. When it saw the next tree, a slender palm, it started grumbling that it should be taller. God heeded the baobab's complaint, but no sooner had the tree grown to reach the palm's tops than it began to covet the flame tree's spectacular flowers. But when it, too, produced flowers, this was still not enough for the baobab. Next, it moaned that, unlike the fig, it had no fruit. This was too much, even for God, and in a rage he wrenched the baobab out of the ground and dashed it back in, headfirst." In a version of this tale, the upside-down baobab was a chance outcome of God's actions. Fed up with the baobab's jealousy and incessant complaining, he uprooted it and flung it far away, where it ended up in an inverted position.

The baobab's upside-down appearance has also been attributed to the malevolence of the hyena, who planted it upside down out of ignorance or spite, to an angry Nigerian hunter called Odede, and to the devil. According to *The Encyclopedia Britannica* (1967), it was an Arab legend that attributed the baobab's inverted appearance to the devil. Similarly, Watson (2007:9) reported that the "Giriama on the Kenyan coast say that the sweltering devil lay down to rest in the shade of a baobab tree. Enraged to find it leafless, as the tree is for nearly two thirds of the year, the devil tore it out of the ground and replanted it upside down." Curiously, the tales indicate that God and the devil inverted the baobab for similar reasons: because it did not give water in the first case and because it did not give shade in the second. The baobab's association with the concept of the devil is, no doubt, a reflection in part of the influence of the Islamic and Christian traditions in Africa. Assogbadjo and Loo (2011:8) reported that there "are many legends about the tree and although in many areas mature trees are protected and highly valued, people in southern Benin consider the species diabolical and eliminate seedlings and saplings from their fields."

It is noteworthy that upside-down appearance is the only explanation put forward in the literature to account for the baobab being named the upside-down tree. There is more to the baobab being called the upside-down tree, however, than mere appearance. Like tree shape and face-down flower, the baobab's significance within the framework of the religious oppositional symbolism characteristic of all cultures reinforces the widespread traditional view that the baobab is an inverted tree. That is, the baobab, unlike normal trees, is upside down symbolically, which places it on the female side. This includes not only its size, shape, soft wood, hollow trunk, fruit shape, and drink color, as earlier noted, but also its association with water as a spring rain indicator, planting cue, and tree cistern; and it includes its links to the darkness of night and a variety of nocturnal floral associates that feed on the flower and its nectar.

Dangers Linked to the Baobabs as a Hub of Life

The baobab's rich ecology, especially its active nightlife and dangerous fauna, is also responsible for its association with harmful spirits, malicious individuals, and ritual specialists. Parrinder (1969:56) noted, for example, that "Great trees such as the African oak, mahogany, silk cotton, or baobab are obvious haunts of spiritual power," and reported that "some trees are dangerous, or may be considered the abode of witches, especially if great bats cluster in their branches and emerge to flap across the sky in the evening dusks." According to Layser (2001:2), baobab stories like "most folk stories contain an element of truth or purpose." The baobab stories he presented included the widely reported flower-picking taboo, a hollow baobab haunted by a ghostly python, and the baobab as a personified being that imprisoned four beautiful "maidens" in its hollow trunk because it was jealous. Lyser wrote:

> Casually picking baobab flowers would deprive the people of the edible and useful fruits, and snakes frequently do den in the hollow trunks of aged trees. At times, at night, the trees seem to "hiss with spirits" as the fluttering fruit bats arriving to drink flower nectar, and the scurrying nocturnal bush babies, also attracted to the nectar, can wail eerily.

Caution is required when dealing with the baobab, and the reasons for concern involve more than the baobab's spooky nightlife linked to bats and bushbabies attracted to its nocturnal flowering. As an ecological tree of life, not all the baobab's many visiting or resident associates are benign. Some are among the real dangers of the savanna, such as aggressive bees and poisonous snakes, and especially carnivores resting in the baobab's shade, hiding in its surrounding bushes, or inhabiting its hollow trunk.

The baobab would have been a potentially dangerous environment for ancestral hominins, as it is for contemporary human beings. From Sweeney's (1974:54) account, we get some sense of the harm attributed to the baobab because of dangerous animals and "vague fears" resulting from its association with spirits. Sweeney identified twelve issues that interfered with his effort to collect data on 150 baobab trees in the Sudan in order to understand their association with insects, particularly the cotton stainer bug, which was a problem for cotton growers. The first four issues were problems for nine trees and included staff difficulties and illnesses, natural disasters (including flooding, lightning, and wind), and aggressive bees. The remaining eight issues that plagued the survey team on an "almost weekly basis," were snakes, leopards, lions, scorpions, centipedes, falling fruits, falling branches, and "vague fears of devils or supernatural animals."

In his accounts of the natural history of the Sudan, Sweeney (1969:85) reported that the "wild honey-bees of Africa, although but a variety of the hive-bees of Europe, are notorious, despite their small size, for aggressiveness, but nowhere in Africa have I encountered swarms as vicious as in the Sudan." He tells of encountering these bees when he camped under a baobab tree. Just after dawn, when the air was still "fresh and chill," he went to warm himself in a sunlit spot and noticed about a half dozen bees already buzzing about. By the time he was ready for breakfast, he had to flee. "Bees in countless numbers arrived from two great hives in the tebeldi trees (as baobabs are called in the Sudan), a favourite place for bees' nests." Sweeney tells us in great detail that during the next three hours, he made two attempts to reach his Land Rover and pack up his things, only to be driven away each time.

Although there are no reports that the Hadza share the "vague fears" resulting from the baobab's association with spirits, they do face the reality of the baobab's harmful wildlife. For example, while Sweeney discussed differences in the aggressiveness of various kinds of honey bees, it is noteworthy that the Hadza point to the effects on bees of the kinds of plants from which they gather honey. They told Peterson et al. (2013:85), "All of us, men, women and children know from experience that different flowers not only result in different flavors, but also affect the behavior of bees in different ways. For example, when the *mukangako* (*Acacia tortilis*) blossoms profusely from January to mid-February, it makes the honeybee very aggressive .... Fortunately, *militimoko* (*Acacia mellifera*) follows with an extensive bloom in mid-February through March making the bees passive so that one can pursue honey more safely." The Hadza told Peterson et al. (2013:86): "When harvesting from a hive [high up in a baobab tree], we use smoke to pacify the bees. Nonetheless, bees do sting and while we are used to it, men or boys do sometimes get stung so badly while harvesting high up in a tree that they fall and are injured."

Summary

Although the baobab is generally known as a tree of life, we have seen in this chapter that it is also perceived by some groups as a tree of danger. The idea of danger is especially tied to the baobab's association with death, which is discussed in the next chapter.

Chapter 22
The Baobab and Death

Although some cultures traditionally abandoned the body of someone who has died at the place where the death occurred, ritual disposal of the dead is the norm among human beings, and the practice varies widely from culture to culture based on environment, mode of adaptation, and social organization (Hertz 1960a, b; van Gennep 1960; Huntington and Metcalf 1979; Robben 2017). Some groups lay the body on the ground in a chosen place, either exposed on the surface or covered with rocks or earth to form a mound. Some bury it, burn it, or dispose of it in rivers, lakes, or the sea; or they place it in a tree or on a platform out of the reach of animals. Some use natural containers, such as caves, animal burrows, or hollow trees, and some build structures, such as tombs and vaults.

Death, as represented around the world, involves the disposal of the corpse and the fate of its animating spirit, especially the spirit's passage to the afterlife and its future rebirth. Because of this, the disposal of the dead is deeply intertwined with religious beliefs and practices. The baobab is associated with African mortuary traditions in a variety of ways including a placement site for the dead, tree tomb, burial place, grave planting, and cemetery tree. This chapter discusses the Hadza as an example of the baobab's diverse associations with death, especially as a temporary placement site for the dead. It is possible early hominins had a similar orientation to the baobab compared to what has been documented for settled food producers.

Disposal of the Dead in Hollow Baobab Trees

The most frequent reference to the baobab's mortuary associations is its hollow trunk used as a tomb. But characterizations of the use of the baobab tree can be misleading when the reports of researchers and travelers are stripped of their ethnographic and historical context to become mere generalizations. For example, the frequent assertion without elaboration that hollow baobabs are used as tombs. Writing of "the drier parts of Africa," Dalziel (1937:115) noted that in "some cases

J. Rashford, *Baobab*, https://doi.org/10.1007/978-3-031-26470-2_22

the hollow trunk has served as a drying chamber, in which a suspended body (denied burial) can be conveniently mummified." There are also those who refer to "the negroes" (Smith 1882:37) and to "the natives" (Blanks 1952:532) using baobab chambers for human remains. According to Sidibe et al. (2002:23), hollow baobabs are used as tombs in West Africa, especially in Senegal, and also in Zimbabwe, and they noted that the baobab is the only tree "in Africa preserved as repositories for the ancestors and hence have spiritual power over the community's welfare."

While it is true that hollow baobabs are sometimes used for disposing of the dead, it is an exaggeration to attribute this practice to "the negroes," or to people "in the drier parts of Africa." In fact, baobab tombs have only been reported for Zimbabwe, Tanzania, Mali, Senegal, and Ghana. There is no evidence, nor is there reason to believe, that baobab entombment could ever have been the general practice of an entire population. There would not be enough hollow baobabs in all of Africa to support this method of disposal of the dead, even if each baobab "tomb" contained more than one corpse. Moreover, using the baobab as a disposal chamber would likely cause people to avoid the trees, which would be counterproductive, given the baobab's multipurpose value. The hollow baobab is not generally known for its use as an ancestral tomb as Sidibe et al. (2002) have suggested, but for its use to protect the healthy from the sick, the fertility of the earth, and the bones of chiefs from enemies in times of war.

"In Tanzania," according to Wickens and Lowe (2008:52), "the hollow trunks of some baobab trees are used to inter those who have died from certain diseases." Citing Pern (1982) and Velton (2000), they wrote, "The Dogan [of Mali who live in villages at the base of the Bandiagara escarpment] inter lepers in hollow baobabs and seal the openings with clay and straw to prevent infection." Wickens and Lowe went on to point out that the Dogan "do not eat the leaves and fruits from hollow baobabs for fear of becoming infected."

One of the best-known traditions of the hollow baobab used as a living tomb occurred in Senegal where a person's status in life was a factor in the disposal of their corpse. Griots are the hereditary musicians, poets, and storytellers of West Africa whose role as historian and genealogist has been central to their performance as praise singers, and to their service as ritual specialists and advisors to influential families among the political, economic, and military elite (Tamari 1991; Wise 2006). It is the griots of the Serer people of Senegal in particular who have been identified as traditionally entombed in the hollow trunk of old baobab trees, which some authors have identified as baobab vaults (Blanks 1952:532), baobab cemeteries (Mauny 1955:75), baobab burial chambers (Ross 2012:11), baobab sepulchers (Anonymous 2005), and baobab mausolea (Anonymous 2014).

Mauny (1955), who published a widely cited account of this practice, noted that the explanation for it has remained the same from the earliest reports to the present: it is said that if griots were buried in fields, the fields would cease to produce because the land would become sterile. According to Mauny, the first person to mention this custom was Alvarez d'Almada in 1594, and Mauny indicated that there were similar reports in 1637, 1668, 1732, 1822, 1853, and 1855. Mauny's (1955:75) account comes from the unpublished summer school notebooks of the students of William

Ponty School, who responded to a written assignment on the theme of traditional funeral rites.

> After … death one proceeds with funeral preparations which are similar to the ones done for other castes: The deceased is dressed in his best clothes, wrapped in a cloth, then 'tied with leaves of Romer in such a manner that a solid package is formed'. He then is placed on a stretcher and carried to the Baobab cemetery by two hefty young men, and accompanied by the crowd of griots (men and women) from the village.
>
> The hollow Baobab which is chosen bears an opening obstructed by thorns to prevent hyenas and jackals from entering. When the procession arrives in front of the Baobab, the young people push and shove one another for the privilege of being the first to enter. They are armed and carry on a wild struggle for this purpose, egged on by the screams from women urging them to display their bravery and recalling their past exploits. Finally, one of them succeeds: from now on, he will do his best with the help of his weapons to keep others from entering. From the outside, the hopeful candidates fight with the winner who is inside. The battle stops only when a second candidate beats the winner: both of them then carry the body into the tree amidst applause and singing from the audience. With no other ceremony, they lay the dead man on top of the previous bodies. The two winners are considered as heroes by the young women, while the other young men who were beaten during this struggle 'have lost their honor and tarnished the family name. They will encounter difficulties in finding a bride ….

Elsewhere in West Africa, baobab tombs have served a different purpose. According to the Danish physician Peter Thonning, who in 1795 traveled to what was then Danish Guinea (now part of modern Ghana), "the inhabitants of Ashanti, Akim, [and] Aquapim always bury their principal dead secretly and often in … [the baobab] tree especially in times of war, when they fear that the enemy will discover the body and keep the bones on his drums as a sign of victory and of his enemy's disgrace. The inhabitants declare that the body is dried without putrefaction in this tree" (cited in Hepper 1976:31).

Baobab Cemeteries

The baobab is known for its association with cemeteries. According to Ross (2008:138), "baobabs and cemeteries have been inextricably linked in West Africa since ancient times. Serer cemeteries especially appear as extensive baobab forests." Ross further noted that "cemeteries which have been abandoned for centuries can still be discerned in the landscape by observing the configuration of baobabs." There are, however, important conceptual differences between the baobab as a burial place, grave planting, and cemetery tree. A true baobab cemetery is one with graves intentionally placed under and around the tree. There are probably also cases where the baobab is only incidentally associated with a cemetery; it either existed before the cemetery, or it sprang or was planted as a landscape tree after the cemetery was established. Tombs and cemeteries are not the only connections the baobab has to death. The burial place baobab is a single tree or a grove of trees that have been selected as a grave site of a particular individual, while a grave planting means baobab trees are sometimes grown at the head and foot of graves. This practice, along with the baobab used as a burial place and cemetery tree, has been reported for coastal East Africa in association with Swahili culture and its Islamic traditions (Watson 2007; Wickens and Lowe 2008), and for the Siddi of India (The reDiscovery Project 2018); baobab grave planting is also mentioned in the folklore of Zambia (de Vos 1968:119).

The Baobab and Hadza Mortuary Traditions

The Hadza have never been reported to use the baobab as a burial place, grave planting, or cemetery tree because they did not traditionally inter the dead. The Hadza deal with their dead by abandoning the location where someone died, by shallow burial in dug graves and anteater holes (Woodburn 1982b:189), and by the use of placement sites of spiritual significance, including the baobab (Wells 2011:205–206). Citing Woodburn (1982b), Marlowe (2004:695), noted that the "Hadza leave corpses out for hyenas to eat, or if the deceased is an older person, they may push his or her hut down on top of them and set it ablaze, then move away." Woodburn's discussion (1982b) of Hadza mortuary traditions was based on fieldwork done in the 1950s, and we learn things were already changing. Marlowe (2010:65) also pointed to change.

> More and more these days, they bury a body because they know the government does not approve of their leaving the dead out to be eaten. This may also be partly due to the influence of the occasional missionary. According to Woodburn (1982b), however, it was also common in the past to make a shallow grave and to leave the epeme ostrich feathers on the grave or to break a decorated gourd used for holding fat over the grave.

Woodburn and Marlowe presented a view of the Hadza treatment of the dead that made no mention of the baobab tree or of secondary burial. The geneticist Spencer Wells (2011:205–206) offered a different perspective in which the baobab was a preferred temporary placement site for the dead.

> One evening, back at camp, I asked Julius how the Hadzabe dealt with their dead – burial or, perhaps, cremation? He smiled and shook his head. "No," he explained, "we don't bury the bodies or burn them. We leave them out in the open in a sacred place, near a baobab tree, and let the animals eat them. Usually it is the hyena. There is a small ceremony performed by the elders, and we leave food for the passage to the afterlife. We return a few months later to find the bones, and then hide them in the bush. The land we live in is full of the remains of our ancestors, and it connects us to them.

While the Hadza do not have baobab tombs, burial places, or cemeteries, Well's account suggests they practice a form of "secondary burial" that sometimes involves the baobab. Judging from Hadza traditions, the value of the baobab as a life source outweighs its use as a final resting place. The baobab is a permanent location for the living, but only a temporary location for the dead. As the birthing place of celebrated ancestors, the baobab is a place of life preferentially conceived of as a womb, rather than a place of death, conceived of as a tomb or burial place. Given the baobab's many uses on the one hand and the need to separate the world of the living from that of the dead on the other, it would be impractical to use hollow baobabs as tombs and the baobab's immediate surrounding of baobab trees as graveyards. Symbolically, it would seem the baobab's contribution to life outweighs its use as a place for disposing of the dead. This is probably why the Hadza only temporarily place the body "near the baobab" rather than under, on, or inside the tree, and why the bones are eventually collected and disposed of away from the baobab (and presumably from other spiritually significant sites as well).

Based on what his informant Julius said, Wells (2011:206) went on to speculate about its meaning as it related to the Hadza conception of their environment; in doing so, he made no mention of the baobab, writing instead:

> The simplicity of [Julius's] ... answer is what struck me most. The entirety of the Hadzabes' land is a burial ground, with no separate place for the dead to spend their afterlives. They continue on in the landscape after they are dead, and this is part of the reason that the Hadzabe feel such a close connection to their territory. Land, to the Hadzabe, is not just about something that provides subsistence, though of course that is important – it's also a tangible connection with their ancestors. The land and the people are inseparable, and this connection is the essence of their *mythos*. To attempt to separate them or to exploit the land in some way would be an insult to the entire Hadzabe worldview. It was and is unthinkable.

What Julius told Wells and what Wells understood are not exactly the same, however. In this work, burial specifically means putting the body in the ground, and as we have seen, this is related to the baobab as a burial place, grave planting, and cemetery tree. This means the "entirety" of the Hadza territory is not a burial ground since it appears the Hadza did not traditionally bury their dead in the sense of internment. It is also inappropriate to identify "the entirety" of the Hadza territory as a burial ground as Wells was told the dead were placed on the ground in a place of religious significance, such as "near a baobab," and later the bones were collected and hidden. This suggests the baobab's link to death and the disposal of the dead is the Hadza link to their ancestors.

Honey and Baobab for the Ancestors

A recently developed link between the Hadza and the baobab and death involves the southern Hadza with ties to the Isanzu. As earlier noted, the baobab is the tree that allows us to characterize the mosaic savanna of the Hadza as a land of milk and honey into which the Hadza are born. The following account by Woodburn (1982b:182) points to the ritual importance of honey and baobab fruit. In addition to being among the Hadza's most valued weaning foods, baobab and honey are presented here as appropriate gifts for ancestors:

> A very small minority of Hadza with close connections with Isanzu, a neighbouring Bantu-speaking people, believe themselves to be possessed, following a severe illness, by *alungube* (the Hadza form of the Isanzu word for 'spirits of the dead'). Men possessed by

> *alungube* occasionally run naked into the night and are believed to be guided by these spirits to select particular pieces of plant or wood or stone as medicines. Sometimes they believe themselves to be possessed by specific named ancestors who have to be placated by gifts of honey mixed with baobab pulp at an *epeme* dance. I was told that all this is a comparatively recent importation and that in the past Hadza were not possessed by *alungube*.

Summary

The baobab has long been known for its association with death, and the Hadza provide an example of the baobab as a place of both life and death. From the baobab and death, we now turn to the baobab as a preferred camping site for the Hadza; that is, their preferred place for the daily activities of their residential life.

Part VII
The Hadza and Baobab Regeneration

Chapter 23
The Baobab and Hadza Central-Place Residential Camps

The baobab's size, distribution, and ecology contribute to its value as a resource environment for human beings, and this includes a place for camping. In the heat of the tropical savanna, the welcome shade of the baobab has made it a settlement site whose uses have ranged from a place to momentarily rest or relax to a temporary or permanent site for residential communities. People across Africa gather at baobab trees to socialize, tell stories, adjudicate disputes, conduct community affairs, stage festive events, and engage in religious activities related to the baobab status as a cosmic tree linked to shrines and altars (Ross 2008). The baobab's importance in the siting and relocating of Hadza residential camps is what makes it their central-place tree. Against this background, we can speculate that the baobab must also have been a place where early hominins gathered and settled.

The intriguing association between the baobab and the settlement environment of villages and urban areas has generated a significant discussion in the literature about which came first, the settlement or the baobab. The question is not as straightforward as it first appears, however. It is increasingly clear that the answer is more complex and is ultimately related to the nature of anthropogenic environments associated with human incidental and intentional influences on the reproductive success of plants and animals. This chapter examines the links between the baobab as a resource-rich environment and the nature, siting, and movement of Hadza central-place residential camps. The sections of this chapter discuss the baobab and Hadza camps with respect baobab camping environment, living space, and bedding place; the link between the baobab and dwellings, including the use of hollow baobabs as dwellings and the making of baobab branch-framed houses; the reasons the Hadza site and relocate their camps near the baobab; the important connection between the baobab, rocky places, and Hadza camps; and the baobab's contribution to the seasonal size and character of Hadza camps. In light of these considerations, it is likely that the baobab in rocky places would have played a part in the development of hominin central-place foraging.

J. Rashford, *Baobab*, https://doi.org/10.1007/978-3-031-26470-2_23

The Baobab and Hadza Camps

Traditionally, as a distinct ethno-linguistic group comprised of regional populations, the Hadza did not have a higher-level social organization above their residential groups (Woodburn 1968a, b). Blurton Jones (2016) reported a mean of twenty-one individuals, and Marlowe (2010), a mean of thirty individuals, based on a survey of fifty-three camps ranging in population from 6 to 139 individuals. Hadza households may consist of individuals or friends, but more often – and most importantly – they are the living spaces of families and other kinds of kin-based groupings. These central-place residential groups are a key aspect of Hadza forager adaptation, and their strategic siting and relocation provides efficient access to the variety of resources the Hadza require through the year for the success of their forager way of life. Individuals or groups are free to move between camps, and the fluid nature of between-camp movements has been especially identified with conflict avoidance, socializing visitation, gambling opportunities, and hunting success (Woodburn 1964, 1968a; Marlowe 2010; Blurton Jones 2016). In his influential article on the stability and flexibility of Hadza central-place communities in which he criticized the application of the concept of "band" to the Hadza, Woodburn (1968b:103) wrote:

> Hadza residential groupings are open, flexible, and highly variable in composition. They have no institutionalized leadership and, indeed, no corporate identity. They do not own territory and clear-cut jurally defined modes of affiliation of individuals to residential groupings do not exist. The use of the term "band," with its connotations of territorial ownership, leadership, corporateness, and fixed membership is inappropriate for Hadza residential entities and I prefer to use the term "camp," meaning simply the set of person who happen to be living together at one place at one time (Woodburn 1964).

Baobab Camping Environment

An efficiently sited camp is one situated in a protected and comfortable environment with good access to essential resources, including house-building material, grinding stones, and firewood, and nothing indicates the intimacy of the relationship between the Hadza and the baobab more than the fact that they, like the Kung of the Kalahari (Lee 1979:478–479), often establish their camps where there are baobab trees. The Hadza choose to settle near baobab trees for various reasons and varying lengths of time. This is particularly true of fruiting, honey-producing, and water-storing baobabs. We have seen that baobab trees are of strategic significance for the Hadza with respect to the efficient access to resources, especially in the dry time when the tree is in fruit and water is scarce. According to Tanaka (1969:41), baobab trees in the Hadza environment are most common in "mountain areas where many rocks lie exposed" (Tanaka 1969:41). In ecological discourse, the natural environment specific to a plant or animal is often identified as its home, habitat, niche, range, or territory. In this sense, the home of the Hadza, as earlier noted, is the rocky wooded hills of their mosaic savanna, where there are what researchers have described as baobab forests or baobab woodlands. But home for the Hadza is not only their preferred camping environment; it is also their preferred campsite (Hawkes 2016).

Baobab Campsites

In his 1968 article on Hadza ecology, Woodburn was less interested in the reasons the Hadza gave for choosing the places in which they camped than he was in the leisurely nature of their hunter-gatherer way of life, the social composition and fluid membership of their camps, and the factors that governed the relocation of camps. Nevertheless, he wrote that although part of Hadza territory included "open grass plains," they "never build camps there." Camps, he reported, were "invariably sited among trees or rocks and, by preference, among both" (1968a:50). While there are no doubt times when the Hadza prefer to locate their camps in rocky elevated places with any kind of tree cover, Woodburn (1970:8) reported two years after his 1968 article on Hadza ecology: "Baobab trees are so important in their economy that the Hadza often camp among them." Cooper (1949) and Woodburn (1970) published excellent photographs of the Hadza camping in baobab groves.

Baobab Living Spaces

Home is not only a choice of camping environment and campsite, however. Setting up a new camp for the Hadza involves the choice of place within the camp in which living spaces are to be located. Living spaces are the dwelling places of the individuals and groups that make up Hadza residential communities. The Hadza living space is chosen for its comfort and safety. Its essential components are a bedding place; a hearth for cooking, warmth, light, cheer, and security; domestic utensils, especially containers; a place for the short-term supplies of water and food, and for the storage of foraging gear, clothing, and personal effects such as are carried in pouches or worn on the body. As the place of domestic activities, living spaces are the defining feature of Hadza residential communities whose spatial arrangement also includes general and gendered areas for communal life, special activities, refuse disposal, and latrines (Kroll and Price 1991; O'Connell et al. 1991). The living space of the Hadza may also contain a built shelter depending on the seasons. Living spaces are often in the open in the dry time with shelter being constructed during the annual rains.

Baobab and Bedding

Finally, home is where one sleeps (Fruth and Hohmann 1994; Fruth et al. 2018; Samson et al. 2017a). Whatever their form, baobab dwellings as living spaces are first and foremost bedding places with a hearth. Considered as a whole, bedding formally defined presupposes an open or sheltered living space and includes sleeping foundation, mattress, lining, cover, and pillow or headrest. Bedding in some instances could be considered more important than shelter, and this is just as evident for gorillas and chimpanzees as it is for humans. Plant-woven bedding in the canopies of trees or on the ground is probably one of the earliest material products of apes. Gorillas and chimps do not often build shelters but routinely have bedding. Apes sleep in the protective canopy of trees on strong limbs that serve as sleeping foundations, and whose leafy branches they intertwine to make their container-shaped mattresses, which are sometimes lined with leafy twigs. It is understandable that for hominins whose material dependence requires such things as handiness and the container-carrying of a variety of objects, the technique of intertwining flexible branches, vines, and sapling stems, and the interlocking of stiff, dry, woody branches would ultimately come to be associated with a great many things. The interlacing of leafy branches in ape nest-building probably developed along with the use of large leaves to make containers, wipers, and wrapping. For hominins, these developments likely preceded the use of plants for other kinds of containers, such as baskets and net bags, and the use of plant fibers for cordage, the refinement of which, combined with weaving, would lead to the production of fabrics.

The Hadza, as we have noted, have open-air living spaces in the dry time, which means their branch-framed houses are a seasonally variable feature of their living environment. But they always have bedding. While the Hadza are known to use impala sleeping mats, Wickens and Lowe (2008:118) have reported that "The Bushmen of southern Africa make fibrous baobab matting, which is sometimes used as a blanket or kaross (Stow 1905; Braun 1929)." It is likely that baobab mats were also used for sleeping.

Baobab and Dwellings

Hollow Baobabs

From all that has already been said, it is clear that old baobabs becoming hollow with age, producing a spacious interior widely used by people wherever the tree is found, is one of the most remarkable characteristics of the tree. The ecological importance and human uses of these baobab chambers have already been noted, including the fact that Hadza mythic ancestors were thought to live in a hollow baobab. The concern here is only with the hollow baobab used as a short- or long-term dwelling that includes bedding and, sometimes, a fireplace. "Dwelling" is here defined simply as a structure in which one lives or sleeps, and although the hollow baobab dwelling is routinely mentioned in the literature (e.g., Dalziel 1937:115; Owen 1970:26–27; Guy 1971; Sweeney 1974:55; Wickens 1978:30, 1982:190; Wickens and Lowe 2008:106), it is not with the same rich details associated with the discussion of the baobab reservoir and tomb. It is noteworthy that Battuta did not mention seeing a single baobab dwelling in his travels in West Africa in the fourteenth century. Given Sweeney's extensive experience in the Sudan and elsewhere in Africa as both a baobab enthusiastic and researcher, it is surprising that he encountered only one baobab dwelling (Sweeney 1974:55). He wrote, "Near Teisi I found a '*house*' in a baobab with a wooden door and square window, while the floor had been closely paved with large flat stones. I was some distance from the nearest habitation, unoccupied, and we never discovered the owner."

According to Wickens and Lowe (2008:106, 107), however, it "is quite common to see hollow [baobab] trees" used as dwellings. They cite Livingstone, who mentioned such baobab dwellings in Mozambique and reported "one large enough to permit 20–30 men to lie down and sleep inside (Livingstone 1962)." Wickens and Lowe also noted that "In Limpopo Province of South Africa hollow baobabs have been used as temporary dwellings." This use of the hollow baobab has also been reported for the San (Guy 1971) and Hadza. According to the Meleka Foundation, which claims "one of the largest translated film archives of the Hadza," meleka in the Hadza language is an old hollow baobab used as a shelter or dwelling. Readal (2022) also presented a photograph they identified as the "Hadza baobab tree house in Tanzania," which they credited to Creative Commons. It shows the bulbous base of a hollow baobab with a door-sized entrance cut into it. At the entrance are three boys, the oldest of which has a small bow and arrows. Hung on the outside of the tree are two skulls of game animals and the Hadza string instrument, and leaning against the tree is the bow and arrows of an adult hunter.

Baobab Branch-Framed Houses

"Hut" is an inadequate name for the quickly built and readily abandoned branch-framed house of foragers, since hut is defined in dictionaries as a small, crude house identified as synonymous with cabin, cottage, shanty, shed, and hovel. The Hadza house is indeed small and can rightly be described as plain, rude, or crude in comparison to permanent shelters in any society today. But house size is not the most important consideration in thinking about the significance of shelter in forager adaptation of early hominins. What is being emphasized here is the nature of the Hadza house with respect to both its form, covering, function, and content, and its construction requirements involving readily available plant material for framing branches and covering. The baobab is also used in constructing the traditional Hadza house because of its strong flexible branches and sapling stems, its renewable bark, and its readily obtainable wood sheets. The Hadza house is a temporary dome-shaped or conical structure of intertwined or interlocking branches whose covering, though often made of grass, also includes big leaves, leafy branches, bark, and wood sheets. It is primarily used for nighttime sleeping (with respect to bedding area and materials and to sleeping fire) and in the daytime for rest, relaxation, and napping. It is also used when someone is sick, as well as for privacy, and for storing portable belongings. Daylight hours at home, however, are mostly spent outdoors in the living spaces or in communal areas. Given the importance of the branch-framed house as an essential component of Hadza residential communities, it is surprising that there is no systematic documentation of the materials used in their construction and some uncertainty about exactly how they are built. Because the Hadza routinely move their camps over relatively short periods of time, their ephemeral branch house, like that of other tropical foragers, is easily constructed, requires minimal maintenance, and is readily abandoned, with a new one built when they relocate.

While the Hadza house is indeed small and ephemeral, it is not as "roughly built" as often reported. The house is far from being thrown together. Its construction, like that of dwellings of other tropical foragers, requires a knowledge of the seasonally available framing and covering plant materials. The houses are either branch-framed, that is, made with pliable branches set in the ground with their tops woven together to make a dome-shaped structure; or, they are stick-framed, made with large dry sticks with branches that are set in the ground and interlocked at the top to form a more conical frame. In either case, they are covered with grass or other plant material, or both. As savanna and desert foragers, the Hadza and San probably choose between making stick- or branch-framed houses depending on the seasons, and things such as the weather and the procurement costs of locally available plant material. It is inadequate to characterize the small, simple, ephemeral Hadza house (like those of other nomadic tropical foragers) as a "grass hut," since this characterization obscures the variety of materials used to cover the frame. The most common covering is indeed grass, and in this specific context, we can speak of the branch-framed grass house. But there are also the branch-framed houses covered with such things as large leaves, leafy branches, bark, and wood sheets. For example, in a blog recounting a visit to Yaeda Valley posted in *Science*, Gibbons (2018:702) wrote:

> The Hadza are particularly concerned about Datoga pastoralists who let their cattle graze on grass and drink from water holes on Hadza land year-round. In one Hadza camp, a woman named Tutu pointed to her people's huts. Their tree-branch frames were draped with clothes and bark instead of the traditional grass thatch. "The cows eat all the grass," she explained.

The branches, sapling stems, bark, and wood of the baobab are selected because of their availability and special properties. The Hadza use pliable baobab branches and sapling stems (along with the branches and sapling stems of other conveniently available trees and shrubs) because they are strong and highly flexible. Baobab sheets are used because they are a natural and easily obtainable product of the baobab's decay. Branch-framing (whether involving dry branching sticks or freshly obtained flexible branches or sapling stems) is the true distinguishing feature of this kind of house structure because it is a constant; the covering, however, is variable. This kind of dwelling described for the San, Hadza, and Mbuti is best identified as a branch-framed house.

Writing of the Kua foragers of southeast Botswana, Van der Post and Taylor (1984:48–49) reported that men gather the house-building materials for their branch-framed house. They start by digging a circle of holes in which they set the bases of the stiff branching sticks of trees and shrubs whose rigid tops are interlocked to form a somewhat conical rather than rounded frame. This is different, of course, from house frames woven with the pliable branches and sapling stems of living shrubs and trees illustrated by Marlowe (2010), Blurton Jones (2016:108), and others. Samson et al. (2017b:95) reported that the "most common domiciles were grass huts constructed by women; several women work together for up to two hours, gathering flexible branches that they bend into an inverted U shape and stick the ends into the ground…. Once braced in the ground, they crisscrossed several branches to form an upside-down bowl-shaped structure that is then stuffed with grass." But what Samson et al. (2017b:95) illustrated is clearly not a house frame made from intertwined pliable branches, but one made by interlocking the branching ends of dry sticks. And contrary to Marlowe's idea (2010), the Hadza branch-framed house Samson et al. (2017b:95) described is not generally constructed by bending flexible branches into a U-shape and setting both ends of these branches into the ground. What is commonly illustrated in the literature is the thicker end of branches and sapling stems fixed in the ground and their upper leafy ends interwoven. Finally, the shape of the stick-framed leaf house is more conical rather than dome shaped.

The kind of shelter associated with Hadza campsites is highly variable. For example, during the November-April rains, especially the March-April autumn peak in annual rainfall, the Hadza sleep in branch-framed houses. But to cope with the peak of the annual rains in autumn, they move their camps to rock shelters. Curiously, it seems that for the Kung, at least, the effort put into building branch-frame houses is dependent more on the length of their intended stay at a particular campsite than just the simple fact of rainfall. Lee (1979:32–33) noted that "A !Kung rainy season hut can be constructed in a day." But water-tethering in the dry time means camp duration in the dry time is longer. Consequently, we learn that "A dry season hut, to last for several months, takes three or four days to build" and that in the "dry season" the "dome of the hut is left open to catch the warming sun." While the Hadza build substantial houses during the annual rains or move to rock shelters, in much of the May-October dry time they sleep in the open. Part of the variability of shelters associated with Hadza campsites is that rock shelters are more common in some regions of Hadzabeland than in others. For example, Marlowe (2010:95) has noted that "In certain areas, such as the hills of Sipunga and in Dunduiya to the west of Lake Eyasi, camps are sometimes made in rock shelters." We also learn that "In the Sipunga area, the Hadza often live where dry grass is less plentiful and where rain is more plentiful. There, they make huts that are covered with large strips of bark and big leaves of various kinds, rather than grass."

The "large strips of bark" Marlowe referred to are likely sheets of baobab wood. Fitzpatrick (2018:14–15) provided a photograph described as depicting "a simple hut composed of a wooden frame covered with grass and bark." Here, also, the bark, judging from the photograph, appears to be sheets of baobab wood. An article titled

"Hadzabe Huts of Baobab, Sisal and Grass" was posted on the website of Nomad Architecture by Gordon Clarke (2017) in which "Hadzabe house building" was described. The process was filmed and put online by The Institute of Nomadic Architecture, which focuses on research involving temporary and traditional buildings around the world. The film documents the use of branches and wood sheets in Hadza house building. Clarke (2017) indicated that of the many woody species suitable for house framing, "young baobabs are preferred although many species are suitable. These are "planted" into holes in the ground and brought together at the top where the branches are skillfully woven together. No ties are required." The Hadza use of baobab materials in their houses has received little attention in the literature, but it appears to be important. Clarke (2017) also noted that the Hadza house is covered with a variety of material, according to availability, and he wrote "In lowland areas, near the lake, they will be thatched with long grasses, in the forests, grass, liana, and even the separated layers of rotting [baobab] tree trunks are used."

The use of baobab wood in house building has also been reported for other parts of Africa and for Madagascar. The moist, spongy wood of the African baobab and of other *Adansonia* species consists of easily separated concentric rings. According to Wickens and Lowe (2008:114), "the Malagasy collect them from withered, dead or healthy [baobab] trees, dry them in the sun, and sell them in the local markets for roofing." They noted that "Wood from fire-killed *A. rubrostipa* in the vicinity of Morondava is used in a similar manner to that of *A. grandidieri* to make thatching tiles (Baum 1995b)." We also learn that "In south-west Madagascar, sapwood (*voroke*) and bark of *A. rubrostipa* was used for the walls and roofing of the 1.5m high traditional rectangular huts (*tranoboroke*, *tongalibaty*) of the Mikea, and is still used for huts (*baibo*) in the cultivated fields (Stiles 1998; Philipson 2002, personal communication)." Wickens and Lowe (2008: xvii) provided two photographs of baobab wood used for roofing and siding. One shows a large area with sheets of *Adansonia rubrostipa* drying in the sun, and the other shows a building with a roof made from an unidentified *Adansonia* species.

The Baobab and the Reasons for the Hadza Relocating Their Camps

The Hadza move their camps within their territory from one known camping environment to another. As often noted, there are many reasons for the Hadza's choice of camping sites and for the size, social life, duration, and movement of their camps. The general view, however, is that the Hadza relocate their camps for protection, enhanced foraging, the joy of being on the move, the pleasure of socializing, and the benefits of traditional and current commercial trade (Woodburn 1968b, Marlowe 2010, Blurton Jones 2016).

Protection

Security has likely been an important influence on the evolved capacity for the camping-based gendered division of labor characteristic of hominins, including immediate return foragers like the Hadza. In the early twentieth century, security was noted as an important consideration in the relocation movements of Hadza camps. As Woodburn (1968a, b) pointed out, hilly places with rocks and baobabs must be unobtrusive in the sense of off-the-beaten-track of strangers and aggressive neighbors. The Hadza have been subject to interspecific competition with neighboring groups, the activities of European colonists in the nineteenth and twentieth centuries, and current interactions with government officials, missionaries, scholars, and tourists. The Hadza also relocate their camps to protect themselves from inclement weather, harmful animals, competing neighbors, social conflict, and places where there has been serious illness or death. The deterioration of their branch-framed house is also a contributing factor to Hadza camp relocation.

Wellness, broadly concerived, is concerned with such things as prosperity, health, ease, comfort, and happiness, and it is often associated with protection. Wellness is an ever-present concern for the Hadza as it is for all people. Of course, the concern with wellness could also be stated negatively as avoiding situations associated with deprivation, illness, hardship, discomfort, and sadness. Obviously, wellness concerns would have influenced the movement and siting of camps for early hominins as it does today for the Hadza, and camp relocation to ensure wellness includes such things as the plesure of being on the move, leaving behind a place of sickness or death, and the fresh start that comes with living in a new place. Although Marlowe (2010:41) thought the "main constraint on camp locations is water," he wrote "camps are often up on a hill where there is a breeze, a view, and good shade. Camps are not located in the middle of very wet areas with many mosquitoes. Around the springs in Mangola, for example, camps are up on the nearby hills, which even during the rainy season are much drier and windier and consequently free of mosquitoes."

As earlier noted, the grass dwellings of the Hadza are minimally constructed during the dry time and are often reduced to wind breaks. Sometimes the Hadza sleep in the open, especially when they are harvesting berries. But roof leaks and wet floors become a problem during the main November-December spring rains and the March-April autumn rains. At these times, some Hadza camp in places with plentiful grass and branches for sturdy branch-framed houses in well-drained locations, while others move their camps to rock shelters in the hills. During periods of heavy rainfall, the accumulating waste of the camp – excreta, garbage, and swept debris – produces an increase in insect pests and, as Woodburn has noted (1972:204), "lice, mosquitos, white ants, etc. – multiply and become troublesome." The Hadza, we learn, "are less disturbed by these conditions than might be expected, but they do play some part in encouraging frequent movements, especially in the wet season when insects are at their worst and dirt is more obtrusive."

Protection from prying eyes is hard to achieve given the intimacy of the forager way of life. No doubt, protection for the Hadza, especially in the large camps of the dry time, must also include some measure of privacy. Marlowe (2010) has reported, for example, attempts by Hadza hunters to hide honey for family consumption. Presumably, in the smaller, more kin-oriented camps of autumn, privacy would be less of a consideration.

Resource Acquisition

Resource availability is the fundamental reason for Hadza camp moves. The Hadza relocate their camps to enhance foraging both in response to the seasonal changes in the availability of essential resources and to the depletion of resources around existing camps. As a resource-rich environment, the multipurpose baobab is an important factor not only in the siting of Hadza overnight and residential camps, but also in Hadza mobility, including camp relocation, resource forays, inter-camp travel, and the fusing and fissioning of camps. It is well documented in the literature on mobile foragers that the longer the duration of their camp, the greater the local resource depletion, measured in both availability and procurement costs. Procurement costs are especially greater because of the ever-increasing distance of resource forays (Kelly 1983). Consider, for example, the account of Cooper (1949:9), a British game ranger between 1945 and 1947 who spent ten days with the Hadza during the dry time of year on two separate occasions. He observed that during the long fruiting period, baobab camps were "occupied a week or ten days till local supplies of baobab fruit are exhausted." He went on,

> I have seen the women out collecting the seed pods from the baobab trees, which provide their principal source of sustenance for about five months of the year [which is the end of autumn in June to the early spring on October]. The cream of tartar pulp, together with the seeds, are pounded on a flat stone till reduced to the consistency of meal. This is then mixed with water and cooked very slightly. When the supplies of baobab fruit fail [from the mid-spring of October to mid-autumn in May which is the rainy time of the year], the Kindiga are dependent on certain roots, of which I am unfortunately unable to give the species. I have not seen them collecting these, since each of my two visits coincided with the height of the baobab season.

According to Kelly (1995), Hadza women forage two to six hours per day in the dry time for an average foraging time of four hours. They forage four to eight hours per day in the wet time for an average of six hours. We can conclude that a foraging time of five hours is toward the upper limits in the dry time and the lower limits in the wet time. In the dry time, especially during the annual peak harvest in the winter dry season, Hadza camping sites, mobility patterns, and resource forays are determined by the seasonal availability of the baobab. I would hypothesize that in the early period of the dry time, especially in the winter dry season when most baobab fruits fall from the tree, foraging time would be closer to two hours. At the height of the dry time, from the end of winter in September to early spring in October, foraging time would likely be closer to six hours. At a distance of six hours of travel time, we can assume that baobab forays would be more intensive, as evidenced by field processing, where only fruit pulp and seeds are brought back to camp.

Copper's account tallies with the general view that the baobab harvest is the dry time from the end of autumn in early June to early spring in October, but the peak harvest period is from the end of autumn in June through the start of spring in late September, with supplies declining from the early spring of October to the early autumn of April. For this reason, the supply of baobab fruits at the height of the September-October dry time is highly variable. This explains why the fieldwork of Hawkes et al. (1995:694) recorded that baobab fruit was plentiful in the "late dry season" of 1986 and scarce in the "late dry season" of 1988. This meant the choice the Hadza faced was tubers close to camp and berries at some distance from camp. The end of the September-October period is the turning point when the Hadza leave the baobab-clad rocky areas for the floodplain areas, where they are closer to the berries they will harvest from mid-spring through early autumn. It is likely that as the cool, windy, winter dry season comes to an end with the approach of the spring rains, increasingly intensive baobab harvesting forays become necessary. The Hadza, therefore, are more likely to engage in the field processing of baobab fruits at this time, with only pulp with embedded seeds taken back to camp.

In addition to fruit, the Hadza site and move their camps to take advantage of other baobab resources, especially honey and water (Woodburn 1968a, b; Marlowe 2010). In this regard, as earlier noted, the Hadza move camps because all baobab water has been consumed or to take advantage of baobab trees with water, particularly after the autumn rains, when seasonal runoff is in decline and pools are drying up. They also move camps to improve their hunting success, to gather roots and fruit other than the baobab, and to acquire the diverse raw materials needed for their hunter-gatherer lifestyle, especially foraging equipment, firewood, and the flexible branches and plant-covering material used in constructing their dwellings. Residential moves also occur to take advantage of socializing opportunities including initiation rites, gambling, and singing and dancing, and in association with inspirational life, especially religious activities, and storytelling. Woodburn (1968b:106) noted that the Hadza enjoy being on the move and these days they increasingly move their camps to take advantage of opportunities for commercial exchange.

Trade

The movement of Hadza camps to seek opportunities for trade probably has a long history, considering the importance of metal and gourd containers in their practical and inspirational life. Today, camps move to pursue exchange possibilities reflect an increasing commercial dimension to Hadza life. For example, recent accounts make a distinction between "Hadza bush camps" and "Hadza tourist camps." In an article that discussed the destructive impact of farmers, herder, scholars, missionaries, tourists, and state officials that "threaten to wipe out some of the world's last hunter-gatherers," Gibbons (2018:703) noted that "To survive some Hadza take handouts of maize flour from missionaries or trade meat and honey for flour to make porridge. Or they head to one of a dozen "tourist camps" in the Mangola region, where they earn money by re-enacting their traditional ways." Gibbons (2018:704) also noted that researchers are well aware of the influence of their studies on the development of Hadza tourism:

> In the roughly 3 weeks that ecological anthropologists Haruna Yatsuka of Nihon University in Mishima, Japan, was in a tourist camp in Mangola in 2013, 40 tourist parties came from 19 nations. The tourists began arriving at 6 a.m. and watched the Hadza hunt (for show – they seldom got meat when with tourists), dig up tubers, or perform dances. In one camp, Hadza wore baboon skins, which is not their traditional dress but fits tourist expectations, Leach says. The Hadza got money by selling souvenirs such as bead bracelets, or from tips. "Tourism now brings income to the Hadza and has had a tremendous effect on their livelihood, diet, residence, and nomadic patters," Yatsuka says.

Some Hadza camps move to be close to farmers during the May-June autumn harvest. Woodburn (1972:203) notes: "Sometimes also neighbours' harvests attract Hadza camps to the area to feast off cereal obtained by begging." Farmers also engage the Hadza to protect crops, and they use the opportunity for ambush hunting.

The Baobab and Rocky Places

Natural stone is highly variable in size, shape, composition, and durability, and occurs in ready-to-use forms that are put to a variety of uses. We have noted that for a number of reasons including food, water, privacy, security, shelter, and overall well-being, the Hadza prefer to camp in tree-shaded rocky hills where resource-rich baobab trees abound. Rocky hills provide the lithic technology necessary to process the baobab fruit pulp and seeds and the stones necessary for resting cooking pots to make baobab porridge. Stone use as an aspect of Hadza material culture should be broadly understood to also include accessing water from rock seeps and rock pools including those that form in shallow depressions of boulders. Marlowe (2010:77–78) mentioned the Hadza throwing stones as a weapon, and they probably have occasionally done so also to reap fruits. In addition to those used in cooking fires, stones of domestic life include seats and sharpening stones. There is also the inspirational significance of stone, such as its association with ornaments, rock paintings, and smoking pipes. One of the earliest and most important stone implements of the archaeological record is the stone chopper. Its importance in hominin evolution is evident in Oldowan stone technology of choppers and flakes identified with the evolution of the genus *Homo* around 2.5 million years ago. Chopper and flakes have been associated not only with cutting meat and processing skins, but also with cutting, scraping, and boring plant materials.

Hadza domestic life is often portrayed in photographs of women and children sitting in a circle atop a huge boulder or rock slab pounding baobab fruit pulp and seeds to make the highly nutritious baobab flour. Rocks and baobabs go together. This includes both in-place rocks and loose stones. For example, "Hammerstone" is an archaeological term frequently used in discussing the use of percussion fracturing to produce flake tools, such as blades and projectile points, and core tools, such as the hand axe. Marlowe (2010) has adopted this term in his discussion of the Hadza. For clarity, however, I distinguish between hammerstone and anvilstone, grindstone and flatstone, and pestlestone and mortarstone. The stone pestle-and-mortar is not used by the Hadza. And instead of the hammerstone and anvilstone, I would place emphasis on the grindstone and flatstone when discussing the Hadza, with the recognition that the grindstone and flatstone are also used as a hammerstone and anvilstone. While angular or round stones can be used as hammerstones, Marlowe has pointed out that the preferred hammerstones of the Hadza are smooth, round, hand-held stones that the Hadza collect and transport, although they are often found in place from earlier camps. As a general term, "flatstone" has been adopted here to include natural grinding slabs and in-place natural flatstones such as surface rocks and the top of large boulders.

Processing the fruit of the baobab involves breaking into its indehiscent woody fruit pod, extracting the hard, dry pulp in which the seeds are tightly embedded, pounding the pulp to separate it from the seeds and to dehull the seeds, and grinding the pulp and seeds to make flours, which are used separately and together. As Marlowe (2010:79) rightly pointed out, the Hadza do not need a hammerstone to break the hard woody fruit shell of the baobab. They can be stomped on it, or they can be smashed against a tree or rock, or against the hard ground typical of the dry time of the year when the tree is in fruit. But the Hadza do need a large cobblestone for use as a grinder, which also serves as a hammerstone. It is noteworthy that chimps use throwing stones as weapons, stone hammer and anvil (as well as other things used as hammer and anvil), seed-extracting stones and twigs, and stone wedges. Unlike Pleistocene *Australopithecus* some 1.5 million years ago, however, they do not use a grindstone/flatstone for processing the seeds of the baobab and palm. As earlier noted, O'Connell et al. (1991:65–68), in their paper titled "The Interpretation of Archaeological Spatial Patterning," described Hadza camps as having "household, communal, and special activity areas":

> *Special activity areas* included bedrock grinding facilities, and defecation areas. The former were large, relatively flat-topped outcrops of granite or schist, whose surfaces were marked by small pecked and polished patches, 30 to 50 cm diameter, where baobab and other hard seeds were ground. Small handstones were frequently found in association with such features. All residential base camps had these facilities, which suggest their presence was a determinant of camp location. Defecation areas were located at the margins of camp, often forming a discontinuous ring around it.

Marlowe (2010:79) has also pointed to the centrality of such rocky sites in the Hadza economy. "Camps," he noted, "are almost always located where there are big, exposed smooth rocks that women can sit on and use as anvils." Reaffirming the point made by O'Connell et al. (1991), Marlowe continued by noting that the rocky substrate in Hadza life "is something archaeologists might keep in mind when surveying possible sites."

Marlowe (2010:74–78) also identified the presence of "rock art" (p. 18) in the Hadza environment and the importance of "rock shelters" (pp. 95–96) and other stone structures in Hadza life. Based on Wendell Oswalt's (1973) system for describing, naming, and classifying tool kits, and under the heading of Hadza tools, Marlowe (2010:74–78) described a stone structure "artifact" identified as a "tended facility" used, like "grass," "branches," and "leaves," to conceal hunters.

Where humans differ from apes and monkeys with respect to the baobab and other resources such as water, honey, fiber, and roots is that more intensive use of equipment enables greater access to these resources. Chimps and baboons eat baobab fruit pulp, but they cannot access the seeds as humans do. Chimps climb the baobab for honey but find it difficult to open the hive. These primates also have greater difficulty in obtaining tree water or digging deep roots. The more intensive resource acquisition of humans based on technology, which results in such things as greater access to baobab seeds, water, and honey, means greater reproductive success as evidenced by population increase and widespread eventually global migration.

Given the shade, water, food, and other resources the baobab provides, and the importance of stones in the processing of baobab fruit, fruit pulp, and seeds, it is understandable that baobab-clad rocky hills, with a view and a good breeze to drive away mosquitoes, are the preferred camping site of the Hadza, especially in the cool winter dry season when the baobab harvest is at its peak. I agree with Marlowe (2010:41): "The placement of Hadza camps merits study by paleoanthropologists and archaeologists interested in the most likely places to search for fossil and archaeological remains. If central places were being used, and I suggest they were from very early on … they should have been located in places like those the Hadza choose for their camps." Baobab trees growing in the rocky hills are key to understanding the dry-time survival of the Hadza in their mosaic savanna environment, and this means the placement of Hadza camps also merits study by ethnobotanists and ethnobiologist generally.

In an article broadly concerned with Hadza land use patterns, especially camping sites, Mabulla (2003:33) presented an ethnoarchaeological examination of the use of rock shelters that included "factors affecting rock shelter choice, times of occupation, group size, activities carried out, and discard patterns." Mabulla's explicit aim was to use "the Hadzabe pattern as a model for Middle Stone Age … and Late Stone Age … foragers in the Eyasi Basin." Unfortunately, he made no effort to address the distribution and use of the baobab tree in relation to rock shelters. Yet we learn that on the floors of six of the forty-two rock shelters he examined, he found "remains" of baobab fruit, bark~~s,~~ and seeds as well as "cereals (e.g., corn, millet), and animal bones" (2003:42). And in his discussion of Hadza camps "near Masiedha-Sipunga hills and their rocky outcrops," Mabulla wrote: "In 1993 we found one abandoned base camp that had been occupied during the 1992 rainy season. This … camp was composed of ten shelters. It was located near rock shelters of the hills in an area rich in wild plant foods …. Baobab trees surrounded this abandoned base camp."

The rocky landscape in which baobab trees are especially concentrated and the exceptional resource-rich environment they provide suggest they would have influenced early hominins in their choice of camping sites and their reasons for relocating camps. The baobab would indeed have been a multi-dimensional tree of life in the development of the forager way of life that was key to the evolution of hominins. Support for this line of thought is evident today in the simple fact that human beings worldwide continue to be deeply attracted to the baobab, even if only for its inspirational value as a manifest tree of life.

The Seasonal Size and Characteristics of Hadza Camps

There is no doubt that rainfall seasonality exerts a powerful influence on Hadza life both directly, given water's many essential uses, and indirectly through its influence on the terrestrial environment, especially on the growth and reproduction of plants and animals. According to Woodburn's wet season/dry season framework, Hadza camps are small, dispersed, and gender-integrated in the "wet season" compared with the "dry season," and as earlier indicated, Woodburn's view is widely represented in the literature (e.g., Marlowe and Berbesque 2009; Marlowe 2010; Blurton Jones 2016:15). Marlowe (2010:40), citing Woodburn (1964), attributed the fact that "Hadza camps tend to be larger in the dry season" mostly to water scarcity, and this is the point emphasized by those who frame their discussion in terms of the wet season/dry season polarity. According to Marlowe (2010:41), "In the rainy season, drinking water often comes from pools of water on top of big flat rocks and from running streams," but from the mid-autumn of May through the October end of the dry time, life changes for the Hadza. We learn that "One important reason for moving camp is that the drinking water is drying up."

However, changes in camp size between the wet and dry times of the year are probably not as significant as changes associated with the contrast between the March-April autumn rains, when resources are dispersed, and the May-February resource concentrations of winter, spring, and summer. Based on the seasonal perspective presented in this work, it is far more accurate to say Hadza camps are generally small, more widely dispersed, and most gender-integrated during the peak annual rains of autumn, when the seasonal emphasis of Hadza foraging is on honey and small game for men, and roots for women. For the Hadza, it is the seasonal changes in the concentration and dispersion of resources such as water, game, and plant foods that is important, not simply the seasonal changes involving resource abundance and scarcity or, for that matter, changes in the seasonal availability of water. The most important kinds of resource concentrations are baobab fruits in the cool winter dry season, large game from water-hole ambush hunting during the hottest time of the year, the September-October height of the dry time, and the November-February height of the annual harvest of *Cordia* berries through the spring rains and into the summer dry season. If camp size were determined only by wet or dry periods of the year, and if large camps occurred only in the dry time, then during the spring rains Hadza camps should be small. But they are not because of the abundance of the berry harvest. Woodburn (1968a:104) wrote:

> At the peak of the dry season, when water is scarce and when the berries of *Cordia gharaf.* Ehrenb. and *Salvadora persica* L. ripen in huge quantities near to these few sources of water, the majority of the inhabitants in each of these areas may be found concentrated in a few large camps within perhaps two or three miles or less of each other and close to the berries and water supplies of their area.

Smaller, more mobile camps are especially associated with the autumn rains, when resources are dispersed. Resource foray distances are longer at this time of year, and socializing, especially related to religious activities, is much reduced. We know that foraging time is important to the Hadza, for as Marlowe (2010:40) reported, women expressed their desire to move when local resource depletion made their foraging distance too great. The fact that foraging time is longer in autumn means the Hadza are in a sense putting out a greater effort at this time of year. Moreover, the smaller camps of the autumn rains make sense also, since the chief resource then is honey, which is available only in relatively small quantities. Honey is better shared without conflict when families forage together in small camps. Controlling honey is especially important, for as Woodburn (1968a, b) reported, Hadza wives use the honey their husbands acquire as a trade good.

Some researchers report a significant difference in the quality of social life between large and small camps, and they emphasize that large camps engender more bickering and sex segregation. They think that during the annual rains, when water again becomes widely available, the Hadza are glad to leave behind their difficulties for the more enjoyable experience of living in smaller groups. But large and small camps both have advantages and disadvantages. The view that bickering is more prevalent in the large camps of the dry time, while calm prevails in the small camps of the rains, overlooks the fact that large camps also offer increased opportunities for social activities, including gambling, the Epeme Dance, and the initiation of girls; this view also overlooks the fact that at the spring beginning of the annual rains, the Hadza live in large camps amidst their *Cordia* berry patches. It is the resource dispersion of the March-April autumn rains that requires the Hadza to live in smaller and more separated camps and to depend more on the hunting of small animals and the gathering of honey and roots.

Summary

The focus of this chapter on the baobab's association with Hadza central-place residential camps is yet another example, among many, of how the baobab enters into the Hadza forager way of life. The baobab is not only a part of the Hadza's ideal camping environment and their preferred campsite, it also figures in intriguing ways in their living spaces, bedding, and branch-framed houses. The link between the baobab and Hadza residential life is yet another factor that could have been significant in prehistory.

Chapter 24
Hadza Influence on Baobab Regeneration

It was earlier suggested that the baobab's usefulness made hominins largely responsible for its dispersal throughout the dry lowlands of tropical and subtropical Africa, and in wet areas and high elevations where it would not normally occur. While it is reasonable to assume that the baobab supported our ancestors over the course of hominin evolution, it is not as well recognized that we and our ancestors may have been just as important to the success of the baobab, for an equally long period of time. The human/baobab relation is a mutually beneficial one. To say, as Harlan (1976:11) did, that the "baobab is not man dependent, but man can sometimes be baobab dependent" obscures a key point: in some situations, the baobab is man-dependent because of the benefits of human incidental encouragement and intentional assistance, the latter being the precise theoretical definition of cultivation. This co-evolving relation – of humans settling at naturally occurring baobab trees and the unwitting aid and cultivation provided by humans in their settlement environment – accounts for the presence of the baobab in the landscape of foragers, farmers, herders, and urbanites.

A number of studies of the baobab/settlement association show baobab regeneration is most successful in villages and their immediate surroundings, as opposed to other anthropogenic environments and most natural areas. Baobab parklands, especially those with concentrated baobab groves, stands, and woodlands, have long been regarded as evidence of a previous human settlement. So obvious is this correlation that instead of arguments about the baobab being linked or not to anthropogenic environments, the discussion has turned instead to understanding the precise ways in which baobab regeneration is determined by human activity and their built environment. In this context, the following question is frequently asked: Do Africans normally establish villages where there are existing baobab trees, or do baobab trees spring up after villages have been established? This chapter discusses the likely influence of central-place foraging on the regeneration of the baobab in the Hadza environment, especially their routinely used residential areas. This discussion of the Hadza is in the context of a long history of speculations regarding the relationship between the presence of the baobab in the environment and past and present human

J. Rashford, *Baobab*, https://doi.org/10.1007/978-3-031-26470-2_24

settlements. The baobab/settlement association includes the concept of settlement vegetation, the fate of young baobab trees in such vegetation, and the positive influences of Hadza foraging on the successful reproduction and spread of the baobab.

Baobab First or Settlement First?

Munyebvu (2015:27) wrote in her account of baobab populations in the Omusati region of Namibia, "Several studies have come up with varying conclusions about the factors that affect the population structure and abundance of the tree species near and around human settlements but there is no definite factor to explain such patterns." Despite this assertion, definite factors have been offered to explain the presence of the baobab in the human environment. The least helpful approach is to assume there are only two mutually exclusive alternatives: either the settlement was first, followed by the baobab, or the baobab was first, followed by the settlement. This is how Assogbadjo et al. (2005:51) seemed to frame the issue when they wrote, "It remains debatable whether man comes because of the baobab or man was responsible for dispersing the seeds around villages." This rigid, either/or perspective is without sound empirical basis. In fact, both situations occur, and overlooking this obscures the complex nature of the settlement/baobab association that current field research is bringing to light. Human settlements are established near the baobab, as is clear with foragers like the Hadza and San, for example; but there is no doubt that baobabs become established because of the existence of human settlements, including those of the Hadza and San.

In his comment on West Africa in which he emphasized climatic difference, Allison (1962:248) responded to the "tree first" or "settlement first" issue by accepting that both occurred. He argued, however, that in the drier Sudanian savanna zone, settlements follow the baobab, while in the wetter Guinea Zone, the baobab follows settlements. Allison's (1962) view of the baobab following settlements suggests people would be less inclined to plant the baobab in wet areas, where natural regeneration would likely be higher than in dry areas. Recent studies indicate that the baobab is incidentally dispersed by humans and cultivated in both the Sudanian and Guinea Zone, but the irony is that the tree is more valuable to people in the Sudanian savanna than it is to those in wetter Guinea Zone.

Harlan (1976:11) accepted that both the "tree first" and "settlement first" scenarios occurred, but said he did not know the comparative occurrence of the two. What he did emphasize, however, was a settlement-first perspective characterizing baobab regeneration in the human environment as resulting from the incidental, harvest-based dispersal of seeds (hereafter referred to simply as incidental harvest dispersal). Harlan (1992:64–65) wrote the following of the genus *Adansonia* in his account of cultivated plants that were "intermediate" between being completely wild or fully domesticated:

> There is a class of plants that came to be closely associated with man, but without evident genetic modifications. One example is the baobab (Adansonia spp.), a tree widely distributed through the savannas of Africa, South Asia, and northern Australia. The plant can become very large and is useful in many ways…. In Africa there is a fairly close association between village sites and stands of baobab…. Certainly, if fruits are repeatedly brought to a village it would seem that new seedlings would inevitably become established, and since they are recognized and protected, stands would develop.

The baobab stands associated with villages that Harlan described were actually cultivated, since they were "recognized" and intentionally "protected." Intentional assistance is the definition of cultivation, even if it only involves protection.

Wickens (1982), like Harlan, recognized the possibility that both situations occurred. He acknowledged that settlement first was the prevailing view, but also expressed his strong conviction that in some cases human settlement followed the baobab. For the baobab-first thesis to make sense, however, the baobab would have to be naturally occurring in natural areas. Unfortunately, neither Wickens (1982) nor Wickens and Lowe (2008) discussed the relation between the baobab and the settlement of foragers. Nevertheless, for those like Allison, Harlan, and Wickens, who accept both perspectives, the difficult task has always been to distinguish when a particular settlement/baobab association, whether involving foragers or food producers, was an expression of the "tree-fist" or "settlement-first" scenario.

It is clear that the discussion of humans settling at the baobab and the baobab springing up in human settlements has meant different things to different authors. But instead of the either/or approach, or the perspective that both occurred simultaneously, Sidibe et al. (2002) gave the whole subject a historical interpretation, claiming that in the past it was "tree first," but now it is "settlement first," because baobab regeneration is essentially the result of cultivation. Ultimately, however, all three perspectives – the either/or approach, the argument for simultaneous occurrence, and historical sequencing – are concerned with the reason for the familiar presence of the baobab in the human environment. What is significant in this discussion, however, is that those who argue for baobab first or settlement first share the same widespread assumption that the presence of the baobab on the African savanna, especially at high densities, is associated with human settlement. This chapter discusses the likely influence of the Hadza as foragers on the reproductive success of the baobab in their environment, especially in residential areas.

Settlement Vegetation

In considering the baobab in Hadza foraging as a window on its significance in hominin evolution, one of the most interesting questions that arises is the way baobab regeneration is associated with human beings. This question is ultimately about the nature of human settlements and the kinds of settlement vegetation to which they give rise. Settlement vegetation identifies all the plants that constitute the botanic dimension of any human environment whose presence is determined both by natural occurrence and by the positive and negative selective pressures of human activities and the built environment with which these activities are associated (Sauer 1947, 1952, 1956; Anderson 1954, 1956, 1960; Lathrap 1977; Rindos 1980, 1984; Rashford 1991b, 1994c). Wherever the development of settlement vegetation occurs, the human response to wild plants in their environment, including those that are directly or indirectly human-incidentally dispersed, is particularly important. These wild plants may be rare or common. The common wild plants can be further distinguished as being familiar, weedy, weeds, or invasives. Positively selected wild plants, those that are only incidental beneficiaries of hunter-gatherer activities – especially benefits resulting from harvesting fruit and dispersing seeds by refuse disposal and defecation – would be familiar settlement volunteers, or they might even be weedy. But they would not be weeds or invasives, which are properly categories of agricultural landscapes.

Whether rare or common, wild plants are destroyed when they are harmful or when they interfere with human activity. This is particularly true of weeds that successfully compete in cultivated landscapes and invasives that take over various areas of the settlement environment as well as natural landscapes. They are overlooked or unattended when they spring up in marginal areas; they are accepted when they spring up in active areas of settlement where they do not interfere; and they are positively favored by intentional preservation and protection and by other forms of cultivation when they are perceived as useful in some way (Kimber 1966, 1973, 1978). Intentional assistance is essentially the cultivation of crops, and transplanting, seed planting, and vegetative propagation are among its most important expressions. Settlement vegetation is a key component of the origin of agriculture, but the origin of agriculture, defined as deliberate assistance given to plants, is not synonymous with the origin of settlement vegetation. Long before cultivation and animal husbandry, hunter-gatherers were inadvertently creating settlement vegetation. The Hadza, like all foragers, create settlement vegetation by the positive and negative consequences of their activities on the wild plants associated with their residential environment, paths, and areas in which they hunt and gather. In light of the long-distance trade that accompanied the rise of agriculture and the making of our modern world system (Brockway 2002), settlement vegetation for most people around the world today identifies all wild and domesticated plants, both native and exotic, that are found in their environment and whose presence is determined by the combined outcome of natural and human influences, incidental and intentional (de Candolle 1886; Vavilov 1926).

The Fate of Young Baobabs

A regenerating baobab population is one in which young trees are present in size classes that include seedlings, saplings, and small trees, but there has long been a discussion in the literature about the absence of young baobabs throughout its range. Recent studies in West, East, and Southern Africa show that differences in natural areas and in types of land use do have a significant impact on the distribution of the different size classes of baobabs that represent the structure of a baobab population. Based on size class distribution, we can judge a population to be thriving when there are many young trees present; this has been variously identified as successful regeneration, recruitment, or rejuvenation. We can judge a population to be holding on when there is low recruitment, represented by mostly mature trees; and we can judge a population to be declining when there are no young trees. Baobab trees are most vulnerable when they are young, especially as seedlings, and an abundance of seedlings coupled with the absence of saplings and small trees is an indication of low seedling survival. For some, the reported absence of young baobabs in the landscape was evidence that regeneration was poor or not occurring; this raised concerns about the survival of local baobab populations and about the long-term survival of the species (Romero et al. 2014; Swanepoel 1993; Assogbadjo et al. 2005; Edkins et al. 2007). For scholars like Ndoro (2013:34), however, the absence of young baobabs was not a cause for concern, since "a recent study carried out in South Africa by Venter and Witkowski (2010) pointed out that for long-lived species, such as the baobab tree, recruitment and mortality might be episodic events, as the baobab population in the area was stable and had healthy numbers of mature baobab trees."

Some have attempted to explain the seeming paucity of young baobabs in the environment by arguing that young baobabs are indeed present but unrecognized, especially by visitors to Africa who have written about the tree based on limited experiential knowledge of the landscape. Baobab seedlings and small saplings bear little resemblance to mature trees. This is so not only because of the obvious difference in size, but especially because these young baobabs have simple leaves rather than the palmately compound leaves typical of older trees. "Unrecognized seedlings" as the basis for the many reports of the absence of young baobabs in the environment is not persuasive, however, since even writers with significant experience in Africa have reported the paucity or complete absence of young baobabs. For example, Sweeney (1974:56) wrote:

> One curiosity of the baobab, remarked upon by many travelers, is the apparent absence of seedlings and young trees…. After some months of work on the association of insects with the trees the same thought occurred to me. Why had I never seen a really young one? The smallest I had found by then was taller than a man. I searched locally without success and came to the conclusion that goats, those ultra-destructive creatures, had eaten them all. Eventually, high on a *jebel*, I discovered a tiny baobab; after that I found a dozen or more seedlings scattered in places well away from the goat pastures.

Recent studies suggest that it is not flowering, fruiting, seed viability, germination, or seedling establishment that most significantly influence baobab regeneration, but seedling survival, and the survival of saplings and small trees. According to Venter and Witkowski (2013), baobab regeneration was not "seed limited" in terms of seed availability and dispersal, but "microsite limited," especially with respect to rainfall. And while young baobabs are destroyed by drought and fire, there is growing evidence that the most significant causes of their mortality are the predation by herbivorous arthropods and mammals, the latter including elephants and wild and domesticated ungulates, which also trample young plants; root-eating baboons and human beings; and disease (see also Guy (1982) and Venter and Venter (1996)). As Sweeney (1974:56) indicated, it is especially the livestock of African mixed farmers, agropastoralists, and pastoralists that explains why young baobabs become scarce in settlement environments and why the protection of young baobabs has been a crucial factor in the cultivation of the species throughout its range. The correlation between the presence of young baobabs in the landscape and the absence of wild herbivores and livestock is supported by the simple fact that young baobabs are frequently encountered on rocky outcrops and hillsides associated with baboons as likely dispersal agents rather than in more exposed situations, such as floodplains.

Foragers and Baobab Regeneration

The results of current research show that land use has a significant influence on baobab regeneration, and this includes the difference between relatively natural areas and intensively used human environments. But these studies have only focused on the landscapes of farmers and herders and on natural areas such as plains, rocky elevations, nature reserves, and parks. Because they overlook the relationship between foragers and baobab regeneration, they are largely devoid of evolutionary content. As such, they provide little insight into the long prehistory of interaction with the baobab by hominins as camping foragers rather than as villagers. But it is precisely in the context of an evolutionary view that the Hadza/baobab and San/baobab relations become ethnographically and theoretically important. These foragers of the mosaic African savanna offer a window on how our hominin ancestors might have interacted with the baobab as humanity's ancestral tree of life. Human beings have exerted a variety of positive and negative selective pressures on the reproduction of the baobab; it is the net outcome of these influences that indicates the overall impact of human activity on baobab regeneration.

Darwin's (1871:167) view on the origin of agriculture seems to minimize the influence of the hunter-gatherer way of life on the dispersal of plants. He speculated that nomadic foragers could not develop agriculture because sedentary life was a pre-condition for such development:

> Nomadic habits, whether over wide plains, or through the dense forests of the tropics, or along the shores of the sea, have in every case been highly detrimental. Whilst observing the barbarous inhabitants of Tierra del Fuego, it struck me that the possession of some property, a fixed abode, and the union of many families under a chief, were the indispensable requisites for civilisation. Such habits almost necessitate the cultivation of the ground; and the first steps in cultivation would probably result, as I have elsewhere shewn [Darwin 1868:309] from some such accident as the seeds of a fruit-tree falling on a heap of refuse and producing an unusually fine variety.

However, Darwin (1871:167) thought the problem of the transition from foraging to farming and herding was "at present much too difficult to be solved." Instead of Darwin's characterization of prehistoric foragers, which seems to suggest that they were perpetual pioneering foragers, it could be argued that human beings have been home-range foragers, even when they occupy a new environment for the first time. This means a significant anthropogenic environment is not based solely on a fixed abode, as Darwin suggested, but on the strategic seasonal camping of prehistoric foragers in their familiar home range, especially during the Mesolithic period leading up to the end of human global dispersal as foragers and the worldwide origin of agriculture (Cohen 1977, 2002). The archaeological evidence over the past twenty thousand years for the period identified as the Broad Spectrum Revolution suggests home-range foraging would be especially likely to develop in response to the social and environmental circumscription and intraspecific competition that accompanied the rise of agriculture (Carneiro 1970).

As nomadic home-range foragers, the Hadza's settlement activities impact the dispersal of baobab seeds and their reactivation, sprouting, and seedling establishment. It is particularly important to note that Hadza settlement activities also impact the survival of baobab seedlings, saplings, and small trees. In fact, the survival of young baobabs is an important factor in the regeneration of the baobab in the human environment. The distinguishing feature of foraging as a mode of adaptation is that foragers do not intentionally assist plants and animals. It is now clear after almost one hundred years of hunter-gatherer studies that foragers like the Hadza are expert ecologists. They have an intimate knowledge of the useful, harmful, or otherwise noteworthy plants and animals of their environment, which they mark with names. Since the Hadza do not grow crops or raise livestock, we can assume that they do not intentionally assist the baobabs of their environment by transplanting or seed planting. In fact, there is no evidence that they deliberately assist baobab trees in any way, except perhaps in the form of minimal intentional preservation.

The areas of their home range to which foragers seasonally return would favor some wild plants, and foragers like the Hadza and San have probably been promoting the regeneration of the baobab in their environment by incidental encouragement and intentional preservation. There are two major forms of incidental encouragement foragers can be theorized to have on baobab regeneration in and around their camps, along their paths and in their hunting and gathering areas. In addition to the incidental harvest dispersal already described, incidental encouragement also includes advantageous growing conditions produced by the disturbed soil of the Hadza residential environment, as well as the enhanced soil fertility produced by the refuse disposal of their dwellings, community areas, and special activity areas, especially the ash from their many hearths and the nutrients deposits of their defecation. Incidentally advantageous growing conditions are also produced by irrigation, given the centrality of water in human life. We can assume, therefore, that over a long period of time the fertility of the soil in camping areas would be enriched and better watered compared with areas where the Hadza do not routinely camp; this would provide a niche for baobab recruitment. Dhillion and Gustad (2004:94) presented a comparable situation for Cinzana, Mali, where baobab seeds were discarded in refuse areas just beyond village boundaries. In their study of local management practices that influenced baobab regeneration, they wrote, "As the seedling density in plots adjacent to the [boundary] wall was, in some cases, significantly higher than recordings further away, these numbers were omitted from calculations of cropland seedling density. Seedlings adjacent to the wall were more appropriately part of the village areas and the cropland proper tended to begin away from the village wall."

In addition to incidental harvest dispersal and incidental advantageous growing conditions, there is yet another way in which the Hadza use of the baobab in house building might impact the regeneration of the baobab, and it involves the vegetative propagation of the species. Traditionally, the baobab develops in the human environment from tended and untended volunteers and from seed planting. But in light of the growing commercial uses of the baobab, especially the economic value of its fruit and leaves, various asexual techniques for propagating the species have been reported in the literature. These studies have shown that the baobab can be propagated vegetatively by the familiar techniques of cuttings, air layering, budding, grafting, tissue culture, and micropropagation (Assogbadjo et al. 2008; Gebauer et al. 2016; Maranz 2009).

Baobab sapling poles are in effect a kind of cutting that would especially be used during the rains when the Hadza are most in need of shelter. We have already noted the use of baobab branches and sapling stems to frame the Hadza branch house. In the process, holes are dug in which the branches and stems are set and the soil around them compacted, sometimes with stones. Baobabs are known to reproduce from cuttings. One cannot but wonder if these cuttings "'planted' into holes in the ground" (Clarke 2018) do not sometimes take root and grow into multistem baobab trees. It would be interesting to learn from the Hadza if they have ever observed the baobab poles of their branch-framed homes growing into new trees. It would also be interesting to survey past residential sites to see if such incidental propagation can be detected in the landscape.

Sidibe et al. (2002:52) reason that "since pruned branches frequently sprout young leaves when minimal conditions are present [a phenomenon I myself have often observed]," it must therefore be the case that the tree can be easily grown from cuttings. The establishment of baobab trees from baobab poles seems unlikely, however, when we consider the familiar plants of tropical home gardens that are routinely vegetatively propagated, especially by cuttings. They include, for example, the highly adaptable and productive cassava (*Manihot esculenta*); "quick stick" or "grow stake" (*Gliricidia sepium*) used to make a living fence and as an ornamental; various figs grown also for living fences and for fruit, fodder, and shade; and several *Spondias* plums, most notably the June plum (*Sonpdias dulcis*). The baobab is clearly not among this group. "However," wrote Wickens and Lowe (2008:242), "Okafor (1980) noted that vegetative propagation was practiced in West Africa, although it was more usual to plant young seedlings." There are in fact no reports in the literature of using baobab cuttings as a primary mode of propagation. Wickens and Lowe (2008:242) reported that "some attempts at vegetative propagation of *A. digitata* in South Africa from truncheons and cuttings have proved unsuccessful (Villiers 1951; Stapleton 1955; Hines and Eckman 1993)." In my own fieldwork, I was often asked if the baobab could be grown from cuttings. I recorded several attempts to grow the baobab from cuttings and none were successful. The approach in the literature is to indicate that the baobab "may" be propagated by cuttings, not that this is anywhere routine.

In addition to incidental harvest dispersal, incidental advantageous growing conditions, and the slight possibility of incidentally "planted" branch and stem cuttings, the other two forms of incidental encouragement foragers provide the baobab are the unwitting protection of young baobabs by hunting and scaring-away herbivores and seed predators; and intentional preservation involving not stepping on baobab seedlings and small saplings and not eating their leaves or roots.

Hadza incidental dispersal of seeds would account for young baobabs in camping areas and elsewhere in their environment. However, it must be considered that the Hadza are also committed seed predators. Because the Hadza harvest and process the fruit pulp and seed of the baobab for food, we can assume that only a small number of seeds would be incidentally dispersed in and around their camps. It is noteworthy that the Hadza eat both seeds collected from the fruits they gather and seeds collected from baboon dung. Consequently, the Hadza have a negative impact on baobab regeneration insofar as they reduce the number of seeds returned to the environment. Baboons are significant dispersal agents for the baobab. The Hadza also reduce the number of seeds returned to the environment when they kill baboons, one of their favorite prey. Hunting baboons protects the baobab in both seed production and seedling survival, but it also limits the positive role of baboons as seed dispersers, especially in rocky uplands away from large herbivores from where the baobab seeds of mature baboon-dispersed trees are later washed to lower elevations.

Despite being baobab seed predators, the Hadza likely spread seeds in their nutrient-rich camping environment that are the result of incidental harvest dispersal involving seeds that are lost, spat out, or occasionally discarded as refuse; or, perhaps, seeds defecated from accidental ingestion. The Hadza eat the fruit in-camp and on the move, and on these occasions, there are no reports that the seeds are always eaten or saved. If discarded, these seeds could be a source of baobab regeneration. It must be noted here, however, that to whatever extent baobab seeds have been incidentally dispersed in human environments worldwide, the baobab has never been described as a weedy tree like such tropical species as the guava (*Psidium guajava* Griseb.), tamarind (*Tamarindus indica* L.), mango (*Mangifera indica* L.), and guinep (*Melicoccus bijugatus* Jacq.). This is so even though a single mature baobab can produce hundreds of fruits, each containing in some cases hundreds of seeds. Spat-out seeds produce the weedy tamarind, but this particular mode of seed dispersal does not appear to be significant for the baobab. My observations of the spread of the baobab in the Americas suggest it does not spring up readily from fruit fall or from incidentally discarded seeds. Nevertheless, over centuries, even a small success rate in Hadza incidental dispersal of the baobab in combination with the baobab's longevity would be significant in ensuring an association between forager camping areas and the baobab.

Incidental and intentional protection are important ways in which human beings support the regeneration of the baobab, both as foragers and as food producers. Hadza incidental protection could in part account for the survival of young trees to maturity. Incidental dispersal and favorable growing conditions might not be as important in the Hadza influence on baobab regeneration as the incidental protection of young baobabs from wild herbivores, which become shy or scarce around Hadza settlements because of hunting. The hunting of herbivores likely allows some young baobabs to escape predation. This is especially true of the hunting of the baboon, which feeds on immature baobab fruits and the taproot of young baobabs.

Baboons have a complex relation to baobab regeneration. Like the Hadza, baboons are both baobab seed predators and seed dispersers, but they are only seed predators when they feed on immature fruits. It is particularly noteworthy, as Venter (2012:1.6) has reported that "baboons (*Papio ursinus* and *P. Anubis*) are known to feed on immature baobab fruit early in the wet season, when there are few other fruit available, and this is thought to have a negative effect on seed production (Lieberman et al. 1979; Pochron 2005; Kunze and Linsenmair 2007)." Feeding on the immature fruits destroys developing embryos or disperses seeds before they are capable of germinating. The impact of baboon fruit predation on baobab regeneration should not be underestimated, as the following report of Venter and Witkowski (2011:637) makes clear:

> Predation of immature baobab fruit by baboons, although not quantified, has been observed elsewhere in Africa (Wickens 1982; Pochron 2005; Kunz and Linsenmair 2007; Watson 2007). In our study fruit production was reduced by 58% and 85% in areas where baboons were found, in contrast to 0% and 1% in areas where baboons were scarce. As baboons eat fruit that do not contain mature seed they contribute to seed destruction rather than dispersal. This shows the dramatic effect baboon predation can have on final mature fruit yield and therefore the presence of baboons needs to be taken into account when assessing areas for fruit harvesting and recruitment.

According to Venter (2012:9.3), the yield per hectare of baobab fruit was "substantially higher" in the settlement environment of villages and fields than in "natural landscapes," due to "the combination of significantly higher densities of trees… and higher fruit production per tree." She also reported, "Extremely low yields from Rocky Outcrops were directly attributed to baboon predation." Venter (2012: 6.21–6.22) continued:

> Baboons are known to reduce fruit production by a least 85% on the ridges and within nature reserves in the study region (Venter and Witkowski 2011). Furthermore, unpublished data from Skelmwater (long term growth monitoring plot near Musina, South Africa) show that predation of baobab fruit by baboons has resulted in multiple crop failures. Fruit predation may also be a contributing factor to poor recruitment in the Kruger National Park (KNP) and Mana Pools National Park, Zimbabwe, where an absence of saplings has been noted (Swanepoel 1993; Hofmeyr 2003). No other fruit predator (other than baboons) has been found in ongoing (2007–2011) fruit production surveys. Once fruit mature they drop off the trees and are mostly collected by villagers and a few by baboons, thus the main dispersal agents are people who live in the area (SM Venter, unpublished data).

For the Hadza, minimal intentional preservation of baobab, unlike incidental protection, would involve not knowingly trampling on or eating the taproot of baobab seedlings and small saplings. Forgoing baobab taproot as a food source is significant when we consider that the Hadza are greatly dependent on some seven species of roots, and it is noteworthy that in West Africa the young saplings in fallow farmland and pastures far from villages are harvested for their root.

The outcome of all these kinds of positive and negative selective influences has likely determined the overall impact of Hadza activities on baobab regeneration in their settlement environment. This impact is identified here as incidental encouragement; early hominins would have provided some protection for the baobab in their foraging environment.

In considering the baobab in the landscape of the Hadza, we could naively say their camping-based settlements follow the baobab, since the tree is one of their preferred camping sites. But over centuries, old baobabs might have been the result of earlier Hadza incidental encouragement. If the Hadza camp next to a baobab that resulted from an earlier settlement, we can say the baobab followed Hadza settlement and Hadza settlement also followed the baobab. In the absence of historical records, the extent to which Hadza camps follow the baobab or the baobab follow Hadza camps cannot be stated without ambiguity. What is clear, however, is the interdependence of Hadza and baobab.

Origin of Agriculture and Baobab Cultivation

Humans settling at the baobab and the baobab springing up in human settlements are not mutually exclusive possibilities. The fundamental point is that both scenarios are ultimately concerned with accounting for the relationship between human beings and the baobab, especially the human influence on baobab regeneration. But the reason for the baobab's presence in human environments can only be genuinely understood by addressing two important questions: How readily does the baobab become established by incidental encouragement and intentional preservation by foragers, as discussed earlier? And to what extent is the African baobab a cultivated tree, defined as a recipient of intentional assistance? This distinction between incidental encouragement and deliberate assistance highlights the importance of baobab settlement volunteers that are not only accepted in place but also transplanted. As important as this distinction is, however, it often leads to the misleading impression that intentionally assisting volunteers, unlike seed-planted baobabs, is something other than cultivation. This overlooks or underplays the fundamental importance of both activities as aspects of baobab cultivation.

Simitu and Oginosako (2002:14) provide an example of this perspective in their socio-economic survey of the baobab and tamarind in Kitui, Kenya. Referring to the baobab in particular, they wrote:

> The trees exist both on the farm and wild. Trees on farm existed prior to land ownership. Some trees date back to more than 100 years. Other small trees are growing on some farms. Establishment is thus by natural regeneration and nobody ever planted a baobab tree. They, however, take care of the trees by clearing bush around the tree and protecting the small tree from destruction by domestic or wild animals.

Along similar lines, Sidibé and Williams (2002:23) have noted that "although the baobab tree is not commonly planted, some farmers practice assisted regeneration: they protect seedlings from fire and animals; they water them or even transplant them." But "assisted regeneration," insofar as it involves protection, land clearance, soil preparation, transplanting, seed planting, fertilization, irrigation, and weeding, is cultivation.

Several early accounts describe the baobab as a cultivated tree. For example, Hutchinson et al. (1958:334) reported that West Africans "often planted [the baobab] near villages, mainly in the drier region." Palmer and Pitman (1961:233) noted: "When the natives of tropical Africa move from one area to another they often take baobab seed with them, which they plant at once in their new home." According to Sidibe et al. (2002:23), "The association of baobab with the farmed parklands or savannah… is a deliberate association with the agricultural environment because of the tree uses, and regeneration may well depend on trees being deliberately planted near settlements." Unfortunately, these authors did not specify exactly what they meant by "planting," and Pitman's idea of Africans seed-planting baobab trees as the primary method of regeneration has received little support in the literature.

That the planting of baobab seeds appears to be less common than the presence of unattended, preserved, and cultivated volunteers suggests a high incidence of self-sown baobabs in the human environment; the idea seems to be that planting baobab seeds is associated with conditions where volunteer baobab seedlings are rare (often because of domesticated animals). For example, in their study of the population, management, and use of the baobab in eastern Burkina Faso, Schumann et al. (2012:271) reported that the strongest agreement among respondents was that they intentionally preserved volunteer baobabs, mostly on cropland. "Nevertheless," wrote Schumann et al. (2012:271), "many informants also declared that they do not protect baobab trees at all…. Planting, sowing, or transplanting of baobab trees were never mentioned."

Assogbadjo and Loo (2011:9) gave a different view of Burkina Faso from the one presented by Schumann et al. (2012:271). "In Burkina Faso and Nigeria," they explained, "rural people plant the trees around their houses or on their farms. In other countries the tree is not often planted from seed, but seedlings are transplanted to field edges or close to houses. Young trees require a high degree of protection, including fencing to exclude livestock grazing." There is also documentation of baobab transplanting and seed planting in Senegal (Gifford 1974), Mali (Sidibe et al. 2002; Dhillion and Gustad 2004), and Nigeria (Assogbadjo and Loo 2011). In their study of local management practices that influence the viability of the baobab in different land-use types in Cinzana, Mali, Dhillion and Gustad (2004:92) reported that in some villages, farmers intentionally protected and transplanted baobab volunteers. They wrote, "Some seeds are consumed, but there is reason to believe that fruit harvesting for pulp yields an excess of seeds, which are returned to the cropland when the household refuse is spread onto the fields." The seedlings of these refuse-disposed seeds are unattended or intentionally preserved where they spontaneously occur, or they are transplanted, often to the residential environment and its nearby farms and pastures. The account presented by Dhillion and Gustad (2004:92) points to the importance of preserving volunteer baobabs and of baobab cultivation:

> Sixty-nine persons were asked, using targeted interviews, about how young, naturally established trees are protected by villagers. The most common answer was "general surveillance" (42 respondents) which, in practice, means that the owner notifies his neighbors of the presence of a young tree that he does not wat to be damaged. Special care should be taken when plowing (seven respondents) or using fire (four respondents). Some mentioned investments in physical barriers to prevent browsing (12 respondents) or seedling maintenance practices (8 respondents) like irrigation cleaning and preparing a basin-shaped bed for water collection. Five person mentioned pruning or cutting surrounding vegetation.

While they indicate the importance of baobab preservation in the human environment, Dhillion and Gustad (2004:92) leave no doubt that the baobab is cultivated in Mali:

> Villagers (total n = 89) more often protected trees they had sowed or transplanted (72%, n = 64) than natural germinations (49%, n = 43). They were also prepared to use more intensive methods like building barriers to prevent browsing (85%, n = 79 for cultivated individuals versus 17%, n = 15 for natural germinations). Seedlings or young trees that are discovered in fallow vegetation are commonly transplanted to a place nearer to the village, either a cropping field or protected garden. When specifically asked (targeted questions) 25 of 57 respondents said that seedlings in fallows were susceptible to being removed by humans. Seedling damage due to browsing (six respondents) or fire (two respondents) were mentioned less frequently (Dhillion and Gustad 2004:92).

Reports of baobab cultivation come not only from West Africa but also from the Sudan, East Africa, and Southern Africa. In their review article on the baobab in Sudan and Kenya, Gebauer and colleagues (2016:389–390) gave the impression that the baobab in residential areas is only a preserved volunteer.

> For Sudan, our observations indicated that baobabs are integral parts of human settlements. In the town of El Obeid they grow in urban gardens (Gebauer and Egbert 2005), are signs of market places and planted as avenue trees (Fig. 13). Wiehle et al. (2014) studied 61 homegardens in four villages of the Nuba Mountains and found *A. digitata* in 46% of the gardens with a total of 74 individuals. With baobab fruits collected from the wild by the gardeners, seeds are extracted from the nearby woodlands in the Nuba Mountains, transferred to homesteads, disposed in the gardens and often spontaneously germinate in place. In the village of Sama, in South Kordofan, Goenster et al. (2011) found baobab trees in homegardens, where they were not actively planted, but had been retained during land clearance for crop planting.

But this emphasis on preservation is hardly convincing when we consider the account of Wiehle et al. (2014:1500), for example, who noted that the "baobab is of particular importance for local food and nutrition security of local communities, including agroforestry systems of the Nuba Mountains, where baobab was the third most abundant (16%) indigenous fruit tree species."

Johansson (1999:14) also reported the cultivation of the baobab in Tanzania, and her account suggests that it is possible to have too many baobabs in the settlement environment. That farmers value the baobab for its many uses does not mean that every baobab volunteer in their residential community and farmland will be welcomed, or even ignored. How farmers respond to these volunteers will obviously depend on the size of their farms, the number of volunteers, the number of mature baobabs already present, the ecological influence of the baobab on crops, and the cultivation of subsistence crops verses commercial crops.

Johansson (1999:14) interviewed five elders of five villages in the Kondoa Eroded Area of the Irangi Hills, Tanzania, and compiled the results into composite answers representing the views of the group. The elders said there were many farms with baobab trees. In their five villages, twenty-two baobabs died in the seventeen years from 1978 to 1997, yet in the same period, twenty-six were planted. Asked about the value of their baobab trees, they said that in addition to fruit, the leaves were eaten as vegetables, the seeds used for making liquor, and the bark for making ropes and brooms. The elders also reported that people wanted baobabs in their farmlands for their agronomic value. Baobabs in farmland "keep moisture in the soil, the soil under the tree is fertile, they act as windbreaks, the shade under the tree is good for the cattle, and sweet potatoes grow well under the baobab tree" (Johansson 1999:14). However, when asked why there were so few young baobabs, the elders said they were destroyed by wild animals, cattle, and dry-time fires. They were also destroyed by people because they "take up a big area, the shade is not good for cash crops, [and] they hide birds which are harmful." And when old baobabs "die, or are removed, they are not replaced because they grow slowly and take a big space" (Johansson 1999:14).

Not surprisingly, the baobab is also cultivated in Southern Africa. Of Northern Venda, South Africa, Venter (2012:7.16) wrote:

> Respondents said that baobabs in homesteads and fields were owned by the individuals on whose land they grew, and trees outside these areas were not owned by anybody. Only 3% had planted a baobab, all of them at their homes. All respondents felt it was not necessary to plant baobab trees, but that if they were given a tree they would plant it either at their home (67%) or in their field (37%).

In her study of the baobab in the Omusati Region of Namibia, Munyebvu (2015:27) reported that "sixty seven percent… of the respondents from Onesi constituency (described as a "communal area") mentioned that they plant baobabs compared to 27% in Outapi constituency." In more rural Onesi, the baobab was more intensively cultivated than in the more urban Outapi. Onesi respondents transplanted the baobab to get trees where they wanted them in relation to their residencies and nearby fields. However, twenty percent said young baobabs were destroyed in active fields. The difference between Onesi and Outapi is the value they place on the baobab, the consequence of this for the dispersal of refuse seeds in near-village farmland, and the acceptance, preservation, and cultivation of volunteers. Munyebvu (2015:87) wrote:

> According to Duvall (2007), Manika-speaking people of West Africa discard baobab seeds around villages where they germinate and tend to be concentrated. This is a likely factor to have resulted in the villagers from Onesi constituency having more baobabs within their homesteads. More than 50% of the homesteads in Onesi constituency had more than two baobabs within their settlements together with the surrounding fields as the discarded seeds within the fenced areas recruited.

Far from being merely an unattended settlement volunteer in the landscapes of food producers, the baobab should rightly be considered a cultivated tree in Africa dating back to the dawn of agriculture, and its cultivation has been intensifying ever since with the dramatic growth of the human population, and now also with the rapidly developing global trade in baobab products. That the baobab is more a transplanted volunteer than seed-planted is not as significant as the fact that in both cases it is intentionally assisted. Many regard "protection" as prior to cultivation and so it is if it is incidental protection. But if cultivation means deliberate assistance, then intentional protection is an aspect of cultivation, albeit one characterized here as the minimal cultivation of volunteers. This means discussions of baobab "management practices," "incipient domestication," and "assisted regeneration" are all expressions of baobab cultivation. It is in this sense that the baobab can justifiably be theorized to be one of Africa's oldest cultivated plants. Seed-planting seems to develop only where volunteers are absent or rare or there is intensive use of the tree.

Summary

Africans have probably been exerting a variety of positive and negative pressures on the baobab's regeneration for a very long time, and the overall outcome of these pressures indicates the true impact of human activity on this regeneration. The transition from the incidental encouragement of the baobab characteristic of foragers to intentional assistance by settled food producers can also be viewed as the expansion of incidental encouragement to include increasingly intensive assistance. Incidental encouragement continues; nothing makes this clearer than the cultivation of volunteer baobabs, especially cultivation that involves transplanting these volunteers. The baobab has likely been incidentally encouraged by human activity since the dawn of the branching hominin tree and cultivated in Africa since the dawn of agriculture. And there is no doubt that its cultivation has intensified with the growth of the human population as well as global social circumscription and intraspecific competition (Carneiro 1970). The need for more intensive cultivation is further driven by the growing demand for baobab products on the world market and efforts to promote the economic development of local communities.

Chapter 25
The Hadza Baobab Retreat

Darwin's figurative representation of the branching nature of the evolutionary process as the Great Tree of Life is now a well-established metaphor in the biological sciences. But the tree-of-life concept has not been significant in ethnographic studies or studies of human evolution. This book has featured an analytical approach to the idea of the tree of life. We find at least informal support for the practicality of this approach in publications and on the Internet, where, as we have seen, the baobab is widely identified not only as *a* tree of life, but as *the* tree of life. It is remarkable indeed that with the exception of being an evergreen the baobab is a tree of life in every sense in which this concept is used today. We have seen that it is a life-manifesting given its extraordinary size, shape, tenacity, longevity, ecology, and utility. Its importance as a resource environment for many species justifies the names life-giving tree of life and ecological tree of life. For human beings, the concept of the baobab as a life-giving tree includes its multipurpose use, its contribution as a staple, and, especially its value as a hunger-season food source. All of these "gifts" of the baobab are associated with cultural narratives of human creation, fertility, and the acquisition of vitality. And the extent to which the baobab is used to model the universe of sky and earth—a tree microcosm representing the tree-like macrocosm—it is also a life-representing tree of life. This study has sought to lay the groundwork for identifying the baobab as the Hadza's tree of life, and thus, the evidence suggests, as humanity's ancestral tree of life. There is every reason to theorize that the resource-rich baobab environment would have been just as important to ancestral hominin foragers as it has been to the Hadza and other Africans throughout its range. This chapter considers the idea that the fate of the Hadza will in part be determined by the outcome of their relationship to their baobab trees.

J. Rashford, *Baobab*, https://doi.org/10.1007/978-3-031-26470-2_25

The Hadza All-Purpose Baobab

To the extent that contemporary studies have dealt with the significance of the baobab in Hadza life, the focus has largely been on the food value of the fruit's pulp and seeds. This work has sought to show that for the Hadza, the paramount value of the baobab is found in its status as a resource environment deeply intertwined with their practical and inspirational activities. Since the Hadza model the prehistoric forager adaptation, it is in this sense that the baobab can aptly be described as humanity's tree of life. This book has reviewed the sources that document the Hadza multipurpose use of the baobab. In the process, it has raised as many questions as it has attempted to answer. Some of the most important answers will come only with future fieldwork. As one example, people elsewhere in Africa who live in intimate association with the baobab are known to differentiate (by some distinguishing feature or features) certain individual baobabs and kinds of baobabs from baobabs in general. It is plausible to assume that the Hadza, too, differentiate some baobab from others. We know the Hadza mark water-storing baobabs because they have names for them. But there must also be baobabs that are famous among them for their size, shape, tenacity, longevity, and fruitfulness, their highly desirable fruit pulp and seeds, their historical or religious importance, or other reasons yet to be identified.

This account of the Hadza/baobab connection and its significance for thinking about African prehistory is only a beginning. There is still much to be learned about the Hadza use of this remarkable tree in such important areas as health and religion. Only systematic ethnobotanical and ethnobiological studies of Hadza and San foraging in the landscape of human evolution, pursued in a thoroughly interdisciplinary way, will reveal the full sense in which the baobab is indeed humanity's ancestral tree of life. Nevertheless, the picture that emerges from the Hadza literature places the seemingly all-purpose baobab at the center of their successful central-place, tool-based, hunter-gatherer way of life – an adaptation to Africa's mosaic savanna as the landscape of our evolutionary development. When the tree-of-life status of the baobab is unrecognized, it is easy to overlook the tree's fundamental importance to the Hadza, to offer a naïve view of the Hadza heartland, and to miss what the Hadza themselves say is important to their current survival as a people.

An example of a limited perspective on the tree-of-life baobab was Stephanie McCrummen's June 10, 2007, article in the *Washington Post* titled "50,000 Years of Resilience May Not Save Tribe." McCrummen discussed changes that were threatening to end the Hadza hunter-gatherer way of life, especially efforts underway by "a United Arab Emirates royal family" to use their homeland "as a personal safari playground."

> Their fear is based on a similar agreement the government struck years ago with another company that resulted in dozens of Hadzabe men being arrested for hunting on tribal land. Three of the men died of illness in the bewildering environment of prison, cut off from the open world, their daily hunting and their diet of herbs, roots and honey. Three others died soon after being released.

In her description of all that was essential to Hadza life, it is surprising that McCrummen did not include the baobab. Instead of the rocky baobab-clad hills as the heartland of the Hadza homeland, she described the Hadza as living in the Yaeda Valley and "deep in the scruffy hills surrounding this valley." In the case of the UAE royal family, she reported in her newspaper article:

> A recent meeting in the Yaeda Valley on the issue ended with several Hadzabe shouting at Tanzanian government officials for ignoring them. One of the men was later charged with disruptive behavior and jailed for several days. Two others who have spoken against the deal said they have been threatened with arrest and are now on the run, moving from hut to hut to elude police.

One of the wanted men was Gonga Petro, whom McCrummen interviewed in his camp. "It was morning in his settlement," and she noted "the four straw huts nearly invisible amid waist-high grass, thorny bushes and thick-trunked baobab trees." In contrast to her lack of appreciation of the significance of the baobab to the Hadza, there is the view she attributed to Gonga Petro:

> "It's very important to go to work and hunt, but now, you can just walk from morning to night and if you're lucky, you might come back with a dik-dik," he sighed, referring to an animal that is embarrassingly small for someone who once slew two zebras, an antelope and a buffalo in a single day. "But there's always an alternative. The baobab. Together with the herbs."

The Fate of the Hadza Is the Fate of Their Baobab Trees

Today, competition for land, water, and wildlife is a serious threat to the Hadza's traditional hunter-gatherer way of life – the competition comes from the encroachment of neighboring herders, cultivators, and agropastoralists, and from an influx of commercial interests and their associated immigrants. The baobab has likely suffered as parts of the natural savanna have been converted to grazing lands. This conversion is one significant impact of the Hadza's pastoral and agropastoral neighbors on the baobab population that is not generally recognized, compared with the alteration of the landscape by scaring away game, the deterioration of water holes for foragers and game, and the cutting of woody species for fuel and construction. The Hadza as hunters eliminate wild herbivores, but livestock is protected by their owners. Not only does livestock trample and browse young baobabs, small baobabs are also cut down to provide livestock feed. In this natural grazing land and in distant fallow land used as pastures, baobab regeneration is most affected (as it is in natural and protected areas with high elephant or baboon populations).

For most authors, the things that limit baobab regeneration throughout its range and that raise questions about the long-term survival of the species include environmental conditions such as poor rainfall, drought, floods, and fire; insects that destroy seedlings, and the trampling and browsing of livestock and wild herbivores, especially elephants and baboons; destructive human activities, especially land clearance for commercial agriculture, large-scale construction projects, and urban expansion; the unsustainable harvesting of fruit, leaves, roots, and bark; and now the global commodification of the baobab which contributes to the reduction of seeds in the environment, making them unavailable for recruitment. In the absence of commercially established baobab plantations, the growing trade in baobab products, some of which are now exported around the world, raises concern for the traditional livelihood of local populations that depend on the tree for its many resources. It also raises questions about the conservation of the diverse wildlife populations that depend on the baobab.

Cultural survival for the Hadza means survival of their foraging adaptation, and this may increasingly depend on the continued presence of baobab trees in their environment and their continued control of the baobab harvest. It is remarkable indeed that the Hadza's hunter-gatherer way of life today is similar to that reported by the earliest written accounts from the end of the nineteenth century and early twentieth century. Crucial to this continuity is that they still depend on their baobab. With the baobab going global, the Hadza might soon face the potentially devastating competition resulting from the demands of the world market. The necessary link between the baobab and survival of the Hadza's traditional way of life is recognized in the scientific literature. In concluding their discussion on the nutritional composition of different kinds of honey and of the most important wild plant foods of the Hadza, Murray et al. (2001:10) wrote:

> Baobab seeds are a good source of protein and, due largely to their high fat content, their energy content is equivalent to that from honey. Although the seeds are inadequate in three essential amino acids, minimal amounts of meat would complete these requirements. In addition, the baobab seeds are a more dependable energy source than honey. Women consistently returned with dozens of baobab fruits or with significant quantities of seeds over the majority of months of the year either through direct fruit collection or through collection of seeds in baboon dung piles. Honey is available in quantity only for a few months each year, as is true for non-baobab fruits. Overall, this suggests that baobab is an important, probably preferred, plant food for the Hadza even though other plants are eaten in significant quantities.

In the final sentence of their paper, Murray et al. (2001:10) concluded, "Maintaining areas with healthy baobab groves is important for the future of [the] Hadza and of other people living in agriculturally marginal regions."

Summary

The fate of the baobab in the Eastern Rift Valley is indeed the fate of the Hadza. Given the accelerating transformation of the natural savanna landscape into farms, pastures, and ever-expanding settlements, there are those who think the opportunity to do studies of Hadza baobab use might be running out. There are no indications, however, that the baobabs of the greater Eyasi region are threatened. Nevertheless, we can take the following poetic view of Matthiessen (1972:234) as a cautionary note: "Perhaps the greatest baobabs were already full grown when man made red rock paintings at Darashagan. Today young baobabs are killed by fires, set by the strangers who clear the country for their herds and gardens, and the tree where man was born is dying in Hadza Land." Who knows? Perhaps it is not merely poetry after all.

References

Abdillahi, Maoulida Mohamed, Cyrille Cornu, Raïma Fadul, Michel Charpentier, Roger Edmond, Bakolimalala Rakouth, Jean-Michel Leong, Jean-Michel Leong Pock Tsy, and Pascal Danthu. 2019. The baobabs of the Comoro Islands: Some biogeographical factors towards the protection and conservation of a neglected asset. *Tropical Ecology* 60: 311–325.

Abiodun, R. 1994. Understanding Yoruba art and aesthetics: The concept of Ase. *Africans Arts* 27 (3): 68–78, 102–103.

Acheson, James M. 1981. Anthropology of fishing. *Annual Review of Anthropology* 10: 275–316.

ADAEHP. 1944. *A Dictionary of American English on Historical Principles*, ed. Sir William A. Craigie and James R. Hulbert. University of Chicago Press.

Adam, J.G. 1962. Le baobab (*Adansonia digitata* L.). *Notes Africaines* 94: 33–44.

Adams, R. McC. 1974. Anthropological perspectives on ancient trade. *Current Anthropology* 15 (3): 239–258.

Adanson, M. 1757. *Histoire Naturelle du Sénégal*. Paris: C.J.-B. Bauche.

———. 1771. Description d'un arbre nouveau genre appelé Baobab, observé au Sénégal. *Histoire et Mémoires de l'Académie royale des sciences (Paris)* 1791: 218–243.

Adedayo, M.R., F.F. Olayemi, and E.I. Bamishaiye. 2011. Proximate and mineral composition of a local drink made from baobab fruit (*Adansonia digitata*) pulp. *Advances in Bioresearch* 2 (2): 82–85.

African Wildlife Foundation. 2021. *Bat*. https://www.awf.org/wildlife-conservation/bat. Downloaded 24 Apr 2021 (9.44 pm).

Agova, Mia. 2015. *Editorial politics and significance assessment procedures: A comparative case study of two Hadza necklaces*. Academia: YJRC7. https://www.academia.edu/6198074/. Accessed 18 June 2020.

Aitken, R.B. 1951. The tree that time forgot. *Natural History* 60 (10): 456–462.

Allen, A.A. 1961. *The book of bird life: A study of birds in their native haunts*. Princeton: D. Van Nostrand Company, Inc.

Alcock, J. 1972. The evolution of the use of tools by feeding animals. *Evolution* 26 (3): 464–473.

Allan, W. 1965. *The African husbandman*. London: Oliver and Boyd.

Allchin, B. 1957. Australian stone industries, past and present. *Journal of the Royal Anthropological Institute of Great Britain and Ireland* 87: 115–136.

Allison, P.A. 1962. Historical inferences to be Drawn from the effect of human settlement on the vegetation of Africa. *The Journal of African History* 3 (2): 241–249.

Allport, S. 2000. *The primal feast: Food, sex, foraging, and love*. Lincoln: Harmony Books.

Altman, Nathaniel. 2000. *Sacred trees: Spirituality, wisdom and well-being*. New York: Sterling Publishing Co.

J. Rashford, *Baobab*, https://doi.org/10.1007/978-3-031-26470-2

Altmann, Stuart A. 2009. Fallback foods, eclectic omnivores, and the packaging problem. *American Journal of Physical Anthropology* 140: 615–629.
Alvard, Michael S. 2003. The adaptive nature of culture. *Evolutionary Anthropology* 12: 136–149.
Amanda, Stone, Massey Abby, Theobald Molly, Styslinger Matt, Kane Dan, Kandy Dan, Tung Alex, Adekoya Abisola, Madan Janeen, and Advert Elena. 2011. In *Africa's indigenous crops*, ed. L. Mastny. Washington, DC: World Watch Institute.
Ambrose, Stanley H. 2001. Paleolithic technology and human evolution. *Science* 291: 1748–1753.
Anderson, E. 1954. *Plants man and life*. London: A. Melrose.
Anderson, Edgar. 1956. Man as a maker of new plants and plant communities. In Volume 2: *Man's role in changing the face of the earth*, ed. W.L. Thomas, 763–777. Chicago: University of Chicago Press.
Anderson, Edgar. 1960. The evolution of domestication. In *Evolution after Darwin*, Volume 2: The evolution of man, culture, and society, ed. Sol Tax, 67–84. Chicago: University of Chicago Press.
Anderson, Richard L. 1990. *Calliope's sisters: A comparative study of philosophies of art*. Englewood Cliffs: Prentice Hall.
Anderson, Kit. 2003. *Nature, culture, and big old trees: Live oaks and ceibas in the landscapes of Louisiana and Guatemala*. Austin: University of Texas Press.
Anderson, C. Leigh, Travis Reynolds, Joshua D. Merfeld, and Pierre Biscaye. 2017. Relating seasonal hunger and prevention and coping strategies: A panel analysis of Malawian farm households. *The Journal of Development Studies* 54 (10): 1737–1755. https://doi.org/10.1080/00220388.2017.1371296.
Andriantsaralaza, Seheno, Elysée N. Rasoamanana, C. Onja Razanamaro, Jean-Michel Leong Elgoyen, Pock Tsy, Perle Ramavovololona, Bakolimalala Rakouth, Roger Edmond, and Pascal Danthu. 2016. Vers une compréhension des relations entre les baobabs et quelques éléments des écosystèmes forestiers de l'ouest de Madagascar: pollinisateurs, disperseurs et prédateurs. In *Recherche interdisciplinaire pour le développement durable et la biodiversité des espaces ruraux malgaches*. FSP Parrur.
Annegers, J.F. 1973. Seasonal food shortages in West Africa. *Ecology of Food and Nutrition* 2 (4): 251–257.
Anonymous. 1876. The baobab as a fiber plant. *Scientific American* 34 (16): 240.
——— 1890. *Cultus arborum*. Privately printed. Available online through Kessinger Publishing.
———. 1949. The society's activities – African Affairs. *African Affairs* 48 (191): 160.
———. 2004. Rain plays hide and seek as the people grope for food. *The Nation*, July 20, Nairobi, Kenya.
———. 2005. *Les Griots*. http://www.senegal-online.com/anglais/histoire/griots.htm. Retrieved 1 Dec 2005.
———. 2014. *Griot*. Wikipedia. Downloaded 2 June 2014.
———. 2018a. *Local juices: Bissap, Bouye, Gingembre*. https://eatyourworld.com/.
———. 2018b. *Traditional baobab drink – Aduna*. Newsletter, Aduna Ltd. https://aduna.com/blogs/recipes/traditional-baobab-drink.
Apicella, Coren L., Alyssa N. Crittenden, and Victoria A. Tobolsky. 2017. Huntergatherer males are more risk-seeking than females, even in late childhood. *Evolution and Human Behavior* 38 (5): 592–603.
Archibald, Sally, A. Carla Staver, and Simon A. Levin. 2012. Evolution of human-driven fire regimes in Africa. *Proceedings of the National Academy of Sciences of the United States of America* 109 (3): 847–852. http://www.jstor.org/stable/23077115.
Ardrey, R. 1961. *African genesis: A personal investigation into the animal origins and nature of man*. New York: Atheneum.
———. 1966. *The territorial imperative: A personal inquiry into the animal origins of property and nations*. New York: Atheneum.
———. 1971. *The social contract*. New York: Doubleday.
Armstrong, Patrick. 1977. Baobabs: Remnant of Gondwanaland? *New Scientist* 73 (1036): 212–213.

———. 1979. The history, natural history and distribution of Adansonia: A plant genus of the Indian Ocean littoral. In *The Indian Ocean in Focus. International Conference on Indian Ocean Studies, Perth, Western Australia 1979*, Environment and resources, section 1, 1–21. Perth: Perth Building Society.

Armstrong, P. 1983. The disjunct distribution of the genus *Adansonia* L. *The National Geographical Journal of India* 29: 142–163.

Arum, Gilbert. 1989. *Baobab*. Nairobi: Norwegian Agency for International Development (NORAD).

Assogbadjo, A.E. 2008. *Indigenous knowledge, genetic diversity and domestication of baobab tree (Adansonia digitata L.) in Benin (West Africa)*, Final technical report. The Rufford Small Grants for Nature Conservation.

Assogbadjo, A.E., and J. Loo. 2011. *Adansonia digitata, African baobab: Conservation and sustainable use of genetic resources of priority food tree species in sub-Saharan Africa*. Rome: Biodiversity International.

Assogbadjo, A.E., B. Sinsin, T.T.C. Codjia, and P. Van Damme. 2005. Ecological diversity and pulp, seed and kernel production of the baobab (*Adansonia digitata*) in Benin. *Belgium Journal of Botany* 138: 47–56.

Assogbadjo, Achille Ephrem, R. Glele Kakal, F.J. Chadare, L. Thomson, T. Sinsin, B. Sinsin, and P. van Damme. 2008. Folk classification, perception and preferences of baobab products in West Africa: Consequences for species conservation and improvement. *Economic Botany* 62: 74–84.

Assogbadjo, Achille Ephrem, Romain Glele Kakai, Tina Kyndt, and Brice Sinsin. 2010. Conservation genetics of baobab (*Adansonia digitata* L.) in the parklands agroforestry systems of Benin (West Africa). *Notulae Botanicae Horti Agrobotanici Cluj-Napoca* 38 (2): 136–140.

Attwell, L., K. Kovarovic, and J. Kendal. 2015. Fire in the Plio-Pleistocene: The functions of hominin fire use, and the mechanistic, developmental and evolutionary consequences. *Journal of Anthropological Sciences* 93: 1–20. https://doi.org/10.4436/JASS.93006. Epub 2015 Mar 19.

Aubréville, A. 1975. Essais de géophylétique de Bombacacée. *Adansonia Sér* 2 (15): 57–64.

Austin, Daniel. 2004. *Florida ethnobotany*. New York: CRC Press.

Awori, A. 1989. Forward. In *G. Arum, Adansonia digitata*, Indigenous trees training series, ed. Elizabeth Obel-lawson. Kengo.

Bagshawe, F.J. 1925. The peoples of the happy valley (East Africa): The aboriginal races of Kondoa Irangi. Part 2: The Kangeju. *Journal of the African Society* 24 (94): 117–130.

Bailey, T.N. 1993. *The African leopard*. New York: Columbia University Press.

Balfour, E. 1885. *The cyclopaedia of India*. London: Bernard Quaritch.

Balick, Michael J. 1988. *The palm-tree of life: Biology, utilization and conservation*, Advances in economic botany, Vol. 6. New York: New York Botanical Garden.

Bankole, Olufunmilayo M. 2006. The antimicrobial effects of the cream of tartar ('KUKA') of baobab fruit, on fermented milk ('NONO'). *Journal of Food Technology* 3 (2): 173–176.

Barnard, Alan, and Jonathan Spencer. 1996. *Encyclopedia of social and cultural anthropology*. London/New York: Routledge/Taylor and Francis Group.

Barnard, Alan, and James Woodburn. 1988. Property, power and ideology in hunter-gathering societies: An introduction. In *Hunters and gatherers: Volume 2: Property, power and ideology*, ed. Tim Ingold, David Riches, and James Woodburn, 4–31. Oxford: Berg.

Barnes, R.F.W. 1985. Woodland changes in Ruaha National Park (Tanzania) between 1976 and 1982. *African Journal of Ecology* 23 (4): 215–221.

Bartholomew, G.A., and J.A. Birdsell. 1953. Ecology and the protohominids. *American Anthropology* 55: 481–498.

Bash, Barbara. 1989. *Tree of life: The world of the African baobab*. Hong Kong: Self-Published.

Battuta, Ibn. 1969. *Ibn Battuta: Travels in Asia and Africa 1325–1354*. New York: Augustus M. Kelly.

Baum, D.A. 1995a. The comparative pollination and floral biology of baobabs (*Adansonia* Bombacaceae). *Annals of the Missouri Botanical Garden* 82 (2): 322–348.

———. 1995b. A systematic revision of *Adansonia* (Bombacaceae). *Annals of the Missouri Botanical Garden* 82 (3): 440–471.

———. 2003. Bombacaceae, *Adansonia*, baobab, bozy, fony, renala, ringy, za. In *The natural history of Madagascar*, ed. S.M. Goodman and J.P. Benstead, 339–342. Chicago: University of Chicago Press.

Baum, David A., and Stachy D. Smith. 2013. *Tree thinking: An introduction to phylogenetic biology*. Greenwood Village: Roberts and company Publishers.

Baum, D.A., R.L. Small, and J.F. Wendel. 1998. Biogeography and floral evolution of baobabs (*Adansonia*, Bombacaceae) as inferred from multiple data sets. *Systematic Biology* 47 (2): 181–207.

Baum, David A., Stacey DeWitt Smith, Alan Yen, William S. Alverson, Reto Nyffeler, Barbara A. Whitlock, and Rebecca L. Oldham. 2004. Phylogenetic relationships of Malvatheca (Bombacoideae and Malvoideae; Malvaceae sensu lato) as inferred from plastid DNA sequences. *American Journal of Botany* 91: 1863–1871.

Baumann. 1897. Bevolkerung des bayerischen Schwabens in ihrer gesc ichtlichen Aufeinanderfolge. *Beitrage zur Anthropologie* 12: 105–126.

Beale, Colin M., Colin J. Courtney, Thomas A. Mustaphi, Sally Archibald Morrison, T. Michael Anderson, Andrew P. Dobson, Jason E. Donaldson, Gareth P. Hempson, James Probert, and Catherine L. Parr. 2018. Pyrodiversity interacts with rainfall to increase bird and mammal richness in African savannas. *Ecology Letters* 21 (4): 557–567. https://doi.org/10.1111/ele.12921.

Beals, Ralph L., and Harry Joijer. 1965. *An introduction to anthropology*. New York: The MacMillan Company.

Beech, W.H. 1913. The sacred fig-tree of the A-Kikuyu of East Africa. *Man* 13 (3): 4–6.

Behjati-Ardakani, Zohreh, Mohammad Mehdi Akhondi, Homa Mahmoodzadeh, and Seyed Hasan Hosseini. 2016. An evaluation of the historical importance of fertility and its reflection in ancient mythology. *Journal of Reproductive Infertility* 17 (1): 2–9.

Bell, Duran. 2001. Comments. *Current Anthropology* 42 (5): 696.

Bell, Karen L., Haripriya Rangan, Christian A. Kull, and Daniel J. Murphy. 2015. The history of the introduction of the African baobab (*Adansonia digitata*, Malvaceae: Bombacoideae) in the Indian subcontinent. *Royal Society Open Science* 2 (9): 150370.

Belsky, A.J., R.G. Amundson, J.M. Duxbury, S.J. Riha, A.R. Ali, and S.M. Mwonga. 1989. The effects of trees on their physical, chemical and biological environments in a semi-arid savanna in Kenya. *Journal of Applied Ecology* 26 (3): 1005–1024.

Bender, Renato, Phillip V. Tobias, and Nicole Bender. 2012. The savannah hypotheses: Origin, reception and impact on paleoanthropology. *History and Philosophy of the Life Sciences* 34 (1–2): 147–184.

Bennet, F.J., N.A. Barnicot, and J.C. Woodbury. 1973. Studies on viral, bacterial, rickettsial, and treponemal diseases of the Hadza of Tanzania, and a note on injuries. *Human Biology* 45: 243–272.

Benson, C.W., and J.M. Winterbottom. 1977. The distribution and habitat of Telacanthura ussheri benguellensis. *Bulletin of the British Ornithologists' Club* 97: 46–48.

Berbesque, Julia C., and Frank Marlowe. 2009. Sex differences in food preferences of Hadza hunter-gatherers. *Evolutionary Psychology* 7 (4): 601–616.

Berbesque, J. Colette, Frank W. Marlowe, Ian Pawn, Peter Thompson, Guy Johnson, and Audax Z.P. Mabulla. 2012. Sex differences in Hadza dental wear patterns: A preliminary report. *Human Nature* 23 (3): 270–282.

Berbesque, J. Colette, Brian M. Wood, Alyssa N. Crittenden, Audax Mabullad, and Frank W. Marlowe. 2016. Eat first, share later: Hadza hunter–gatherer men consume more while foraging than in central places. *Evolution and Human Behavior* 37 (4): 281–286.

Bingham, M.G. 1994. Did the Baobabs originate in Madagascar? *Ingens Bulletin* 10: 12.

Bird, Rebecca Bliege, and Douglas W. Bird. 2008. Why women hunt: Risk and contemporary foraging in a Western Desert aboriginal community. *Current Anthropology* 49 (4): 655–693.

Bird, R.B., and B.F. Codding. 2015. The sexual division of labor. In *Emerging trends in the social and behavioral sciences: An interdisciplinary, searchable, and linkable resource*, ed. R.A. Scott and S.M. Kosslyn, 1–15. Wiley.

Bird-David, Nurit. 1992. Beyond the hunting and gathering mode of subsistence: Culture sensitive observations on the Nayaka and other modern hunter-gatherers. *Man (New Series)* 27 (1): 19–44.

Biro, D., N. Inoue-Nakamura, R. Tonooka, G. Yamakoshi, C. Sousa, and T. Matsuzawa. 2003. Cultural innovation and transmission of tool use in wild chimpanzees: Evidence from field experiments. *Animal Cognition* 6: 213–223.

Blanks, H. 1952. The monstrous baobab. *Wild Life, Melbourne* 15: 528–532.

Blasco, Ruth. 2016. Humankind and the avian world: Zooarchaeological evidence for inferring behavioural evolutionary signatures. *Quaternary International* 421: 1–5.

Bleek, Dorothea F. 1931. The Hadzapi or Watindiga of Tanganyika territory. *Africa* 4 (3): 273–286.

Blench, R.M. 2007. The intertwined history of the silk-cotton and baobab. In *Fields of change; progress in African ethnobotany*, ed. Cappers RTJ, 1–20. Gröningen: Barkhuis and Gröningen University.

Blier, Suzanne Preston. 1987. *The anatomy of architecture: Ontology and metaphor in Batammaliba architectural expression*. Chicago/London: The University of Chicago Press.

Blunt, H.S. 1923. Tebeldis. *Sudan Notes and Records* 6 (1): 114–117.

Blurton Jones, N.B., and F. Marlowe. 2002. Selection for delayed maturity: Does it take 20 years to learn to hunt and gather? *Human Nature* 13: 199–238.

Blurton Jones, N.B., K. Hawkes, and J.F. O'Connell. 1989. Studying costs of children in two foraging societies: Implications for schedules of reproduction. In *Comparative socioecology of mammals and man*, ed. V. Standon and R. Foley, 365–390. London: Blackwell.

Blurton Jones, N.B. 1993. The lives of hunter-gatherer children: Effects of parental behavior and parental reproductive strategy. In *Juvenile primates: Life history, development, and behavior*, ed. M.E. Pereira and L.A. Fairbanks, 309–326. Oxford: Oxford University Press.

Blurton Jones, N.B., K. Hawkes, and P. Draper. 1993. Differences between Hadza and !Kung children's work: Affluence or practical reason? In *Key issues in hunter-gatherer research*, ed. E.S. Burch and L.J. Ellana, 189–215. Providence: Berg.

Blurton Jones, N.B., K. Hawkes, and J.F. O'Connell. 1996. The global process and local ecology: How should we explain differences between the Hadza and !Kung? In *Cultural diversity among twentieth century foragers*.

Blurton Jones,, N.B., K. Hawkes, and J.F. O'Connell. 1997. Why do Hadza children forage? In *Uniting psychology and biology: Integrative perspectives on human development*, ed. N. Segal, G.E. Weisfeld, and C.C. Weisfeld, 279–313. Washington, DC: American Psychological Association.

Blurton Jones, N.G. 2016. *Demography and evolutionary ecology of Hadza Hunter-Gatherers*, Cambridge studies in biological and evolutionary anthropology. Cambridge: Cambridge University Press.

Blurton Jones, N.B., K. Hawkes, and P. Draper. 1994. Foraging returns of !Kung adults and children: Why didn't !Kung children forager? *Journal of Anthropological Research* 50 (3): 217–248.

Blurton Jones, N.G., Kristen Hawkes, and James F. O'Connell. 2005. Older Hadza men and women as helpers: Residence data. In *Hunter-gatherer childhoods: Evolutionary, developmental, and cultural perspectives*, ed. Barry S. Hewlett and Michael E. Lamb, 214–236. New York: Routledge.

Boesch, Christopher. 2012. *Wild cultures: A comparison between chimpanzee and human cultures*. New York: Cambridge University Press.

Bogin, Barry. 1988. *Patterns of human growth*. Cambridge: Cambridge University Press.

———. 1997. Evolutionary hypotheses for human childhood. *Yearbook of Physical Anthropology* 40: 63–89.

Boinski, Sue, Robert P. Quatrone, and Hilary Swartz. 2001. Substrate and tool use by brown capuchins in Suriname: Ecological contexts and cognitive bases. *American Anthropologist* 102 (4): 741–761.

Bolandier, G., and J. Maquet. 1974. *Dictionary of Black African civilization*. New York: L. Amiel.

Bonta, Mark, Robert Gosford, Dick Eussen, Nathan Ferguson, Erana Loveless, and Maxwell Witwer. 2017. Intentional fire-spreading by "Firehawk" raptors in Northern Australia. *Journal of Ethnobiology* 37 (4): 700–718. https://doi.org/10.2993/0278-0771-37.4.700.

Booth, F.E.M., and G.E. Wickens. 1988. *Non-timber uses of selected arid zone trees and shrubs in Africa*, FAO conservation guide 19, 103–109. Rome: Food and Agriculture Organization of the United Nations.

Borrow, Nik, and Ron Demey. 2014. *Birds of Western Africa*. 2nd ed. Princeton: Princeton University Press.

Bosch, C.H., K. Sié, and B.A. Asafa. 2004. *Adansonia digitata* L. [Internet] record from PROTA4U, ed. G.J.H. Grubben and O.A. Denton. PROTA (Plant Resources of Tropical Africa/Ressources végétales de l'Afrique tropicale), Wageningen. http://www.prota4u.org/search.asp. Accessed 17 Mar 2018.

Bowen, W. Wedgwood. 1932. Angolan birds collected during the second Gray African expedition: 1930. *Proceedings of the Academy of Natural Sciences of Philadelphia* 84: 281–289.

Bowman, D., J. Balch, P. Artaxo, W.J. Bond, J.M. Carlson, M.A. Cochrane, C.M. D'Antonio, R.S. DeFries, J.C. Doyle, S.P. Harrison et al. 2008. *The evolution and diversification of fire in the Earth system*. Manuscript.

Boyes, Rutledge S., and Michael R. Perrin. 2009. Do Meyer's Parrots *Poicephalus meyeri* benefit pollination and seed dispersal of plants in the Okavango Delta, Botswana? *African Journal of Ecology* 48: 769–782.

Boyes, Rutledge S., and Michael R. Perrin 2010. Aerial surveillance by a generalist seed predator: Food resource tracking by Meyer's parrot *Poicephalus meyeri* in the Okavango Delta, Botswana. *Journal of Tropical Ecology* 26 (4): 381–392. https://doi.org/10.1017/S0266467410000210

Braun, K. 1929. Der Affenbrotbaum (*Adansonia digitata* L.) und seine venwendung besonders als faserpflanze in Deutsch-Ostafrica. *Faserforsch* 8: 90–115.

Braun, David R., John W.K. Harris, Naomi E. Levin, Jack T. McCoy, Andy I.R. Herries, Marion K. Bamford, Laura C. Bishop, Brian G. Richmond, Mzalendo Kibunjia, and Richard G. Klein. 2010. Early hominin diet included diverse terrestrial and aquatic animals 1.95 Ma in East Turkana, Kenya. *Proceedings of the National Academy of Sciences* 107 (22): 10002–10007.

Brace, C. L. 1962. Refocusing on the Neanderthal Problem. *American Anthropologist* 64: 729–741.

Brace, C.L. 1995. *The stages of human evolution*. Englewood Cliffs: Prentice-Hall.

Brenan, J.P.M. 1949. *Check-lists of the forest trees and shrubs of the British Empire: No 5. Tanganika territory. Part 2*. Oxford: Forestry Institute.

Brightman, Robert. 1996. The sexual division of foraging labor: Biology, taboo, and gender politics. *Comparative Study of Society and History* 38: 687–729.

Britannica Online Encyclopedia. 2020. *Passeriform*. https://www.critannica.com/print/article/445721.

Britannica, The Editors of Encyclopaedia. 2006. *Palm summary*. Encyclopedia Britannica, 9 Feb 2006. https://www.britannica.com/summary/palm-tree. Accessed 15 March 2023.

———. 2023. *Ficus*. Encyclopedia Britannica, 7 Mar 2023. https://www.britannica.com/plant/Ficus. Accessed 15 March 2023.

Brockway, Lucile. 2002. *Science and colonial expansion: The role of the British Royal Botanic Garden*. New Haven: Yale University Press.

Brooke, R.K. 1971. The eastern and southern populations of the mottled spinetail. *Bulletin of the British Ornithologists' Club* 91. Biostor, Additional Collections. https://archive.org/search?query=mottled+spinetai. Retrieved 3/5/2023.

———. 1984. *South African Red Data Book – Birds*, South African National Scientific Programmes Report No. 97. Pretoria: Foundation for Research Development, Council for Scientific and Industrial Research.

———. 2018. Mottled spinetail (Gevlekte Stekelstert, *Telacanthura ussheri*). In *The Atlas of Southern African birds*, 630. http://sabap2.adu.org.za/docs/sabap1/422.pdf. Accessed 3 Apr 2018.

Brooke, Michael, and Tim Birkhead. 1991. *Cambridge encyclopedia of ornithology*. Cambridge: Cambridge University Press.

Brondízio, Eduardo S. 2008. The Amazonian Caboclo and the Açaí Palm: Forest farmers in the global market. In *Advances in economic botany*, vol. 16. New York: The New York Botanical Garden Press.

Brown, Leslie. 1965. *Africa: A natural history*. New York: Random House.

Brown, Judith K. 1970. A note on the division of labor by sex. *American Anthropologist* 72 (5): 1073–1078.

Brown, Chris. 2015. Meyer's Parrot – An unusual nest site. *Lanioturdus* 48 (2): 25–26.

Bryant, David Murray. 1985. Swallows. In *A dictionary of birds*, ed. Bruce Campbell and Elizabeth Lack, 571–572. Vermillion: Butea Books.

Buchmann, Christine, Sarah Prehsler, Anna Hartl, and Christian R. Vogl. 2010. The importance of baobab (*Adansonia digitata* L.) in rural West African subsistence – Suggestion of a cautionary approach to international market export of baobab fruits. *Ecology of Food and Nutrition* 49 (3): 145–172.

Buchmann, C., S. Prehsler, A. Hartl, and C.R. Vogl. 2010. The importance of baobab (*Adansonia digitata* L.) in rural West African subsistence: Suggestion of a cautionary approach to international market export of baobab fruits. *Ecology of Food and Nutrition* 49: 149–172.

Bunn, H.T., L. Bartram, and E. Kroll. 1988. Variability in bone assemblage formation from Hadza hunting, scavenging, and carcass processing. *Journal of Anthropological Archaeology* 7: 412–457.

Burkill, H.M. 1985. *The useful plants of west tropical Africa*, Vol. 1, families A-D. London: Royal Botanical Garden, Kew.

Burton, F.D. 2009. *Fire: The spark that ignited human evolution*. Albuquerque: University of New Mexico Press.

Burton, Michael L., and Douglas R. White. 1984. Sexual division of labor in agriculture. *American Anthropologist* 86 (3): 568–583.

Burton, Michael L., Lilyan A. Brudner, and Douglas R. White. 1977. A model of the sexual division of labor. *American Ethnologist* 4 (2): 227–252.

Burton-Page, J. 1969. The problem of the introduction of *Adansonia digitata* into India. In *The domestication and exploitation of plants and animals*, ed. P.J. Ucko and G.W. Dimbleby. Chicago/New York: Aldine Atherton, Inc.

Burtt, B.D., and E.J. Salisbury. 1929. A record of fruits and seeds dispersed by mammals and birds from the Singida district of Tanganyika territory. *Journal of Ecology* 17 (2): 351–355.

Cadamosto, A. 1507 [1937]. *The voyages of Cadamosto*. Trans. G.R. Crone. London: Hakluyt Society.

Cahoon, Donald R., Brian J. Stocks, Joel S. Levine, Wesley R. Cofer, and Katherine P. O'Neill. 1992. Seasonal distribution of African savanna fires. *Nature* 359 (29): 812–815.

Caillault, Sebastien, Aziz Ballouche, and Daniel Delahaye. 2015. Where are the 'bad fires' in West African savannas? Rethinking burning management through a space–time analysis in Burkina Faso. *The Geographical Journal* 181 (4): 375–387.

Caius, J.F. 1941. The medicinal and poisonous plants of India: Dammers, guttifers, silkcottons, teas, tutsams, water-peppers. *Journal of the Bombay Natural History Society* 42: 617–639.

Campbell, Bernard G. and James D. Loy 1996. Humankind Emerging. New York: Harper Collins College Publishers.

Campbell, Bernard G., and James D. Loy. 1996. *Humankind Emerging*. 7th ed. HarperCollins College Publishers.

Campbell, Bernard. 1983. *Human ecology*. New York: Aldine Publishing Company.

Campbell, L. 1995. The baobab: the upside down tree. *Ballya* 2: 32–36.

Cann, Rebecca L. and Allan C. Wilson 1982. Models of Human Evolution. Science 217(4557): 303–304. DOI: 10.1126/science.217.4557.303.

Cann, R.L., W.M. Brown, and A.C. Wilson. 1982. Evolution of human mitochondrial DNA: A preliminary report. *Progress in Clinical and Biological Research* 103 (Pt A): 157–165. PMID: 6298804.

Carmody, R.N., and R.W. Wrangham. 2009. The energetic significance of cooking. *Journal of Human Evolution* 57: 379–391. https://doi.org/10.1016/j.jhevol.2009.02.011.

Carneiro, R.L. 1970. A theory of the origin of the state. *Science* 169 (3947): 733–738.

Caro, T.M. 1987. Cheetah mothers' vigilance: Looking out for prey or for predators. *Behavioral Ecology and Sociobiology* 20: 351–361.

Carvalho, S., and W.C. McGrew. 2012. The origins of the Oldowan: Why chimpanzees are still good models for technological evolution in Africa. In *Stone tools and fossil bones*, ed. M. Dominguez-Rodrigo, 201–221. Cambridge: Cambridge University Press.

Carvalho, Susana, Dora Biro, Eugénia Cunha, Kimberley Hockings, William C. McGrew, Brian G. Richmond, and Tetsuro Matsuzawa. 2012. Chimpanzee carrying behaviour and the origins of human bipedality. *Current Biology* 22 (6): 180–181.

Cashel, R. 1995. *The Baobab in fact and fable*. Amanzimtoti (Town), Natal, South Africa. Unpublished.

Chadare, F.J., J.D. Houndhouigan, A.R. Linnemann, M.J.R. Nout, and M.A.J.S. van Boekel. 2008. Indigenous knowledge and processing of *Adansonia digitata* L. food products in Benin. *Ecology of Food and Nutrition* 47: 338–362.

Chadare, F.J., A.R. Linnemann, J.D. Hounhouigan, M.J. Nout, and M.A. Van Boekel. 2009. Baobab food products: A review on their composition and nutritional value. *Critical Reviews in Food Science and Nutrition* 49: 254–274.

Chadare, F.J., J.D. Hounhouigan, M.J.R. Nout, and M.A.J.S. van Boekel. 2013. Ethno-food knowledge of baobab (*Adansonia digitata* L.) and characterization of its traditional fermented novel foods from Benin. *ISHS Acta Horticulturae* 979: 187–197. https://doi.org/10.17660/ActaHortic.2013.979.17.

Chapotin, S., J.H. Razanameharizaka, and N.M. Holbrook. 2006. Baobab trees (*Adansonia*) in Madagascar use stored water to flush new leaves but not to support stomatal opening prior to the rainy season. *New Phytologist* 169 (3): 549–559.

Cheke, R., and C. Mann. 2008. Hunter's sunbird (Chalcomitra hunteri). In *Handbook of the birds of the world alive*, ed. J. del Hoyo, A. Elliott, J. Sargatal, D.A. Christie, and E. de Juana. Barcelona: Lynx Edicions.

Chevalier, A. 1906. Les baobabs (*Adansonia*) del' Afrique continentale. *Bulletin de la Société Botanique de France* 53: 480–496.

Chevallier, D., R. Duponnois, F. Baillon, P. Brossault, J.-M. Grégoire, H. Eva, Y. Le Maho, and S. Massemin. 2010. The importance of roosts for black storks *Ciconia nigra* wintering in West Africa. *Ardea* 98 (1): 91–96.

Childe, V. Gordon. 1941. *Man makes himself*. London: Watts.

Clark, G. 1944. Water in antiquity. *Antiquity* 18: 1–15.

Clarke, Gordon. 2017. Hadzabe huts of baobab, sisal and grass. Institute of Nomadic Architecture. http://artsection.org/hadzabe.html. Retrieved 3/8/2023.

———. 2018. *Hadzabe huts of baobab, sisal and grass*. Institute of Nomadic Architecture. http://artsection.org/hadzabe.html. Retrieved 3/9/2023.

Clarke, Andrew C., Michael K. Burtenshaw, Patricia A. McLenachan, David L. Erickson, and David Penny. 2006. Reconstructing the origins and dispersal of the Polynesian bottle gourd (*Lagenaria siceraria*). *Molecular Biology and Evolution* 23 (5): 893–900.

Clusius, J.C. 1605. *Exoticum libri decem*. Leiden: Ex Officina Plantiniana Raphalengii.

Codding, Brian F., Rebecca Bliege Bird, and Douglas W. Bird. 2011. Provisioning offspring and others: Risk – Energy trade-offs and gender differences in hunter–gatherer foraging strategies. *Proceedings of the Royal Society* 278: 2502–2509.

Coe, M.J., and F.M. Isaac. 1969. Pollination of the baobab (*Adansonia digitata* L.) by the lesser bushbaby (Galagocrassicaudatus E. Geoffroy). *East African Wildlife Journal* 3: 123–331.

Coetzee, C.G. 1963. *The prey of owls in the Kruger National Park as indicated by owl pellets collected during 1960-61*, 115–125. Johannesburg: Medical Ecology Centre, State Department of Health.

Coetzee, Hendri, Werner Nell, and Leon van Rensburg. 2014. An exploration of cultural beliefs and practices across the Southern Ground-Hornbill's range in Africa. *Journal of Ethnobiology and Ethnomedicine* 10 (28): 1–7.

Cohen, Mark Nathan. 1977. *The food crisis in prehistory*. New Haven/London: Yale University Press.

———. 2002. Demographic expansion: Causes and consequences. In *Companion encyclopedia of anthropology*, ed. Tim Ingold, 265–296. London/New York: Routledge.

Collis, J.S. 1954. *The triumph of the tree*. New York: The Viking Press.

Comitas, Laabros. 1964. Occupational multiplicity in rural Jamaica. In *Proceedings of the American ethnological society*, 41–50. Seattle: University of Washington Press.

Compton, S.G., J.T. Wiebes, and C.C. Berg. 1996. The biology of fig trees and their associated animals. *Journal of Biogeography* 23: 405–407.

Condit, I. J. 1969. *Ficus: The exotic species*. Division of Agricultural Sciences, University of California, Berkeley.

Conover, Michael R., and Gregory G. Chasko. 1985. Nuisance Canada Goose problems in the eastern United States. *Wildlife Society Bulletin* 13 (3): 228–233.

Conroy, Glenn C. 1981. Comments. In: The early hominid plant-food niche: Insights from an analysis of plant exploitation by Homo, Pan, and Papio in Eastern and Southern Africa [and comments and reply]. Author(s): Charles R. Peters, Eileen M. O'Brien, Noel T. Boaz, Glenn C. Conroy, Laurie R. Godfrey, Kenji Kawanaka, Adriaan Kortlandt, Toshisada Nishida, Frank E. Poirier And Euclid O. Smith. *Current Anthropology* 22 (2): 127–140.

Constantino, Paul J., and Barth W. Wright. 2009. The importance of fallback foods in primate ecology and evolution. *American Journal of Physical Anthropology* 140: 599–602.

Cook, R. 1974. *The tree of life: Images for the cosmos*. New York: Thames and Hudson.

Cooper, B. 1949. The Kindiga. *Tanganyika Notes and Records* 27: 8–15.

Coppens, Y., F.C. Howell, G.L. Isaac, and R.E.F. Leakey, eds. 1976. *Earliest man and environments in the Lake Rudolf Basin: Stratigraphy, paleoecology, and evolution*. Chicago: University of Chicago Press.

Cordain, L., S.B. Eaton, J.B. Miller, N. Mann, and K. Hill. 2002. The paradoxical nature of hunter-gatherer diets: Meat based, yet non-atherogenic. *European Journal of Clinical Nutrition* 56: S42–S52.

Cornu, Cyrille, Wilfried Ramahafaly, and Pascal Danthu. 2014. *Adansonia madagascariensis*, a marine hydrochory hypothesis. *Bois et Forets des Tropiques* 68 (320): 7–14.

Cotton, C.M. 1996. *Ethnobotany: Principles and applications*. New York: Wiley.

Coursey, D.G. 1973. Hominid evolution and hypogeous plant foods. *Man, New Series* 8 (4): 634–635.

Cowen, D.V. 1952. *Flowering trees and shrubs in India*. Mumbai: Thacker Press.

Cox, Hillary. 2008. The tree of life. *Cultural Survival* 32 (4) https://www.culturalsurvival.org/publications/cultural-%20survival-quarterly/tree-life. Accessed 18 June 2020.

Cracraft, Joel, and Michael J. Donoghue. 2004. *Assembling the tree of life*. New York: Oxford University Press.

Creative Commons. 2022. *A hadzabe tribe child standing in front of a tree house*. File: A typical Hadzabe tribe tree house.jpg. Wikimedia Commons. https://commons.wikimedia.org/wiki/File:A_typical_Hadzabe_tribe_tree_house.jpg. Retrieved 27 Nov 2022.

Creel, S., and N.M. Creel. 1996. Limitation of African wild dogs by competition with larger carnivores. *Conservation Biology* 10: 526–538.

Crittenden, Alyssa N. 2009. *Allomaternal care and juvenile foraging among the Hadza: Implications for the evolution of cooperative breeding in humans*. Dissertation, University of California, San Diego.

———. 2011. The importance of honey consumption in human evolution. *Food and Foodways: Explorations in the History and Culture of Human Nourishment* 19 (4): 257–273.

———. 2016. Ethnobotany in evolutionary perspective: Wild plants in diet composition and daily use among Hadza hunter-gatherers. In *Wild harvest: Plants in the Hominin and Pre-Agrarian human worlds*, ed. Karen Hardy and Lucy Kubiak-Martens. Oxford: Oxbow Books.

Crittenden, A.N., and David A. Zes. 2015. Food sharing among Hadza hunter-gatherer children. *PLoS One* 10 (7): e0131996. https://doi.org/10.1371/journal.pone.0131996. Accessed 18 June 2020.

Crittenden, Alyssa N., Nancy L. Conklin-Brittain, David A. Zes, Margaret J. Schoeninger, and Frank W. Marlowe. 2013. Juvenile foraging among the Hadza: Implications for human life history. *Evolution and Human Behavior* 34: 299–304.

Crompton, Robin Huw, William I. Sellers, and Susannah K.S. Thorpe. 2010. Arboreality, terrestriality and bipedalism. *Philosophical Transactions of the Royal Society* 365: 3301–3314.

Cron, Glynis V., Nisa Karimi, Kelsey Glennon, Chukwudi Udeh, Ed T.F. Witkowski, Sarah Venter, Achille Assogbadjo, and David Baum. 2016. One African baobab species or two? Synonymy of *Adansonia kilima* and *A. digitata*. *Taxon* 65 (5): 1037–1049.

Crowe, Tim. 2016. African gamebirds are the key to understanding global avian evolution. *The Conservation*, February 23. http://theconversation.com/african-gamebirds. Accessed 18 June 2020.

Crowther, Alison, Patrick Faulkner, Mary E. Prendergast, Eréndira M. Quintana, Mark Horton Morales, Edwin Wilmsen, Anna M. Kotarba-Morley, Annalisa Christie, Nik Petek, Ruth Tibesasa, Katerina Douka, Llorenç Picornell-Gelabert, Xavier Carah, and Nicole Boivin. 2016. Coastal subsistence, maritime trade, and the colonization of small offshore islands in Eastern African prehistory. *The Journal of Island and Coastal Archaeology* 11 (2): 211–237.

Dale, Michael. 2018. The sexual selection of hominin bipedalism. *Ideas in Ecology and Evolution* 11 (1): 47–60. https://doi.org/10.4033/iee.2018.11.6.n.

Dahlberg, Frances. 1981. *Woman the gatherer*. New Haven: Yale University Press.

Dalziel, J.M. 1937 [1948]. The useful plants of West Tropical Africa. London: Crown Agents for Oversea Governments and Administrations.

Dart, R.A. 1925. *Australopithecus africanus*, the man-ape of South Africa. *Nature* 115: 195–199.

Dart, R.A. 1953. The predatory transition from ape to man. *International Anthropological Linguistic Review* 1: 201–219.

Dart, R.A., and Dennis Craig. 1959. *Adventures with the missing link*. New York: Harper and Brothers.

Darwin, Charles. 1868. *The variation of animals and plants under domestication*. London: John Murray, Albemarle Street.

———. 1871. *The descent of man, and selection in relation to sex*. London: John Murray.

Darwin, Charles. 1871. *The descent of man, and selection in relation to sex*. London: John Murray.

David, M., J.S. Bowman, Jennifer Balch, Paulo Artaxo, William J. Bond, Mark A. Cochrane, Carla M. D'Antonio, Ruth DeFries, Fay H. Johnston, Jon E. Keeley, Meg A. Krawchuk, Christian A. Jull, Michelle Mack, Max A. Moritz, Stephen Pyne, Christopher I. Roos, Andrew C. Scott, Navjot S. Sodhi, Thomas W. Swetnam, Michelle Mack, Max A. Moritz, Stephen Pyne, Christopher I. Roos, Andrew C. Scott, Navjot S. Sodhi, and Thomas W. Swetnam. 2011. The human dimension of fire regimes on Earth. *Journal of Biogeography* 38: 2223–2236.

Day, M. H. 1986. Bipedalism: Pressures, origins and modes. In: Topics in Primate and Human Evolution, ed. B. Wood, L. Martin and P. Andrews, pp. 188–202. Cambridge: Cambridge University Press.

Davis, T.A., and S.S. Ghosh. 1976. Morphology of *Adansonia digitata*. *Adansonia sér* 2 (15): 471–479.

De Caluwé, E., S. De Smedt, A.E. Assogbadjo, R. Samson, B. Sinsin, and P. Van Damme. 2009. Ethnic differences in use value and use patterns of baobab (*Adansonia digitata* L.) in Northern Benin. *African Journal of Ecology* 47 (3): 433–440.

De Caluwé, Emmy, Kateřina Alamová, and Patrick Van Damme. 2010. *Adansonia digitata* L. – A review of traditional uses, phytochemistry and pharmacology. *Afrika Focus* 23 (1): 11–51.
De Candolle, Alphonse. 1908 (1886). *Origin of cultivated plants*. New York: D. Appleton and Company.
———. 2011. *The origin of cultivated plants*. Cambridge: Cambridge University Press.
de Groote, Sitske, Emmy De Caluwé, and Patrick Van Damme. 2007. Domestication and development of Baobab and Tamarind (DADOBAT). In *Conference on international agricultural research for development*, University of Kassel-Witzenhausen and University of Göttingen, October 9–11, 2007.
De la Torre, Ignacio. 2011. The origins of stone tool technology in Africa: a historical perspective. *The Philosophical Transactions of the Royal Society B* 366 (1567): 1028. https://doi.org/10.1098/rstb.2010.0350.
de Lamarck, J.-B. 2006 [1890]. *Zoological philosophy: An exposition with regard to the natural history of animals*. Rosamond: Bill Huth Publishers.
De Queiroz, Alan. 2005. The resurrection of oceanic dispersal in historical biogeography. *Trends in Ecology & Evolution* 20 (2): 68–73.
de Schlippe, Pierre. 1956. *Shifting cultivation in Africa*. London: Routledge and Paul.
De Smedt, S., K. Alaerts, A.M. Kouyaté, P. Van Damme, G. Potters, and R. Samson. 2011. Phenotypic variation of baobab (*Adansonia digitata* L.) fruit traits in Male. *Agroforestry Systems* 82: 87–97.
de Villiers, P.C. 1951. Die Kremetartboom (*Adansonia digitata*). *Journal of the South African Forestry Association* 29: 9–18.
De Vos, A. 1968. The baobab: Africa's wonder tree. *Canadian Audubon* 30 (4): 117–119.
Dean, W.R.J. 1974. Breeding and distributional notes on some Angolan Birds. *Novitates* X (8): 109–125.
Dean, W.T.J. 2005. House sparrow Passer domesticus. In *Roberts' birds of Southern Africa*, ed. P.A.R. Hockey, W.R.J. Dean, P.G. Ryan, and S. Maree, 1082–1083. Cape Town: John Voelcker Bird Book Fund.
Dean, W.R.J., and I.A.W. MacDonald. 1981. A review of African birds feeding in association with mammals. *Ostrich* 52: 135–155.
Dean, W.R.J., W.R. Siegfried, and I.A.W. MacDonald. 1990. The fallacy, fact, and fate of guiding behavior in the greater honeyguide. *Conservation Biology* 4 (1): 99–101.
Deeble, Mark, and Victoria Stone 2005. *The queen of trees*. Documentary. Flat Dog Productions Limited.
Deikumah, Justus P., Vida Asieduwaa Konadu, and Richard Kwafo. 2015. Bird naming systems by Akan people in Ghana follow scientific nomenclature with potentials for conservation monitoring. *Journal of Ethnobiology and Ethnomedicine* 11: 75. https://doi.org/10.1186/s13002-015-0062-y.
Dellatola, Claire. 1983. The baobab: Home for everyone. *South African Panorama*: 24–29.
Denis-Esdras, Amon Anoh, Ahoulou Assoum Stanislas, Achah Jacques Auguste Alfred Bognan, Sebe Fiba Doriane, Soro Dodiomon, N'guessan Koffi, and Traoré Dossahoua. 2020a. Ethnobotanical knowledge of medicinal values of Loranthaceae used to treat human diseases by local ethnic Agni Sanwi from Aboisso and Maferé in Côte d'Ivoire. *World Journal of Biology Pharmacy and Health Sciences* 04 (01): 039–050.
Denis-Esdras, Anoh Amon, Victoire Annick Koulibaly, and El Amine Hamed Sako. 2020b. Parasitism of Loranthaceae on crops: Case of plantations in rural area on the periphery of Jean Lorougnon Guédé University of Daloa, Central-West Côte d'Ivoire. *World Journal of Advanced Research and Reviews* 07 (02): 155–167.
Dennell, R.W. 1989. Comments. *Current Anthropology* 30 (1): 1–26.
Destombes, Jérôme. 2006. From long-term patterns of seasonal hunger to changing experiences of everyday poverty: Northeastern Ghana *c*. 1930–2000. *The Journal of African History* 47: 181–205.

DeVore, I. 1965. Male dominance and mating behavior in baboons. In *Sex and behavior*, ed. F. Beach, 266–289. New York: Wiley.

Dhillion, Shivcharn S., and Gunnar Gustad. 2004. Local management practices influence the viability of the baobab (*Adansonia digitata* Linn.) in different land use types, Cinzana, Mali. *Agriculture, Ecosystems and Environment* 101: 85–103. https://doi.org/10.1016/S0167-8809(03)00170-1.

Diamond, A.W., and T.E. Lovejoy 1985. *Conservation of tropical forest birds*. Princeton University Press.

Dick, C.W., E. Bermingham, M.R. Lemes, and R. Gribel. 2007. Extreme long-distance dispersal of the lowland tropical rainforest tree *Ceiba pentandra* L. (Malvaceae) in Africa and the Neotropics. *Molecular Ecology* 16: 3039–3049.

Diop, Aïda Gabar, Mama Sakho, Manuel Dornier, Mandy Cisse, and Max Reynes. 2006. Le baobab africain (*Adansonia digitata* L.) principals caractéristiques et utlisations. *Fruits* 61 (01): 55–69.

Done, Chris. 2010. The boab: Australia's isolated Adansonia. *Bois et forets des Tropiques* 306 (4): 17–22.

Dos Santos, J. 1609. Ethiopia oriental (Convento de S. Domingos de Évora, Évora, 1609).

Doxiadis Associates and Doxiadis Ionides Associates Ltd. 1965. Land and water use survey in Kordofan Province of the Republic of the Sudan. *Ekistics* 19 (111): 114–121.

Du Puy, B. 1996. Faunal interactions with the genus *Adansonia* in the Kiridy Forest. *Primate Report* 46 (1): 329–338.

Dubin, Lois Sherr. 1987. *The history of beads*. New York: Harry N. Abrams.

Durkheim, Emile. 1893. *De la division du travail social*. Glencoe: Free Press.

Duvall, Chris S. 2007. Human settlement and baobab distribution in south-western Mali. *Journal of Biogeography* 34 (11): 1947–1961.

Duxoux, E. 1983. Are baobab flowers only pollinated by male straw-coloured fruit bats Eidolon helvum? [Senegal]. *Revue-d'Ecologie (France)* 38 (2): 229–231.

Eby, Stephanie, Anna Mosser, Ali Swanson, Craig Packer, and Mark Ritchie. 2013. The impact of burning on lion Panthera leo habitat choice in an African savanna. *Current Zoology* 59 (3): 335–339.

Edenmyr, Niklas. 2004. The semantics of Hadza gender assignment. *Africa & Asia* 4: 3–19.

Edkins, M.T., L.M. Kruger, K. Harris, and J.J. Midgley. 2007. Baobabs and elephants in Kruger National Park: Nowhere to hide. *African Journal of Ecology* 46 (2): 119–125.

El-Bushra, El-Sayed, and Mohammed Osman El Sammani. 1977. Urban and rural water supplies in the Sudan. *Athens Center of Ekistics* 43 (254): 36–42. https://www.jstor.org/stable/i40144727.

Eliade, Mircea. 1958. *Patterns in comparative religion*. Sheed and Ward.

———. 1991. *Images and symbols: Studies in religious symbolism*. Princeton: Princeton University Press.

Elliott, Clive C.H. 2006. Bird population explosions in agroecosystems – The quelea, *Quelea quelea*, case history. Acta Zoologica Sinica 52 (Supplement): 554–560.

Ellert, Mary Wilkins. 2006. New information on the origins of bottle gourd (*Lagenaria siceraria*). *Desert Plants* 22 (1): 8–17. http://hdl.handle.net/10150/555919.

Ember, M., and C.R. Ember. 1979. Male-female bonding: A cross-species study of mammals and birds. *Behavior Science Research* 14: 37–56.

Ember, Carol R., and Melvin Ember. 1998. *Anthropology: brief introduction*. 3rd ed. New Jersey: Prentice Hall.

Emboden, William A. 1974. *Bizarre plants: Magical, monstrous mythical*. Littlehampton Book Services Ltd.

Encyclopedia Britannica. 2015. The editors of Encyclopedia. "World tree". Encyclopedia Britannica, 8 Nov. 2015, https://www.britannica.com/topic/world-tree. Accessed again on 7 March 2023.

Erickson, David L., Bruce D. Smith, Andrew C. Clarke, Daniel H. Sandweiss, and Noreen Tuross. 2005. An Asian origin for a 10,000-year-old domesticated plant in the Americas. *PNAS* 102 (51): 18315–18320.

Estalrrich, Almudena. 2018. *Division of labor in human evolution*. Résumé. sciencesconf.org:uispp2018:182681.

Estalrrich, Almudena, and Antonio Rosas. 2015. Division of labor by sex and age in Neandertals: An approach through the study of activity-related dental wear. *Journal of Human Evolution* 80: 51–63.

Etkin, W. 1954. Social behavior and the evolution of man's faculties. *American Naturalist* 88: 129–142.

Evans-Pritchard, E. E. (Edward Evan). 1940. *The Nuer: A description of the modes of livelihood and political institutions of a Nilotic People*. Oxford: Clarendon Press.

Fagan, Brian M. 1989. *People of the Earth: An introduction to world prehistory*. Glenview: Scott, Foresman and Company.

Falk, Dean. 2007. Constraints on brain size: The radiator hypothesis. *Evolution of Nervous Systems* 4: 347–353.

FAO. 2012. *Tropical palms – Historical role of palms in human culture*. Produced by the FAO Forestry Department, November 29. http://www.fao.org/docrep/X0451E/X0451e04.htm. FAO Corporate Document Repository.

Feare, Christopher John. 1985. Starling. In *A dictionary of birds*, ed. Bruce Campbell and Elizabeth Lack, 562–563. Vermillion: Buteo Books.

Fedigan, L.M. 1986. The changing role of women in models of human evolution. *Annual Review of Anthropology* 15: 25–66.

Fenner, M. 1980. Some measurements on the water relations of baobab trees. *Biotropica* 12: 205–209.

Fernández, Esteban, and Olaf Jöris. 2008. Personal ornaments in the Early Upper Paleolithic of Western Eurasia: An evaluation of the record. In *Dating the Middle to Upper Palaeolithic boundary across Eurasia*, ed. O. Jöris and D.S. Adler. Proceedings of Session C57, 15th UISPP, Lisbon, Portugal, September 2006. *Eurasian Prehistory* 5(2) (2007): 31–44.

Ferro-Luzzi, A. 1990a. Social and public health issues in adaptation to low energy intakes. *American Journal of Clinical Nutrition* 57: 309–315.

———. 1990b. Seasonal energy stress in marginally nourished rural women: Interpretation and integrated conclusions of a multicentre study in three developing countries. *European Journal of Clinical Nutrition* 44: 41–46.

Ferry, M.-P., M. Gessain, and R. Gessain. 1974. *Ethno-botanique Tenda*. Paris: Documents du Centre de Recherches Anthropologiques du Musée de l'Homme.

Finlayson, Stewart, and Clive Finlayson. 2016. The birdmen of the Pleistocene: On the relationship between Neanderthals and scavenging birds. *Quaternary International* 421: 78–84.

Fisher, Helen E. 1982. *The sex contract: The evolution of human behavior*. New York: William Morrow.

Fitzpatrick, Katherine K. 2018. *Foraging and menstruation in the Hadza of Tanzania*. Dissertation, Department of Archaeology and Anthropology, University of Cambridge.

Fleuret, A. 1980. Nonfood uses of plants in Usumbara. *Economic Botany* 34: 320–333.

Foley, R. 1995. Causes and consequences in human evolution. *The Journal of the Royal Anthropological Institute* 1 (1): 67–86.

Food and Agricultural Organization 1988. Traditional food plants. FAO Food and Nutrition Paper 42. FAO. Rome.

Fortes, M., and S.L. Fortes. 1936. Food in the domestic economy of the Tallensi. *Africa* ix (2): 237–276.

Fosbrooke, H.A. 1956. A Stone Age tribe in Tanganyika. *South African Archaeological Bulletin* 11: 3–9.

Fowler, Catherine S., and Dana Lepofsky. 2011. Traditional resource and environmental management. In *Ethnobiology*, ed. E.N. Anderson, D. Pearsall, E. Hunn, and N. Turner. New York: Wiley.

Frazer, Sir James George. 1976 (1922). *The golden bough*. New York: MacMillan.

Frazer, J. 1976. *The golden bough*. New York: MacMillan.

Freedman, Bob. 2018. *Adansonia digitata*. Famine Foods Database. Purdue University. https://www.purdue.edu/hla/sites/famine-foods?s=baobab. Accessed 18 June 2020.

Friede, H.M. 1953. Trees in rock paintings. *Tree Society of Southern Africa* 5 (3): 8–11, 13.

Fruth, B., and G. Hohmann. 1994. Nests – Living artifacts of recent apes. *Current Anthropology* 35: 310–311.

Fruth, Barbara, Nikki Tagg, and Fiona Stewart. 2018. Sleep and nesting behavior in primates: A review. *The American Journal of Physical Anthropology* 166: 499–509. https://doi.org/10.1002/ajpa.23373.

Fry, Charles Hilary. 1985. Hoopoe. In *A dictionary of birds*, ed. Bruce Campbell and Elizabeth Lack, 289–290. Vermillion: Buteo Books.

Gabaza, Molly, Habtu Shumoy, Maud Muchuweti, Peter Vandamme, and Kathleen Raes. 2018. Baobab fruit pulp and mopane worm as potential functional ingredients to improve the iron and zinc content and bioaccessibility of fermented cereals. *Innovative Food Science and Emerging Technologies* 47: 390–398.

Galat-Luong, A., and G. Galat. 2000. Chimpanzees and baboons drink filtrated water. *Folia Primatologia* 71: 258.

Galvin, K., K. Hawkes, J.A. Maga, J.F. O'Connell, and N. Blurton-Jones. n.d. *The composition of some wild plant foods used by East African hunter-gatherers*. Unpublished manuscript.

Gautier, D. 1996. Ficus (Moraceae) as part of agrarian systems in the Bamileke region (Cameroon). *Economic Botany* 50 (3): 318–326.

Gebauer, J., and G. Egbert. 2005. Comparison of the salt tolerance of the two under-utilised fruit species, baobab (*Adansonia digitata* L.) and tamarind (*Tamarindus indica* L.). *Conference on International Agricultural Research for Development* 2005: 1–4.

Gebauer, J., K. El-Siddig, and G. Ebert. 2002. Baobab (*Adansonia digitata* L.): A review on a multipurpose tree with promising future in the Sudan. *Gartenbauwissenschaft* 67 (4): 155–160.

Gebauer, Jens, Yahia O. Adam, Aida Cunı´ Sanchez, Dietrich Darr, Muneer E.S. Eltahir, Kamal E.M. Fadl, Gabriele Fernsebner, Michael Frei, Tsige-Yohannes Habte, Karl Hammer, Mauricio Hunsche, Henry Johnson, Maha Kordofani, Michael Krawinkel, Florian Kugler, Eike Luedeling, Tarig E. Mahmoud, Anthony Maina, Dagmar Mithöfer, Chimuleke R.Y. Munthali, Georg Noga, Rabea North, Willis O. Owino, Kathleen Prinz, Freda K. Rimberia, Amina Saied, Martin Schüring, Anne Sennhenn, Martin A. Späth, Mohamed E.N. Taha, Andreas Triebel, Florian Wichern, Martin Wiehle, Nicole Wrage-Mönnig, and Katja Kehlenbeck. 2016a. Africa's wooden elephant: The baobab tree (*Adansonia digitata* L.) in Sudan and Kenya: A review. *Genetic Resources and Crop Evolution* 63: 377–399.

Gebauer, Jens, Cory W. Whitney, and John R.S. Tabuti. 2016b. First record of baobab (*Adansonia digitata* L.) in Uganda. *Genetic Resources and Crop Evolution* 63: 755–762.

Geraghty, Nina. 2016. *Vanessa Bristow: A bird's guide to baobab trees*. https://www.baobabfoundation.co.za/vanessa-bristow-birds-guide-baobab-trees/. Accessed 8 Nov 2020.

Gerber, C. 1895. Contribution à l'histoire botanique, thérapeutique et chimique du genre *Adansonia* (Baobab). *Ann Inst Bot-Géol* 2: 1–78.

Giaimo, Cara. 2016. *The surprisingly sticky tale of the Hadza and the honeyguide bird*. Atlas Obscura. https://www.atlasobscura.com/articles/the-surprisingly-sticky-tale-of-the-hadza-and-the-honeyguide-bird. Retrived 18 June 2020.

Gibbons, Ann. 2012. Generation gaps suggest ancient human-ape split. *Science, All News*. https://doi.org/10.1126/article.25782.

———. 2018. Hadza on the brink. *Science* 360 (6390): 700–704. https://doi.org/10.1126/science.360.6390.700.

Giffard, Pierre L. 1974. *L'Arbre dans le paysage sénégalais sylviculture en zone tropicale sèche*. Paris: Centre technique forestier tropical.

Giglio, Stafanie. 2010. *Picture it: Baobab Juice*. Posted in Malawi by stefaniegiglio on 14 August 2010. https://stefaniegiglio.wordpress.com/2010/08/14/baobob-juice/. Retrived 3/24/2023.

Gilbert, Allan S. 1989. Comment. *Current Anthropology* 30 (1): 1–26.

Gill, Rancis B. 1985. Sunbird. In *A dictionary of birds*, ed. Bruce Campbell and Elizabeth Lack. Vermillion: Buteo Books.

Gill, Frank, Mary Heimerdinger Clench, and Oliver L. Austin. 2018. Passeriform. *Encyclopedia Britannica*. Encyclopedia Britannica, Inc. https://www.britannica.com/animal/passeriform. Accessed 1 June 2020.

Gill, F., Mary Heimerdinger Clench, and Oliver L Austin 2023. *Passeriform*. Encyclopedia Britannica, February 10, 2023. https://www.britannica.com/animal/passeriform. Accessed 5 Mar 2023.

Goenster, S., M. Wiehle, K. Kehlenbeck, R. Jamnadass, J. Gebauer, and A. Buerkert. 2011. Indigenous fruit trees in homegardens of the Nuba Mountains, Central Sudan: tree diversity and potential for improving nutrition and income of rural communities. *Acta Horticulturae* 911: 355–364.

Goodall, Jane. 1986. *The chimpanzees of Gombe: Patterns of behavior*. Cambridge: Harvard University Press.

Goody, J. 1993. *The culture of flowers*. Cambridge: University Press.

Gowlett, J.A.J. 2010. Firing up the social brain. In *Social brain, distributed mind*, ed. Robin Dunbar, Clive Gamble, and John Gowlett (London, 2010; online edn, British Academy Scholarship Online, 31 Jan. 2012), https://doi.org/10.5871/bacad/9780197264522.003.0017. Accessed 1 Mar 2023.

———. 2014. Human evolution: Use of fire. In *Encyclopedia of global archaeology*, ed. C. Smith. New York: Springer. https://doi.org/10.1007/978-1-4419-0465-2_679.

———. 2016. The discovery of fire by humans: A long and convoluted process. *Philosophical Transactions of the Royal Society B* 371: 20150164. https://doi.org/10.1098/rstb.2015.0164.

———. 2018. Solving the puzzle of humanity's mastery of fire: Professor John Gowlett talks to the Athens and Macedonian News Agency. Archaeology Newsroom. https://www.archaeology.wiki/. Downloaded March 1, 2023.

Gowlett, J.A.J., and Richard W. Wrangham. 2013. Earliest fire in Africa: Towards the convergence of archaeological evidence and the cooking hypothesis. *Azania Archaeological Research in Africa* 48 (1): 5–30. https://doi.org/10.1080/0067270X.2012.756754.

Gowlett, J.A.J., J.W.K. Harris, D. Walton, and B.A. Wood. 1981. Early archaeological sites, hominid remains and traces of fire from Chesowanja, Kenya. *Nature* 294: 125–129.

Green, David S., Gary J. Roloff, Brian R. Heath, and Kay E. Holekamp. 2014. Temporal dynamics of the responses by African mammals to prescribed fire. *Journal of Wildlife Management* 79: 235–242.

Gruber, A. 1962. Letters to the Editor: Comments on "Food Transport and the Origin of Hominid Bipedalism." American Anthropologist, 64: 605-606.

Gruber, Martin. 2018. Hunters and guides: Multispecies encounters between humans, honeyguide birds and honeybees. *African Study Monographs* 39 (4): 169–187. https://doi.org/10.14989/236670.

Gruenwald, Joerg. 2009. Novel botanical ingredients for beverages. *Clinics in Dermatology* 27: 210–216.

Gruenwald, J., and M. Galizia. 2005. Market brief in the European Union for selected natural ingredients derived from native species. *Adansonia digitata* L., Baobab. In *United Nations conference on trade and development*, UN, Schweiz.

Grzimek, Bernhard. 1970. *Among animals of Africa*. New York: Stein and Day.

Gupta, Sankar Sen. 1980. *Sacred trees across cultures and nations*. Calcutta: Indian Publications.

Gurashi, N.A., and A. Kordofani. 2014. Morphological variation in fruit shapes of *Adansonia digitata* L., from Blue Nile and North Kordofan States, Sudan. *Journal of Forest Products and Industries* 3 (2): 106–111.

Gustad, Gunnar, S. Shivcharn Dhillion, and Daouda Sidibé. 2004. Local use and cultural and economic value of products from trees in the parklands of the municipality of Cinzana, Mali. *Economic Botany* 58 (4): 578–587.

Guy, G.L. 1971. The baobabs: *Adansonia* spp (Bombacaceae). *Journal of the Botanical Society of South Africa* 57: 30–37.

Guy, R.D. 1972. The gorongosa bee tree. *African Wildlife* 26 (4): 160–161.

Guy, P.R. 1982. Baobabs and elephants. *African Journal of Ecology* 20: 215–220.

Haberman, David L. 2010. Faces in the trees. *Journal for the Study of Religion, Nature and Culture* 4 (2): 173–190. https://doi.org/10.1558/jsrnc.v4i2.173.

Hall, B.P., and R.E. Moreau. 1970. *An atlas of speciation in African passerine birds*. London: British Museum (Natural History).

Halperin, Rhoda. 1980. Ecology and mode of production: Seasonal variation in the division of labor by sex among hunter-gatherers. *Journal of Anthropological Research* 36: 379–399.

Hankey, J.A. 2004. Adansonia digitata A. L. (Bombacaceae). Plantzafrica.com 2011. In *CRC handbook of chemistry and physics,* ed. M. William, 92nd ed. M. William, 122. CRC Press, Boca Raton. Retrieved from http://pza.sanbi.org/adansonia-digitata, 155–160. https://doi.org/10.1007/s10341-017-0328-8.

Hardy, A. 1960. Was man more aquatic in the past? *New Scientist* 7: 642–645.

Hardy, D., and R. LaFon. 1982. The baobab: Silent Colossus of the African bush. *Cactus and Succulent Journal* 54: 51–53.

Harlan, Jack R. [1976] 1992. *Crops and man*. Second Edition (First Edition 1976). Madison: American Society of Agronomy, Inc./Crop Science Society of America, Inc.

Harris, David R. 1976. Traditional systems of plant food production and the origins of agriculture in West Africa. In *Origins of African plant domestication*, ed. J.R. Harlan, J.M.J. de Wet, and A.B.L. Stemler, 311–352. The Hague: Mouton.

Harris, David R., ed. 1980a. *Human ecology in Savanna environments*. New York: Academic.

Harris, David. 1980b. Tropical savanna environments: Definition, distribution, diversity and development. In *Human ecology in Savanna environments*, 3–27. New York: Academic.

Harris, B.J., and H.G. Baker. 1959. Pollination of flowers by bats in Ghana. *Nigerian Field* 24: 151–159.

Harrison, S.G., G.B. Masefield, and Michael Wallis. 1969. *The Oxford book of food plants*. London: Oxford University Press.

Harrison, Terry, Michael L. Mbago, and Charles P. Msuya. 1997. New archaeological sites in the Manonga Valley, North-Central Tanzania. *Nyame Akuma, Bulletin of the Society of Africanist Archaeologists* 47: 53–60.

Hart, Donna, and Robert Sussman. 2009. *Man the hunted: Primates, predators and human evolution*. New York: Westview Press.

Haswell, M.R. 1953. *Economics of agriculture in a savannah village*. London: H. M. S. O. Colonial Research Studies, No. 8.

Hawkes, K. 1990. Why do men hunt? Some benefits for risky strategies. In *Risk and uncertainty in tribal and peasant economies*, ed. E. Cashdan, 145–166. Boulder: Westview Press.

———. 1991. Showing off: Tests of a hypothesis about men's foraging goals. *Ethnology and Sociobiology* 12: 29–54.

———. 2016. Ethnoarchaeology and Plio-Pleistocene sites: Some lessons from the Hadza. *Journal of Anthropological Archaeology* 44: 158–165.

Hawkes, K., J.F. O'Connell, and N.G. Blurton Jones. 1989. Hardworking Hadza grandmothers. In *Comparative socioecology of mammals and man*, ed. V. Standen and R. Foley, 341–366. London: Blackwell.

———. 1991. Hunting income patterns among the Hadza: Big game, common goods, foraging goals and the evolution of the human diet. *Philosophical Transactions: Biological Sciences* 334 (1270): 243–250.

———. 1995. Hadza children's foraging: Juvenile dependency, social arrangements, and mobility among hunter-gatherers. *Current Anthropology* 36 (4): 688–700.

———. 1997. Hadza women's time allocation, offspring provisioning, and the evolution of long postmenopausal life spans. *Current Anthropology* 38 (4): 551–577.

———. 2001a. Hunting and nuclear families: Some lessons from the Hadza about men's work. *Current Anthropology* 42 (5): 681–709.

———. 2001b. Hadza meat sharing. *Evolution and Human Behavior* 22: 113–142.

———. 2001c. Is meat the hunter's property? Big game, ownership, and explanations of hunting and sharing. In *Meat-eating and human evolution*, ed. C. Stanford and H. Bunn, 219–236. Oxford: Oxford University Press.

Heath, M.E. 1992. Manis temminckii. *Mammalian Species* 415: 1–5.

Hedrick, V.P. 1972. *Sturtevant's edible plants of the world.* New York: Dover Publications. First published in 1919 by Lyon, Albany for the State of New York as the Department of Agriculture's 27th Annual Report, Vol. 2, Part 11 [Report of the New York Agricultural Experiment Station for the year 1919].

Hellekson, Lyndsay. 2009. High biodiversity in association with the common baobab tree. *Desert Plants* 25 (1): 3–9.

Hema, Sane, and Vinaya Ghate. 2000. Exotic medicinal plants: Antiquity in ayurveda and ethno-medico-botany. *Asian Agri-History* 4: 283–296.

Hepper, F.N. 1976. *The West African herbaria of isert and thonning.* London: Bentham-Moxon Trust, Royal Botanic Gardens, Kew.

Hertz, R. 1960a [1907]. *Death and the right hand.* London: Cohen and West.

———. 1960b. The collective representation of death. In *Death and the right hand.* Trans. R. Needham and C. Needham, 84–86. Aberdeen: Cohen and West.

Herzog, Nicole M. 2015. *Primate behavioral responses to burning as a model for hominin fire use.* PhD dissertation, University of Utah, Salt Lake City.

Hewes, Gordon W. 1961. Food transport and the origin of hominid bipedalism. *American Anthropologist* 63: 687–710.

———. 1964. Hominind bipedalism: Independent evidence for the food-carrying theory. *Science, New Series* 146 (3642): 416–418.

HHES (Highwood Hills Elementary School). 2005. *Baobab, tree of life.* http://www.highwood.spps.org/classroom_Music_with_Mr__Lacina.html. Retrieved on 27 Sept 2005, 01:37:15 GMT.

Hines, Deborah A., and Karlyn Eckman. 1993. *Indigenous multipurpose trees of Tanzania: Uses and economic benefits for people.* Cultural Survival Canada and Development Services Foundation of Tanzania. http://www.fao.org/docrep/X5327e/x5327e0g.htm.

Hobley, C.W. 1922. On baobabs and ruins. *Journal of East African Natural History Society* 5: 75–77.

Hockey, P.A.R., W.R.J. Dean, P.G. Ryan, and S. Maree, eds. 2005. *Roberts' birds of Southern Africa.* Cape Town: John Voelcker Bird Book Fund.

Hoebel, E. Adamson. 1958. *Man in the primitive world.* New York: McGraw-Hill Book Company, Inc.

Hoesch, W. 1937. Ueber das "Honiganzeigen" von Indicator. *Journal fur Ornithologie* 85: 201–205.

Hofmeyr, M.V. 2003. Spatial demography of selected tree species in the Kruger National Park. In *Relation to Elephant Impacts. Progress Report of Research Committee Meeting.* South Africa: University of the Witwatersrand.

Hollis, R. 1963. Reflections on baobabs. *The Nigerian Field* 28 (3): 134–138.

Holmes, Lowell D. 1971. *Anthropology: An introduction.* New York: The Ronald Press Company.

Holt, D.N. 1959. A note on the underground water resources of the protectorate. *The Nyasaland Journal* 12 (2): 60–82.

Hopler, Whitney. 2017. *Baobab: The miraculous tree of life.* https://www.thoughtco.com/baobab-the-miraculous-tree-of-life 124418.

Houston, David Charles. 1985. Vulture (1). In *A dictionary of birds*, ed. Bruce Campbell and Elizabeth Lack, 634–635. Vermillion: Buteo Books.

Humphries, C.J. 1982. Bombacaceae. In *The flowering plants of the world*, ed. V.H. Heywood, 93–94. Upper Saddle River: Prentice Hall.

Hunt, K. D. 1992. Positional behavior of Pan troglodytes in the Mahale Mountains and Gombe Stream National Parks, Tanzania. American Journal of Physical Anthropology 87:83–105.

Hunt, K. D. 1994. The evolution of human bipedality: Ecology and functional morphology. Journal of Human Evolution 26:183–202.

Hunt, Kevin D. 1996. The postural feeding hypothesis: An ecological model for the evolution of bipedalism. *South African Journal of Science* 92: 77–90.

Hunt, K.D., A.J.M. Cleminson, J. Latham, R.I. Weiss, and S. Grimmond. 1999. A partly habituated community of dryhabitat chimpanzees in the Semliki Valley Wildlife Reserve, Uganda. *American Journal of Physical Anthropology* 28 (Supplment): 157.

Hunter, J.M. 1967. Seasonal hunger in a part of the West African Savanna: A survey of bodyweights in Nangodi, North-East Ghana. *Transactions of the Institute of British Geographers* 41: 167–185.

Huntingford, G.W.B. 1954. The political organization of the Dorobo. *Anthropos* 46: 1–48.

Huntington, Richard, and Peter Metcalf. 1979. *Celebrations of death*. Cambridge: Cambridge University Press.

Huss-Ashmore, R., and J.L. Goodman. 1989. Seasonality of work, weight, and body composition for women in highland Lesotho. In *Coping with seasonal constraints MASCA research papers in science and archaeology*, ed. R. Huss-Ashmore with J.J. Curry and R.K. Hitchcock, Vol. 5, 29–44. Philadelphia: University Museum, University of Pennsylvania.

Hussein, Moustafa H. 2001. Beekeeping in Africa. *Apiacta*: 1–2.

Hutchinson, J., J.M. Dalziel, and R.W.J. Keay. 1958. *Floral of West Tropical Africa*. London: Agents for Overseas Government and Administration.

Ichikawa, Mitsuo. 1998. The birds as indicators of the invisible world: Ethno-ornithology of the Mbuti hunter-gatherers. *African Study Monographs* 25 (Supplementary issue): 105–121. https://doi.org/10.14989/68390.

iGo Terra. 2020. *Tanzania bird checklist*. iGoTerra, May 21. Retrieved 12 June 2020.

Imoro, Abdul-Wahab Mbelayim, and Victor Rex Barnes. 2013. Highly-clumped baobab trees (*Adansonia digitata* L.): Its ecology and impact on neighbouring flora. *Journal of Plant Sciences* 1 (4): 81–91.

Imoro, Abdul-Wahab Mbelayim, Victor Rex Barnes, and Isaac Sackey. 2015. Litter production and decomposition of different baobab (*Adansonia digitata* L.) stands in a savanna parkland of Northern Ghana. *Canadian Journal of Pure & Applied Science* 9 (3): 3585–3593.

Ingold, Tim. 1980. *Hunters, pastoralists, and ranchers*. Cambridge: Cambridge University Press.

International Centre for Underutilized Crops (ICUC). 2002. *Fruits for the future: Baobab*. Factsheet No. 4, March.

Irvine, F.R. 1952. Supplementary and emergency food plants of West Africa. *Economic Botany* 6 (1): 23–40.

———. 1961. *Woody plants of Ghana with special reference to their uses*. London: Oxford University Press.

Isaac, G. 1978a. Food sharing and human evolution: Archaeological evidence from the Plio-Pleistocene of East Africa. *Journal of Anthropological Research* 34 (3): 311–325.

———. 1978b. The food sharing behavior of protouman hominids. *Scientific American* 238 (4): 60–180.

Isack, H.A., and H.U. Reyer. 1989. Honeyguides and honey gatherers: Interspecific communication in a symbiotic relationship. *Science* 243 (4896): 1343–1346.

J. H. H. 1926. Gum Arabic. Bulletin of Miscellaneous Information (Royal Botanic Gardens, Kew) 1926 (10), 485–486. Published by: Springer on behalf of Royal Botanic Gardens, Kew. https://www.jstor.org/stable/i382166.

Jablonski, Nina G., and George Chaplin. 1993. Origin of habitual terrestrial bipedalism in the ancestor of Hominidae. *Journal of Human Evolution* 24: 259–280.

Jackson, Simon, and Anabel Maldonado. 2015. Baobab: The tree of life (an ethnopharmacological review). *Herbalgram, November 2015 – January 2016* 108: 42–53.

Jaeger, P. 1945. Épanouissement et pollinisation de la fleur du baobab. *Compte Rendu Hebd. Séances Acad Sci* 220: 369–371.

———. 1950. La vie nocturne de la fleur du Baobab (*Adansonia digitata* L., Bombacacées). *La Nature* 3177: 28–29.
Jaeggi, Adrian V., and Michael Gurven. 2013. Natural cooperators: Food sharing in humans and other primates. *Evolutionary Anthropology* 22 (4): 186–195.
Jahn, S.A.A. 1979. African plants used for the improvement of drinking water. *Curare* 2 (3): 183–199.
Jallow, Saidou Sireh. 1995. Identification of and response to drought by local communities in Fulladu West District, The Gambia. *Singapore Journal of Tropical Geography* 16 (1): 22–41.
James, Steven R. 1989. Hominid use of fire in the Lower and Middle Pleistocene: A review of the evidence. *Current Anthropology* 30 (1): 1–26. https://doi.org/10.1086/203705.
Jeffrey, C. 1978. Cucurbitaceae. In *Flora Zambesiaca*, ed. E. Launert, vol. 4, 414–499. London: Flora Zambesiaca Managing Committee.
Jelliffe, D.B., J. Woodburn, F.J. Bennett, and E.F.P. Jelliffe. 1962. The children of the Hadza. *Journal of Pediatrics* 60 (6): 907–913.
Jenike, M.R. 1988. *Seasonal hunger among tropical Africans: The lese case*. PhD thesis, University of California, Los Angeles.
Jensen, Jan Svejgaard, Jules Bayala, Haby Sanou, Adama Korbo, Anders Raebild, Sié Kambou, Abasse Tougiani, Henri-Noël Bouda, Anders Søndergaard Larsen, and Charles Parkouda. 2011. A research approach supporting domestication of Baobab (*Adansonia digitata* L.) in West Africa. *New Forests* 41: 317–335.
Jerolmack, Colin. 2008. How pigeons became rats: The cultural-spatial logic of problem animals. *Social Problems* 55 (1): 72–94.
Jobes, Gertrude. 1962. *Dictionary of mythology, folklore and symbols*. New York: Scarecrow Press.
Johansson, M. 1999. *The baobab tree in Kondoa Irangi Hills, Tanzania*, Minor field studies no. 74. Uppsala: Swedish University of Agricultural Sciences.
Johnson, Dennis V. 2012. Multi-purpose palms in agroforestry: A classification and assessment. *International Tree Crops Journal* 2 (3–4): 217–244.
Johnston, Bruce. 1958. *The staple food economies of western tropical Africa*. Palo Alto: Stanford University Press.
Jolly, Cliff. 1970. The seed eaters: A new model of hominid differentiation based on a baboon analogy. *Man* 5: 5–26.
Jolly, Clifford J., and Fred Plog. 1987. *Physical anthropology and archeology*. 4th ed. New York: Alfred A. Knopf.
Jones, Owain, and Paul Cloke. 2002. *Tree cultures*. New York: Oxford University Press.
Jones, N.B., and Frank W. Marlowe. n.d. *History and the Hadza*. Unpublished manuscript.
Jordania, Joseph 2011. Why Do People Sing? Music in Human Evolution. Logos : Tbilisi Ivane Javakhishvili State University, Tbilisi [Georgia].
Jordania, Joseph, and Alexander Jordania. 2011. *Why do people sing? Music in human evolution*. Logos. ISBN 9789941401862 (ISBN10: 9941401861).
Joseph, Grant S., Graeme S. Cumming, David H.M. Cumming, Zacheus Mahlangu, Res Altwegg, and Colleen L. Seymour. 2011. Large termitaria act as refugia for tall trees, deadwood and cavity-using birds in a miombo woodland. *Landscape Ecology* 26: 439–448.
Kaare, Bwire, and James Woodburn. 1999. The Hadza of Tanzania. In *The Cambridge encyclopedia of hunters and gatherers*, ed. Richard B. Lee. Cambridge: Cambridge University Press.
Kaboré, Donatien, Hagrétou Sawadogo-Lingani, Bréhima Diawara, Clarisse S. Compaoré, Mamoudou H. Dicko, and Mogens Jakobsen. 2011. A review of baobab (*Adansonia digitata*) products: Effect of processing techniques, medicinal properties and uses. *African Journal of Food Science* 5 (16): 833–844.
Kamatou, G.P.P., I. Vermaak, and A.M. Viljoen. 2011. An updated review of *Adansonia digitata*: A commercially important African tree. *South African Journal of Botany* 77: 908–919.
Kaplan, Hillard, and Kim Hill. 1985. Food sharing among ache foragers: Tests of explanatory hypotheses. *Current Anthropology* 26 (2): 223–246.
Karangi, Matthew M. 2008. Revisiting the roots of Gĩkũyũ culture through the sacred Mũgumo tree. *Journal of African Cultural Studies* 20 (1): 117–132.

Keeley, Jon E., and Philip W. Rundel. 2005. Fire and the Miocene expansion of C4 grasslands. *Ecology Letters* 8 (7): 683–690.

Kelly, R.L. 1983. Hunter-gatherer mobility strategies. *Journal of Anthropological Research* 39 (3): 277–306. https://doi.org/10.1086/jar.39.3.3629672.

———. 1995. *The foraging spectrum*. Washington, DC: Smithsonian Institution Press.

Kempe, Andreas, Christoph Neinhuis, and Thea Lautenschläger. 2018. *Adansonia digitata* and *Adansonia gregorii* fruit shells serve as a protection against high temperatures experienced during wildfires 2018. *Botanical Studies* 59: 7.

Kent, Susan. 1992. The current forager controversy: Real versus ideal views of hunter-gatherers. *Man (New Series)* 27 (1): 45–70. https://doi.org/10.2307/2803594.

Kennedy, G.E. 2005. From the ape's dilemma to the weanling's dilemma: Early weaning and its evolutionary context. *Journal of Human Evolution* 48: 123–145.

Kessler, Cristina, and Walter Lyon Krudop. 2002. *My great-grandmother's gourd*. New York: Orchard Books.

Kew Science (Plants of the World online). 2020. *Adansonia digitata* L. http://powo.science.kew.org/taxon/urn:lsid:ipni.org:names:558628-1. Accessed 18 June 2020.

Kharia, C.P., ed. 2007. *Indian medicinal plants: An illustrated dictionary*. Berlin: Springer.

Kilham, L. 1989. *The aerican crow and the common raven*. College Station: Texas A&M University Press.

Kimbel, W.H., and L.K. Delezene. 2009. "Lucy" redux: A review of research on *Australopithecus afarensis*. *American Journal of Physical Anthropology* 140: 2–48.

Kimber, C. 1966. Dooryard gardens of Martinique. *Yearbook of the Association of Pacific Coast Geographers* 28: 97–118.

———. 1973. Spatial patterning in dooryard gardens: Implications of a Puerto Rican example. *Geographical Review* 63: 6–26.

———. 1978. A folk context for plant domestication: Or the dooryard garden revisited. *Anthropological Journal of Canada* 16: 2–11.

Kimura, Tasuku. 2019. How did humans acquire erect bipedal walking? *Anthropological Science* 127 (1): 1–12.

Kingdon, Jonathan. 2003. *Lowly origin: Where, when and why our ancestors first stood up*. Princeton: Princeton University Press.

Kingston, John D. 2007. Shifting adaptive landscapes: Progress and challenges in reconstructing early hominid environments. *Pearbook of Physical Anthropology* 50: 20–58.

Kirtley, Paul. 2012. *Hadza movie: By the light of a million fires*. https://frontierbushcraft.com/2012/08/31/hadza-movie-by-the-light-of-a-million-fires/. Downloaded March 3, 2023.

Kislev, Mordechai E., Anat Hartmann, and Ofer Bar-Yosef. 2006. Early Domesticated Fig in the Jordan Valley. *Science* 312 (5778): 1372–1374.

Kistler, L., Á. Montenegro, B.D. Smith, J.A. Gifford, R.E. Green, L.A. Newsom, and B. Shapiro. 2014. Transoceanic drift and the domestication of African bottle gourds in the Americas. *Proceedings of the National Academy of Sciences* 111 (8): 2937–2941.

Ko, Kwang Hyun. 2015. Origins of bipedalism. *Brazilian Archives of Biology and Technology* 58 (6): 929–934.

Kohl-Larsen, L. 1958. *Wildbeuter in Ostafrika: Die Tindiga, ein Jager und Sammlervolk*. Berlin: Dietrich Reimer Verlag.

Konner, Melvin. 2004. Hunter-gatherer infancy and childhood: The !Kung and others. In *Hunter-gatherer childhoods: Evolutionary, developmental, and cultural perspectives*, ed. Barry S. Hewlett and M.E. Lamb, 19–64. New Brunswick: Transaction.

Koops, Kathelijne, William C. McGrew, and Tetsuro Matsuzawa. 2010. Do chimpanzees (*Pan troglodytes*) use cleavers and anvils to fracture *Treculia africana* fruits? Preliminary data on a new form of percussive technology. *Primates* 51 (2): 175–178. https://doi.org/10.1007/s10329-009-0178-6. Epub 2009 Dec 8.

Kopij, G. 2015. Avian diversity in an urbanized South African grassland. *Zoology and Ecology* 25 (2): 87–100.

Kopij, Grzegorz. 2019. Population density and structure of birds breeding in an urban habitat dominated by large baobabs (*Adansonia digitata*), Northern Namibia. *Biosystems Diversity* 27 (4): 354–360.

Kraft, Thomas S., Vivek V. Venkataraman, and Nathaniel J. Dominy. 2014. A natural history of tree climbing. *Journal of Human Evolution* 71: 105–118.

Kramer, Karen L. 2018. The cooperative economy of food: Implications for human life history and physiology. *Physiology and Behavior* 193 (Part B): 196–204.

Kranjac-Berisavljevic, Y.I., Balma, and B.Z. Gandaa. 2009. Securing food in the hungry season: The role of the baobab tree. In *Proceedings of the international symposium on underutilized plants for food security, nutrition, income and sustainable development*, Arusha, Tanzania, ed. Jaenicke, et al., Acta Hort, 806. ISHS.

Kroll, Ellen M., and T. Douglas Price. 1991. The interpretation of archaeological spatial patterning. In *Interdisciplinary Contributions to Archaeology (IDCA)*. New York: Springer/Plenum Press.

Kubitzki, K., and C. Bayer. 2002. *Flowering plants-dicotyledons: Malvales, capparales and non-betalain caryophyllales*. Berlin/Heidelberg/New York: Springer.

Kuhn, Steven L., and Mary C. Stiner. 2006. What's a mother to do? The division of labor among neandertals and modern humans in Eurasia. *Current Anthropology* 47 (6): 953–980.

Kull, Christian A., and Paul Laris. 2009. Fire ecology and fire politics in Mali and Madagascar. In *Tropical fire ecology*, 171–226. Berlin/Heidelberg: Springer, Spring Praxis Books.

Kunwar, R.M., and R.W. Bussmann. 2006. Ficus (Fig) species in Nepal: A review of diversity and indigenous uses. *Lyonia* 11 (1): 85–97.

Kunz, K.B., and K.E. Linsenmar. 2007. Changes in baboon feeding behaviour: Maturity dependent fruit and seed size selection within a food plant species. *International Journal of Primatology* 28: 819–835.

Kupika, O.L., S. Kativu, E. Gandiwa, and A. Gumbie. 2014. Impact of African elephants on baobab (*Adansonia digitata* L.) population structure in northern Gonarezhou National Park, Zimbabwe. *Tropical Ecology* 55 (2): 159–166.

Kusimba, Sibel Barut. 2003. *African foragers: Environment, technology, interactions*. New York: Altamira Press.

Kyndt, Tina, Achille E. Assogbadjo, Oliver J. Hardy, Romain Glele Kakaï, Brice Sinsin, Patrick Van Damme, and Godelieve Gheysen. 2009. Spatial genetic structuring of baobab (*Adansonia digitata*, Malvaceae) in the traditional agroforestry systems of West Africa. *American Journal of Botany* 96 (5): 950–957.

Laden, Greg, and Richard Wrangham. 2005. The rise of the hominids as an adaptive shift in fall-back foods: Plant underground storage organs (USOs) and australopith origins. *Journal of Human Evolution* 49: 482–498.

Lamprey, H.F. 1963. Ecological separation of the large mammal species in the Tarangire Game Reserve, Tanganyika. *East Africa Wildlife Journal* 1: 12–92. https://doi.org/10.1111/j.1365-2028.1963.tb00179.x.

Lancaster, Jane B. 1968. On the evolution of tool-using behavior. *American Anthropologist* 70 (1): 56–66.

Lancaster, Jane B., and Phillip Whitten. 1979. Sharing in human evolution. In *Anthropology: Contemporary perspectives (1985)*, ed. David E.K. Hunter and Phillip Whitten, 45–49. Boston: Little, Brown and Company.

Lancaster, Jane B., and Phillip Whitten. 1980. Family matters: Sharing in human evolution. *The Sciences* 20: 10–15.

Lancaster, J.B., and C.S. Lancaster. 1980. The division of labor and the evolution of human sexuality. *Behavioral and Brain Sciences* 3 (2): 193–193.

Langergraber, Kevin E., Kay Prüfer, Carolyn Rowney, Christophe Boesch, Catherine Crockford, Katie Fawcett, Eiji Inoue, Miho Inoue-Muruyama, John C. Mitani, Martin N. Muller, Martha M. Robbins, Grit Schubert, Tara S. Stoinski, Bence Viola, David Watts, Roman M. Wittig, Richard W. Wrangham, Klaus Zuberbühler, Svante Pääb, and Linda Vigilant. 2012. Generation times in wild chimpanzees and gorillas suggest earlier divergence times in great ape and human evolution. *PNAS* 109 (39): 15716–15721. https://doi.org/10.1073/pnas.1211740109.

Langdon, John H. 2005. *The human strategy: An evolutionary perspective on human anatomy*. New York: Oxford University Press.

Lanzara, Paola, and Mariella Pizzetti. 1978. *Simon & Schuster's guide to trees: A field guide to conifers, palms, broadleafs, fruits, flowering trees, and trees of economic importance*. New York: Simon and Schuster Inc.

Laris, Paul. 2002. Burning the seasonal mosaic: Preventative burning strategies in the wooded savanna of Southern Mali. *Human Ecology* 30 (2): 155–186.

Lathrap, Donald W. 1977. Our father the cayman, our mother the gourd: Spinden revisited, or a unitary model for emergence of agriculture in the new world. In *Origins of agriculture*, ed. C.A. Reed, 713–752. The Hague: Mouton Publishers.

Latimer, B. 1991. Locomotor adaptations in *Australopithecus afarensis*: The issue of arboreality. In *Origine(s) de la Bipedie Chez Les Hominides*, ed. B. Senut and Y. Coppens, 169–176. Paris: CNRF.

Latimer, B., and C.O. Lovejoy. 1990a. Hallucal tarsometatarsal joint in *Australopithecus afarensis*. *American Journal of Physical Anthropology* 82: 125–133.

———. 1990b. Metatarsophalangeal joints of *Australopithecus afarensis*. *American Journal of Physical Anthropology* 83: 13–23.

Latimer, B., J.C. Ohman, and C.O. Lovejoy. 1987. Talocrural joint in African hominoids: Implications for *Australopithecus afarensis*. *American Journal of Physical Anthropology* 74: 155–175.

Layser, Earle F. 2001. *Africa's tree of life: Legends, uses, and traditions of the baobab*. The World & I, a Publication of the Washington Times Corporation.

Leach, H.B. 2011. Baobab (*Adansonia digitata* L.) and tamarind (*Tamarindus indica* L.) management strategies in the midst of conflict and change: A dogon case study from Mali. *Human Ecology* 39 (5): 597–612.

Leacock, Eleanor, and Richard B. Lee. 1982. *Politics and history in band societies*. Cambridge: Cambridge University Press.

Leakey, Richard, and Roger Lewin. 1992. *Origins reconsidered: In search of what makes us human*. New York: Doubleday.

Lee, Richard B. 1965. *Subsistence ecology of ! Kung Bushmen*. Ph.D. dissertation, University of California, Berkeley.

Lee, Richard B. 1972. The !Kung Bushmen of Botswana. In *Hunters and gatherers today*, ed. M.G. Bicchieri, 326–368. New York: Holt, Rinehart and Winston.

———. 1972a. The ! Kung Bushmen of Botswana. In *Hunters and gatherers today*, ed. M.G. Bicchieri, 326–368. New York: Holt, Rinehart and Winston.

———. 1972b. !Kung spatial organization: An ecological and historical perspective. *Human Ecology* 1: 125–147.

———. 1979. *The !Kung San: Men, women, and work in a foraging society*. London: Cambridge University Press.

Lee, Richard B., and Irven DeVore, eds. 1968. *Man the hunter*. New York: Aldine Publishing Company.

Lee, R.B., and Irven DeVore. 1976. *Kalahari hunter-gatherers: Studies of the !Kung San and their neighbors*. Cambridge: Harvard University Press.

Lee, R.B., and M. Guenther. 1995. Errors corrected or compounded? A reply to Wilmsen. *Current Anthropology* 36 (2): 298–305.

Lee, Richard B., and Robert K. Hitchcock. 2001. African hunter-gatherers: Survival, history, and the politics of identity. *African Study Monographs* 26 (Suppl): 257–280.

Leeming, David A. 2001. *A dictionary of Asian mythology*. Oxford University Press.

Lehner, E., and J. Lehner. 1962. *Folklore and odysseys of food and medicinal plants*. New York: Tudor Publishing Company.

Lely, H.V. 1925. *The useful trees of Northern Nigeria*. London: Crown Agents for the Colonies.

Lieberman, D., J.B. Hall, M.D. Swaine, and M. Lieberman. 1979. Seed dispersal by baboons in the Shai Hills, Ghana. *Ecology* 60: 65–75.

Lindshield, Stacy, Jessica M. Rothman, Sylvia Ortmann, and Jill D. Pruetz. 2021. Western chimpanzees (*Pan troglodytes verus*) access a nutritionally balanced, high energy, and abundant food, baobab (*Adansonia digitata*) fruit, with extractive foraging and reingestion. *American Journal of Primatology* 83: e23307.

Linton, S. 1971. Women the gatherer: Male bias in anthropology. In *Women in perspective-A guide for cross-cultural studies*, ed. S.E. Jacobs, 9–21. Urbana: Univ. Ill. Press.

Lippi, A. 1704. *Description des Plantes Observeés en Egypte*. Unpublished manuscript.

Lisao, K., C.J. Geldenhuys, and P.W. Chirwa. 2017. Traditional uses and local perspectives on baobab (Adansonia digitata) population structure by selected ethnic groups in northern Namibia. *South African Journal of Botany* 113: 449–456.

Livingstone, Frank B. 1962. Reconstructing man's Pliocene pongid ancestor. *American Anthropologist* 64 (2): 301–305.

Long, R. Charles. 1956. The breeding colonies of large water and marsh birds within the Port Herald District. *The Nyasaland Journal* 9 (2): 29–50.

Longhurst, Richard. 1986. Household food strategies in response to seasonality and famine. *IDS Bulletin* 17 (3): 27–35.

Lorenz, Konrad. 1966. *On aggression*. New York: Harcourt, Brace & World.

———. 1973. *Behind the mirror: A search for a natural history of human knowledge*. New York: Harcourt Brace Jovanovich.

Lovejoy, C.O. 1981. The origin of man. *Science* 211 (4480): 341–350.

Lucas, G.L. 1971. The baobab map project. *Mitteilungen Bot Staatsamml München* 10: 162–164.

Luckert, Martin K., Nonto Nemarundwe, Loren Gibbs, Isla Grundy, Grant K. Hauer, Dzidzaye Maruzane, Sheona Shackleton, and June Sithole. 2014. Contribution of baobab production activities to household livelihoods: A case study in Zimbabwe. In *Bark: Use, management, and commerce in Africa*, ed. Anthony B. Cunningham, Bruce M. Campbell, and Martin K. Luckert. New York: The New York Botanical Garden Press.

Luxford, George. 1842. Notes on the baobab tree, (*Adansonia digitata*). *The Phytologist* XX: 433–455.

Lynn, C.W. 1937. *Agriculture in Mamprusi*. Gold Coast Department of Agriculture Bulletin (Accra). No. 34.

Mabberley, D.J. 1990. *Mabberley's plant-book: A portable dictionary of plants, their classification and uses*. Cambridge: Cambridge University Press.

———. 1997. *The plant-book* (First edition 1990). Cambridge: Cambridge University Press.

Mabulla, Audax. 1996. *Middle and Later Stone Age land-use and lithic technology in the Eyasi Basin, Tanzania*. PhD dissertation, Department of Anthropology, University of Florida, Gainesville.

———. 2000. Strategy for cultural heritage management (CHM) in Africa: A case study. *African Archaeological Review* 17: 211–233.

———. 2003. Archaeological implications of hadzabe foragers land use in the Eyasi Basin, Tanzania. In *East African archaeology: Foragers, potters, smiths, and traders*, ed. Chapurukha M. Kusimba and Sibel B. Kusimba, 33–58. Philadelphia: University of Pennsylvania Museum of Archaeology and Anthropology.

———. 2007. Hunting and foraging in the Eyasi Basin, Northern Tanzania: Past, present and future prospects. *The African Archaeological Review* 24 (1–2): 15–33.

MacDonald, K. 2017. The use of fire and human distribution. *Temperature (Austin)* 4 (2): 153–165. https://doi.org/10.1080/23328940.2017.1284637.eCollection.

Maceda, Marcelino. 1964. *The culture of the Mamanua*. Manila: Catholic Trade School.

Mackey, Wade C., and N.S. Coney. 2000. Cultural evolution and gender roles: A re-affirmation of J.K. Brown's note. *International Journal of Comparative Sociology*. 41 (3–4): 285–298. https://doi.org/10.1163/156851801511765.

Madsen, Andrew. 2000. *The hadzabe of Tanzania: Land and human rights for a hunter-gatherer community*. Copenhagen: IWGIA (International Work Group for Indigenous Affairs), Document No. 98.

Magill, Clayton R., Gail M. Ashley, and Katherine H. Freeman. 2013. Water, plants, and early human habitats in Eastern Africa. *Proceedings of the National Academy of Sciences of the United States of America* 110 (4): 1175–1180.

Maguire, R.A.J. 1931. Masai water-trees. *Man* 31 (July): 142–143.

Mail and Guardian. 2013. *ANC on Mandela: 'The large baobab tree has fallen.'* Staff Reporter, December 6. https://mg.co.za/article/2013-12-05-anc-remembers-mandela-as-large-african-boabab/. Retrieved 2 Apr 2014.

Malinowski, B. 1922. *Argonauts of the Western Pacific*. New York: Dutton.

Mallol, Carolina, F.W. Marlowe, B.M. Wood, and C.C. Porter. 2007. Earth, wind and fire: Ethnoarchaeological signals of Hadza fires. *Journal of Archaeological Science* 34: 2035–2052.

Manica Post. 2017. *Baobab coffee producers think big*. https://www.pressreader.com/zimbabwe/the-manica-post/20170602/281651075068667. Accessed 18 June 2020.

Mann, A.E. 1972. Hominid and cultural origins. *Man* 7: 379–387.

Maranz, Steven. 2009. Tree mortality in the African Sahel indicates an anthropogenic ecosystem displaced by climate change. *Journal of Biogeography* 36 (6): 1181–1193.

Marchant, Linda F., and William C. McGrew. 2005. Percussive technology: Chimpanzee baobab smashing and the evolutionary modelling of hominin knapping. In *Stone knapping: The necessary conditions for a uniquely hominin behavior*, ed. Valentine Roux and Blandine Bril, 341–349. McDonald Institute Monographs.

Marks, Stuart A. 1976. *Large mammals and a brave people: Subsistence hunters in Zambia*. Seattle/London: University of Washington Press.

Markusjais. 2011. Simon Thomsett on the African Crowned Eagle – Part 2. African Raptors. https://www.africanraptors.org/2011/06/.

Marlowe, Frank. 1999. Male care and mating effort among Hadza foragers. *Behavioral Ecology and Sociobiology* 46: 57–64.

Marlowe, Frank W. 2002. Why the Hadza are still hunter-gathers. In *Ethnicity, hunter-gatherers, and the other: Association or assimilation in Africa*, ed. Sue Kent, 247–275. Washington, DC: Smithsonian Institution Press.

———. 2003. A critical period for provisioning by Hadza men: Implications for pair bonding. *Evolution and Human Behavior* 24: 217–229.

———. 2004. The Hadza. In *Encyclopedia of medical anthropology*, 689–696. New York: Kluwer Academic/Plenum Publishers.

———. 2006. Central-place provisioning: The Hadza as an example. In *Feeding ecology in apes and other primates: Ecological, physical and behavioral aspects*, ed. G. Hohmann, M.M. Robbins, and C. Boesch, 359–377. Cambridge: Cambridge University Press.

Marlowe, F.W. 2007. Hunting and gathering: The human sexual division of foraging labor. *Cross-Cultural Research* 41 (2): 170–195.

Marlowe, Frank W. 2009. Sex differences in food preferences of Hadza hunter-gatherers. *Evolutionary Psychology* 7 (4): 601–616.

———. 2010. *The Hadza: Hunter-gatherers of Tanzania*. Berkeley: University of California Press.

Marlowe, Frank W., and Julia C. Berbesque. 2009. Tubers as fallback foods and their impact on Hadza hunter-gatherers. *American Journal of Physical Anthropology* 140: 751–758.

Marlowe, F.W., J.C. Berbesque, B. Wood, A. Crittenden, C. Porter, and A. Mabulla. 2014. Honey, Hadza, hunter-gatherers, and human evolution. *Journal of Human Evolution* 71: 119–128.

Marshall, L. 1976. *The !Kung of Nyae Nyae*. Cambridge: Harvard University Press.

Martins, Cléo, and Roberval Marinho. 2002. *Iroco: o Orixá da Ávore e a Ávore Orixá*. Rio de Janeiro, Brazil: Pallas.

Marzke, Mary W. and M. Steven Shackley 1986. Hominid hand use in the pliocene and pleistocene: Evidence from experimental archaeology and comparative morphology. *Journal of Human Evolutio*, 15(6): 439–460.

Masefield, G.B., Michael Harrison Wallis, S.G. Harrison, and B.E. Nicolson. 1969. *The Oxford book of food plants*. Oxford: Oxford University Press.

Matagi, Samuel Vivian. 2004. A biodiversity assessment of the Flamingo Lakes of Eastern Africa. *Biodiversity* 5 (1): 13–26.
Matthiessen, Peter. 1972. *The tree where man was born*. New York: E. P. Dutton & Co.
———. 2010. *The tree where man was born*. London: Penguin Classics.
Mauny, R. 1955. Baobabs – Cimetières à griots. *Notes Africaines* 65: 72–76.
Maurin, Olivier, T. Jonathan Davies, John E. Burrows, Barnabas H. Daru, Kowiyou Yessoufou, A. Muthama Muasya, Michelle van der Bank, and William J. Bond. Savanna fire and the origins of the 'underground forests' of Africa. *New Phytologist* 204 (1): 201–214. https://doi.org/10.1111/nph.12936. Epub 2014 Jul 17. https://www.ncbi.nlm.nih.gov/pubmed/28680931.
Mauss, M. 1925. *The gift: Forms and functions of exchange in archaic societies*. Glencoe: Free Press.
May, Jacques M. 1965. *The ecology of malnutrition in Middle Africa*. New York: Hafner Publishing Company.
Mbiti, John S. 1970. *African religions and philosophies*. New York: Anchor Books.
McCullough, D.R., and D.E. Ullrey. 1983. Proximate mineral and gross energy composition of white-tailed deer. *Journal of Wildlife Management* 47: 430–441.
McDowell, W. 1981a. *Hadza traditional economy and its prospects for development*. Report to the Rift Valley Project, Ministry of Information and Culture, Dar-es-Salaam, Tanzania.
———. 1981b. *A brief history of Mangola Hadza. Report to the Rift Valley Project, Ministry of Information and Culture, 1993 state of the peoples: A global human rights report on societies in danger*. Boston: Beacon Press.
McDowell, William. 1982. Hadza relations with foreign researchers. Appendix to D. K. Ndagala and S. A. C. Waane, "The effect of research on the hadzabe, a hunting and gathering group of Tanzania". *Review of Ethnology* 8 (11–19): 94–105.
McGregor, Alasdair 2019. How did the iconic boab tree get to Australia? Australian Geographic. https://www.australiangeographic.com.au/topics/science-environment/2019/07/how-did-the-iconic-boab-tree-get-to-australia/. Retrieved 3/9/2023.
McGregor, Alasdair. 2019. How did the iconic boab tree get to Australia? *Australian Geographic Magazine*. Issue 150, May/June 2019.
McGrew, W.C. 1987. Tools to get food: The subsistants of Tasmanian aborigines and Tanzanian chimpanzees compared. *Journal of Anthropological Research* 43 (3): 247–258.
———. 1992. *Chimpanzee material culture: Implications for human evolution*. Cambridge: Cambridge University Press.
———. 2004. *The cultured chimpanzee. Reflections on cultural primatology*. Cambridge: Cambridge University Press.
———. 2013. Is primate tool use special? Chimpanzee and New Caledonian crow compared. *Philosophical Transaction of the Royal Society B* 368: 20120422. https://doi.org/10.1098/rstb.2012.0422.
McGrew 2012. The Cultured Chimpanzee: Nonsense or Breakthrough? Human Ethology Bulletin – Proc. of the XXII. ISHE Conference (2015): 41-52.
McGrew, W.C., C.E.G. Tutin, and P.J. Baldwin. 1979. Chimpanzees, tools, and termites: Cross-cultural comparisons of Senegal, Tanzania, and Rio Muni. *Man* 14 (2): 185–214. New Series.
McGrew, W.C., P.J. Baldwin, L.F. Marchant, J.D. Pruetz, S.E. Scott, and C.E.G. Tutin. 2003. Ethnoarchaeology and elementary technology of unhabituated wild chimpanzees at Assirik, Senegal, West Africa. *Paleoanthropology* 5 (2): 1–20.
Mcwilliam, A.N., and R.A. Cheke. 2004. A review of the impacts of control operations against the red-billed quelea (Quelea quelea) on non-target organisms. *Environmental Conservation* 31 (2): 130–137.
Medler, Michael J. 2011. Speculations about the effects of fire and lava flows on human evolution. *Fire Ecology* 7 (1): 13–23. https://doi.org/10.4996/fireecology.0701013.
Mehlman, Michael James. 1989. *Later quaternary archaeological sequences in northern Tanzania*. Dissertation, University of Illinois at Urbana-Champaign. https://hdl.handle.net/2142/22520.
Menninger, Edwin A. 1967. *Fantastic trees*. New South Wales: Generic Publications.

Mertz, O., A. Lykke, and A. Reenberg. 2001. Importance and seasonality of vegetable consumption and marketing in Burkina Faso. *Economic Botany* 55: 276–289.

Mikusiński, Gregorz, Maciej Gromadzki, and Prezemyslaw Chylarecki. 2001. Population trends of widespread woodland birds in Europe. *Conservation Biology* 15 (1): 208–217.

Miller, Jennifer M., Elizabeth A. Sawchuk, Amy L.R. Reedman, and Pamela R. Willoughby. 2018. Land snail shell beads in the Sub-Saharan archaeological record: When, where, and why? *African Archaeological Review* 35: 347–378.

Miracle, M.P. 1961. Seasonal hunger: A vague concept and an unexplored problem. *Bulletin de l'Institut français d'Afrique noire. Série B, Sciences naturelles* 23: 273–283.

Mitchel, S.R. 1955. Comparison of stone tools of the Tasmanians and Australian aborigines. *Journal of the Royal Anthropological Institute of Great Britain and Ireland* 85: 131–139.

Mithen, Steven. 2010. The domestication of water: Water management in the ancient world and its prehistoric origins in the Jordan Valley. *Philosophical Transactions: Mathematical, Physical and Engineering Sciences* 368(1931); *Water and Society: Past, Present and Future* (28 November 2010): 5249–5274.

Mittre, Vishnu. 1981. Wild plants in Indian folk life – A historical perspective. In *Glimpses of Indian ethnobotany*, ed. S.K. Jain. New Delhi: Oxford/IBH Publishing Company.

Mlingwa, Charles. n.d. *Palearctic-African migratory birds in Tanzania*. Arusha: Tanzania Wildlife Research Institute.

Mogea, J., B. Seibert, and W. Smits. 1991. Multipurpose palms: The sugar palm (*Arenga pinnata* (Wurmb) Merr.). *Agroforestry Systems* 13: 111–129.

Mogg, A.O.D. 1950. Trees in South Africa. *The Journal of the Arboricultural Society of South Africa* 4: 12–14.

Moiselet, Dominique. 1998. *Flora of Senegal: The Baobab*. https://www.senegal-online.com/flore-du-senegal/le-baobab/. Accessed 24 Mar 2023.

Molien, G. 1967. Travels in the interior of Africa to the sources of the Senegal and Gambia performed by command of the French government in the year 1818. 1st edn. Bowditch TE (ed). Cass, London: 1820.

Monteiro, Joachim J. 1875. *Angola and the river Congo*. Vol. 1. London: Macmillan.

Morelli, Federico, Anna Maria Kubicka, Piotr Tryjanowski, and Emma Nelson. 2015. The vulture in the sky and the hominin on the land: Three million years of human-vulture interaction. *Anthrozoös* 28 (3): 449–468.

Morris, Desmond. 1967. *The naked ape: A zoologist's study of the human animal*. New York: McGraw-Hill.

———. 1969. *The human zoo*. New York: McGraw-Hill.

Mpala, Canisius, P. Sibanda, M. Dlamini, and B. Sibanda. 2015. Are quelea birds really a menace? Innovative use of indigenous knowledge systems in the harvesting and utilization of quelea, *Quelea quelea lathamii* in Hwange District of Matabeleland North Province. *International Journal of Agricultural Sciences* 5 (3): 476–486.

Mshigeni, K.E., and L. Hangula. 2001. *Africa's baobab resource: Unlocking their economic potential and unrealized virtues within the framework of the zero waste concept*. Windhock: UNDP/UNOPS Regional Project RAF/99/021, University of Namibia.

Mundy, P. 1992. On the vultures of Africa. In *The vultures of Africa*, ed. P. Mundy, D. Butchart, J. Ledger, and S. Piper, 110–114. Acorn Books and Russel Friedman Books: Randburg and Halfway House (South Africa).

Munyebvu, Faith. 2015. *Abundance, structure and uses of baobab (Adansonia digitata L.) populations in Omusati Region, Namibia*. MA thesis, The University of Namibia.

Murdock, George P. 1937. Comparative data on the division of labor by sex. *Social Forces* 15: 551–553.

———. 1949. *Social structure*. New York: Macmillan.

Murdock, George P., and Caterina Provost. 1973. Factors in the division of labor by sex: A cross-cultural analysis. *Ethnology* 12: 203–225.

Murray, Shaun S., Margaret J. Schoeninger, Henry T. Bunn, Travis R. Pickering, and Judith A. Marlett. 2001. Nutritional composition of some wild plant foods and honey used by Hadza foragers of Tanzania. *Journal of Food Composition and Analysis* 14: 3–13.

MWO. 2005. *Dictionary by Merriam-Webster: America's most-trusted online dictionary.* https://www.merriam-webster.com/.

Nakahashi, Wataru, and Marcus W. Feldman. 2014. Evolution of division of labor: Emergence of different activities among group members. *Journal of Theoretical Biology* 348: 65–79. https://doi.org/10.1016/j.jtbi.2014.01.027. Epub 2014 Jan 28.

Napier, J.R. 1963. The locomotor functions of hominids. In *Classification and human evolution*, ed. Sherwood L. Washburn, 178–190. Chicago: Aldine Publishing Company.

Ndagala, D.K. 1988. Free or doomed? Images of the Hadzabe. In *Hunters and gatherers in the modern context, book of presented papers, volume 11*, Seventh international conference on hunting and gathering societies, CHAGS, ed. L.J. Ellanna, 477–485. Moscow: Institute of Ethnology and Anthropology, Russian Academy of Sciences.

Ndagala, D.K., and S.A.C. Waane. 1982. The effect of research on the hadzabe, a hunting and gathering group of Tanzania. *Review of Ethnology* 8 (11–19): 94–105.

Ndagala, D.K., and N. Zengu. 1989. From the raw to the cooked: Hadzabe perception of their pasts. In *Who needs the past?* ed. R. Layton, 51–54. London: Unwin Hyman.

Ndoro, Onias. 2013. Impact of elephants (*Loxondonta africana* L.) on baobab trees (*Adansonia digitata* L.) in Mana Pools National Park of the Midzambezi Valley Region. Masters thesis, University of Zimbabwe.

Negro, Juan José, Ruth Blasco, Jordi Rosell, and Clive Finlayson. 2016. Potential exploitation of avian resources by fossil hominins: An overview from ethnographic and historical data. *Quaternary International* 421: 6–11.

Newbold, D. 1924a. More notes on tebeldis. *Sudan Notes and Records* 7 (1): 135–137.

———. 1924b. Still more notes on tebeldis (A reply). *Sudan Notes and Records* 7 (2): 117–123.

Newton, L. 1974. Is the baobab tree succulent? *Cactus and Succulents Journal* 36: 57–58.

Newton, Ivy. 2015. *African tree tales: Baobab the tree god planted upside down.* Published by the Author.

Niassy, Saliou, Karamoko Diarra, Abdoul A. Niang, and Alain Coache. 2011. Entomological survey and biodiversity conservation in the Madeleine Island Park of Senegal: *Analeptes trifasciata* (Coleoptera, Cerambycidae), a threat to the insular baobab. *International Journal of Biological and Chemical Sciences* 5 (1): 386–391.

Nickrent, Daniel L. 2011. Santalales (including mistletoes). In *Encyclopedia of life sciences (ELS)*. Chichester: Wiley.

Nicol, Bruce M. 1957. Ascorbic acid content of baobab fruit. *Nature* 180: 287. https://doi.org/10.1038/180287a0.

Nicolas, R.P., and F.J. Nicolas. 1955. Recherches sur la valeur sémantique du mot "baobab". *Notes Africaines* 67: 77–78.

Nicolson, S.W., and P.A. Fleming. 2003. Nectar as food for birds: The physiological consequences of drinking dilute sugar solutions. *Plant Systematics and Evolution* 238: 139–153.

Niemitz, C. 2010. The evolution of the upright posture and gait – A review and a new synthesis. *Naturwissenschaften* 97 (3): 241–263.

Nixon, Roy W. 1951. The date palm: "Tree of life" in the subtropical deserts. *Economic Botany* 5 (3): 274–301.

Nkana, Z.G., and S. Iddi. 1991. *Utilization of baobab (Adansonia digitata) in Kondo District, Central Tanzania.* Morogoro: Sokoine University of Agriculture.

Nkwame, V.M. 2007. Tanzania: Bushmen lives threatened as famine hits. *The Arusha Times*, October 26.

North, Rabea, Dietrich Darr, Jens Gebauer, Dagmar Mithöfer, Willis Omondi Owino, Fredah Karambu Rimberia, Katja Kehlenbeck, Muneer Elyas Siddig, Tarig Elsheikh Mahmoud, Mohamed El Nour Taha, Andreas Triebel, Henry Johnson, Anthony Maina, Mahmoud Fadl El Mula Ahmed, Yahia Omar Adam, Muthali Chimuleke, Elwasila Muktar, Mohamed Elwasila, Tsige-Yohannes Habte, Michael B. Krawinkel, Myori Jeremiah Mbugua, Kavoi Mutuku Muendo, El Amin Sanjak, and Martin Schüring. 2014. Promoting the use of Baobab (*Adansonia digitata* L.) in rural communities in Eastern Africa. In *Conference on international research on food security, natural resource management and rural development organized by the Czech University of Life Sciences, Prague*. Prague, Czech Republic, September 17–19, 2014.

O'Connell, J.F., K. Hawkes, and N. G. Blurton Jones. 1988. Hadza hunting, butchering, and bone transport. *Journal of Anthropological Research* 44 (2): 113–161.

O'Connell, James F., K. Hawkes, and N.G. Blurton Jones. 1988a. Hadza scavenging: Implications for Plio-Pleistocene hominid subsistence. *Current Anthropology* 29 (2): 356–363.

———. 1988b. Hadza hunting, butchering, and bone transport and their archaeological implications. *Journal of Anthropological Research* 44: 113–162.

———. 1990. Reanalysis of large mammal body part transport among the Hadza. *Journal of Archaeological Science* 17: 301–316.

O'Connell, James F., Kristen Hawkes, and N. G. Blurton Jones. 1991. Distribution of refuse-producing activities at Hadza residential base camps: Implications for analyses of archaeological site structure. In *The interpretation of archaeological spatial patterning*, ed. Ellen M. Kroll and T. Douglas Price. New York: Plenum Press.

O'Higgins, Paul, and Sarah Elton. 2007. Walking on trees. *Science* 316: 1292–1294.

O'Meara, Donna. 2003. The Kalahari: The lonely baobab tree offers food, water, medicine, and shelter to the people and animals of the Kalahari. *Faces: People, Places, and Cultures* 19 (6): 6. Cobblestone Publishing Company.

Oakley, Kenneth P. 1959. *Tools makyth man*, Smithsonian report for 1958, 431–445. Washington, DC: Smithsonian Institution.

Obizoba, I.C., and J.U. Anyika. 1994. Nutritive value of baobab milk (gubdi) and mixtures of baobab (*Adansonia digitata* L.) and hungry rice, acha (*Digitaria exilis*) #ours. *Plant Foods for Human Nutrition* 46: 157–165.

Obst, Erich. 1912. Von Mkalama ins Land der Wakindega. *Mitteilungen der geographischen Gesellschaft in Hamburg* 27: 1–45.

Ogbu, J.C. 1973. Seasonal hunger in tropical Africa as a cultural phenomenon. *Africa* 43: 317–332.

Ogden, John C. 1985. Vulture (2). In *A dictionary of birds*, ed. Bruch Campbell and Elizabeth Lack, 634–636. Vermillion: Buteo Books.

Ogen-Odoi, Asaph A., and T.G. Dilworth. 1984. Effects of grassland burning on the savanna hare-predator relationships in Uganda. *African Journal of Ecology* 22 (2): 01–106. https://doi.org/10.1111/j.1365-2028.1984.tb00683.x.

Ohler, J.G. 1984. *Coconut, tree of life*. Rome: Food and Agriculture Organization of the United Nations.

Okafor, J.C. 1980. Trees for food and fodder in the savanna areas of Nigeria. *International Tree Crops Journal* 1 (2–3): 131–141.

Olaitan, Ngozi Innocentia, Mike Ojotu Eke, and Emmanuel Aitiya. 2014. Quality evaluation of coffee-like beverage from baobab (*Adansonia digitata*) seed. *Advance Journal of Food Science and Technology* 6 (9): 1050–1055. https://doi.org/10.19026/ajfst.6.158.

Olasky, M. 2005. *Planners vs searchers: How bad advice hurts poor people*. Washington, DC.: Capital Research Center.

Oldfield, Matthew. 2010. *Hadzabe men hunting Marabou Stork on the shore of Lake Eyasi, Tanzania.* http://cdn.c.photoshelter.com.

———. 2017. *Stork hunting, Tanzania*. http://www.sciencephoto.com/media/150334/view.

Ombuor, Joe. 2004. Face to face with starvation. *The Nation*, August 4.

Onderstall, J. 1979. The flower of the baobab. *Veld and Flora* 65 (3): 76.

Oschadleus, H. Dieter. 2016. Nesting associations between vultures and weavers. *Vulture News* 70: 3–21.

Oswalt, W.H. 1973. *Habitat and technology*. New York: Hold, Rinehart and Winston.

Owen, John C. 1968. *Water storage properties of Adansonia digitata (Baobab)*, The West African archaeological newsletter. Vol. 9, 55–56. Ibadan: Institute of African Studies, University of Ibadan.

———. 1970. The medico-social and cultural significance of *Adansonia digitata* (Baobab) in African communities. *African Notes* 6: 24–36.

———. 1974. A contribution to the ecology of the African baobab. *Savanna* 3 (1): 1–12.

Pakenham, Thomas. 2004. *The remarkable baobab*. New York: W. W. Norton and Company.

Palmer, Eva, and Nora Pitman. 1961. *Trees of South Africa*. Amsterdam: A. A. Bolkema.

Pander, Heike. 2015. *Baobab Fruit Juice – Very tasty! November 15, 2015 blog posted on the website Baobab: Stories about the African tree of life*. http://baobabstories.com/en/baobab-fruit-juice-very-tasty/.

Panger, Melissa A., Susan Perry, Lisa Rose, Julie Gros-Louis, Erin Vogel, Katherine C. Mackinnon, and Mary Baker. 2002. Cross-site differences in foraging behavior of white-faced capuchins (*Cebus capucinus*). *American Journal of Biological Anthropology* 119 (1): 52–66. https://doi.org/10.1002/ajpa.10103.

Panter-Brick, C. 2002. Sexual division of labor: Energetic and evolutionary scenarios. *American Journal of Human Biology* 14: 627–640.

Parker, Christopher Hugh. 2015. *On the evolution of human fire use*. PhD dissertation, Department of Anthropology, University of Utah.

Parker, C.H., E.R. Keefe, N.M. Herzog, J.F. O'connell, and K. Hawkes. 2016. The pyrophilic primate hypothesis. *Evolutionary Anthropology* 25 (2): 54–63. https://doi.org/10.1002/evan.21475.

Parr, M.W. 1924. Still more notes on Tebeldis. *Sudan Notes and Records* 7 (2): 117–119.

Paris, Paul. 2011. Humanizing savanna biogeography: Linking human practices with ecological patterns in a frequently burned savanna of Southern Mali. *Annals of the Association of American Geographers* 101 (5): 1067–1088. https://doi.org/10.1080/00045608.2011.560063

Parrinder, Geoffrey. 1949 [1970]. *West African religion*. New York: Barnes and Noble.

———. 1954 [1970]. *African traditional religion*. Westport: Greenwood Press.

Parrinder, G. 1949 (1970). *West African religion: A study of the beliefs and practices of Akan, Ewe, Yoruba, Ibo, and kindred peoples*. New York: Barnes and Noble.

Patrut, Adrian, Karl F. von Reden, Daniel Lowy, Andries H. Alberts, John Pohlman, Rudolf Wittmann, Dana Gerlach, and Xu. Li. 2007. Radiocarbon dating of a very large African baobab. *Tree Physiology* 27 (11): 1569–1574. https://doi.org/10.1093/treephys/27.11.1569.

Patrut, Adrian, Diana H. Mayne, Karl F. von Reden, Daniel A. Lowy, Robert van Pelt, Ann P. McNichol, Mark L. Roberts, and Dragos Margineanu. 2010. Fire history of a giant African baobab evinced by radiocarbon dating. Proceedings of the 20th International Radiocarbon Conference, edited by A J T Jull. Arizona Board of Regents on behalf of the University of Arizona. *Radiocarbon* 52 (2–3): 717–726.

Pausas, Juli G., and Jon E. Kelley. 2009. A burning story: The role of fire in the history of life. *Bioscience* 59 (7): 593–601. https://doi.org/10.1525/bio.2009.59.7.10.

Pennisi, Elizabeth. 1999. Did cooked tubers spur the evolution of big brains? *Science* 283 (5410): 2004–2005.

Penrith, Mary Louise. 2021. *Experiences with ASF and African wild pigs*. PowerPoint Presentation, Department of Veterinary Tropical Diseases, University of Pretoria. Downloaded 12 Feb 2021. https://www.oie.int/eng/OIE-CICASF_2014/docs/pdf/10_Mary_LPenrith/files/assets/common/downloads/10_Mary_LPenrith.pdf.

Pern, S. 1982. *Masked dancers of West Africa. The Dogon peoples of the wild*. Amsterdam: Time-Life.

Persley, Gabrielle J. 1992. *Replanting the tree of life: Towards an International ATenda for Coconut Palm Research*. Oxon: C-A-B International.

Peters, Charles R. 1979. Toward an ecological model of African Plio-Pleistocene hominid adaptations. *American Anthropologist* 81 (2): 261–278. https://doi.org/10.1525/aa.1979.81.2.02a00010.

Peters, Charles R., and Eileen M. O'Brien. 1981. The early hominid plant-food niche: Insights from an analysis of plant exploitation by *Homo*, *Pan*, and *Papio* in Eastern and Southern Africa. *Current Anthropology* 22 (2): 127–140.

Peters, Dick, Coert Engels, and Soungalo Sarra. 2012. Natural spread of plant viruses by birds. *Journal of Phytopathology* 160 (10): 591–594.

Peterson, Jean Treloggen. 1978. Hunter-gatherer/farmer exchange. *American Anthropologist* 80 (2): 335–351.

Peterson, Daudi, Richard Baalow, and Jon Cox. 2013. *Hadzabe: By the light of a million fires*. Dar-es-Salaam: Mkuki na Nyota Publishers Ltd.

Petignat, Andry, and Louise Jasper. 2012. *Baobabs of the world*. South Africa: Struik Nature.

Pettet, Antony. 1977. Seasonal changes in nectar-feeding by birds at Zaria, Nigeria. *Ibis* 119 (3): 291–308.

Pettigrew, Jack D., Karen L. Bell, Adhil Bhagwandin, Eunice Grinan, Ngalla Jillani, Jean Meyer, Emily Wabuyele, and Claudia E. Vickers. 2012. Morphology, ploidy and molecular phylogenetics reveal a new diploid species from Africa in the baobab genus *Adansonia* (Malvaceae: Bombacoideae). *Taxon* 61 (6): 1240–1250.

Phillips, John. 1959. *Agriculture and ecology in Africa: A study of actual and potential development South of the Sahara*. London: Faber and Faber.

Philpot, J.H. 1897. *The sacred tree in religion and myth*. London: Macmillan and Co. Limited.

Piciformes. 2020. *Lybiidae (African Barbets)*. Piciformes Taxon Advisory Group. http://www.piciformes.org/lybiidae.htm. Accessed 18 June 2020.

Pickering, Charles. 1879. *Chronological history of plants: Man's Record of his own existence illustrated through their names, uses, and companionship*. Boston: Little, Brown, and Company.

Piearce, G.D., G.M. Calvert, C. Sharp, and P. Shaw. 1994. *Sooty baobab: Disease or drought?* Zimbabwe Forestry Commission, Research Paper 6. Harare.

Pielou, E.C. 1952. Notes on the vegetation of the Rukwa Rift Valley, Tanganyika. *The Journal of Ecology* 40 (2): 383–392.

Pienaar, U. de V. 1969. Predator-prey relationships among the larger mammals of Kruger National Park. *Koedoe* 12: 108–176.

Pierce, John D., Jr. 1986. A review of tool use in insects. *The Florida Entomologist* 69 (1): 95–104.

Pipe-Wolferstan, K. 1988. *Traditional food plants*, Food and Nutrition Paper 42. Rome: FAO.

Platt, B.S. 1954. Food and its production. In *The development of tropical and sub-tropical countries: With special reference to Africa*, ed. A. Leslie Banks. London: Arnold.

Plotkin, Mark J. 1988. The outlook for new agricultural and industrial products from the tropcs. In *Biodiversity*, ed. E.O. Wilson, 106–118. Washington, DC: National Academy Press.

Pochron, S.T. 2005. Does relative economic value of food elicit purposeful encounter in the yellow baboon (*Papio hamadryas cynocephalus*) of Ruaha National Park, Tanzania. *Primates* 46: 71–74.

Polanyi, K. 1944. *The great transformation*. New York: Rinehart.

Polanyi, K. 1957. The economy as instituted process. In *Trade and market in the early empires: Economics in history and theory*, ed. K. Polanyi, C.M. Arensberg, and H.W. Pearson, 243–270. Glencoe: Free Press.

———. 1959. *The great transformation: The political and economic origins of our time*. Boston: Beacon Press.

Polunin, N. 1960. *Introduction to plant geography and some related sciences*. London: Longmans Press Ltd.

Porteous, Alexander. 1928. *Forest folklore, mythology and romance*. London: George Allen & Unwin.

———. 1968. *Forest folklore, mythology and romance*. Detroit: Singing Tree Press.

Potts, Richard. 1998. Environmental hypothesis of hominin evolution. *Yearbook of Physical Anthropology* 41: 93–136.

Power, Camilla. 2015. *Hadza gender rituals – epeme and maitoko – Considered as counterparts*, Hunter gatherer research 1.3. Liverpool: Liverpool University Press. https://doi.org/10.3828/hgr.2015.18

Power, Camilla, and Ian Watts. 1997. The woman with the zebra's penis: Gender, mutability and performance. *Journal of the Royal Anthropological Institute* 3 (3): 537–560.

Prance, G.T. 2007. Ethnobotany, the science of survival: A declaration from Kauai. *Economic Botany* 61 (1): 1–2. https://doi.org/10.1007/BF02862367.

Prassack, Kari Alyssa. 2011. Fossil water and marsh birds, as paleoenvironmental indicators at Olduvai Gorge, Tanzania. *American Journal of Physical Anthropology* 144 (Supplement 52): 243.

Pruetz, Jill D., and Nicole M. Herzog. 2017. Savanna chimpanzees at Fongoli, Senegal, Navigate a fire landscape. *Current Anthropology* 58 (516): S337–S350.
Pruetz, J.D., and T.C. LaDuke. 2010. Brief communication: Reaction to fire by savanna chimpanzees (*Pan* troglodytes verus) at Fongoli, Senegal: Conceptualization of "fire behavior" and the case for a chimpanzee model. *American Journal of Physical Anthropology* 141 (4): 646–650. https://doi.org/10.1002/ajpa.21245.
Prussin, L. 1999. Non-Western sacred sites: African models. *Journal of the Society of Architectural Historians* 58 (3): 424–433.
Pyne, S.J. 1995. *World fire. The culture of fire on earth*. Seattle: University of Washington Press.
———. 2001. *Fire: A brief history*. Seattle: University of Washington Press.
Rabi'u, Tukur, and Murtala Rabi'u. 2013. An assessment of multi-purpose use of Adansonia digitata (Baobab tree) for sustainable development in the semi urban fringes of Dutsinma Katsina State Nigeria. *Academic Research International* 4 (1): 486–494.
Radcliffe Brown, A.R. 1964. *The Andaman Islanders*. New York: Free Press.
Raebild, Anders, Anders Søndergård Larsen, Jan Svejgaard Jensen, Moussa Ouedraogo, Sitske De Groote, Patrick Van Damme, Jules Bayala, Boukary Ousmane Diallo, Haby Sanou, Antoine Kalinganire, and Erik Dahl Kjaer. 2011. Advances in domestication of indigenous fruit trees in the West African Sahel. *New Forests* 41: 297–315.
Raji, A.G., and A.O. Adeyemi. 2018. *Thin layer drying kinetics and quality of African baobab leaf (Adansonia digitata)*. www.researchgate.net. Accessed 14 Jan 2019.
Randhawa, M.S. 1965. *Flowering trees*. New Delhi: National Book Trust India.
Rangan, Haripriya, Karen L. Bell, David A. Baum, Rachael Fowler, Patrick McConvell, Thomas Saunders, Stef Spronck, Christian A. Kull, and Daniel J. Murphy. 2015. New genetic and linguistic analyses show ancient human influence on baobab evolution and distribution in Australia. *PLoS One* 10 (4): 20119758. https://doi.org/10.1371/journal.pone.0119758.
Rashford, John. 1997a. *An explanation for tamarind and baobab trees growing together in Africa and the Caribbean: The case of St. Croix*, Virgin Islands agriculture and food fair, bulletin number 11, 22–26. University of the Virgin Islands.
———. 1997b. Africa's baobab tree in Jamaica: A further comment. *Jamaica Journal* 26 (2): 51–58.
———. 2019. Human beings and baobab and tamarind trees in the Caribbean: Ethnographic, historical and evolutionary perspectives on an ancient African memetic complex. Elsa Goveia Memorial Lecture, Department of History and Archaeology, University of the West Indies, Mona, Jamaica, March 12, 2019.
———. 2022. A historically contextualized account of the baobab trees (*Adansonia digitata* L.) of Tobago. *Economic Botany* 76 (2): 127–157.
———. 1985. The cotton tree and the spiritual realm in Jamaica. *Jamaica Journal* 18 (1): 49–57.
———. 1987a. The search for Africa's baobab tree in Jamaica. *Jamaica Journal* 20 (2): 3–11.
———. 1987b. The baobab tree and seasonal hunger in Africa: The case of the San. *The Botswana Society* 19: 57–68.
——— 1987b. The Baobab Tree and Seasonal Hunger in Africa: The Case of the San. The Botswana Society, 19:57–68.
———. 1991. *The Grove Place baobab tree*, Virgin Islands agriculture and food fair, bulletin number 5, 65–69. University of the Virgin Islands.
———. 1992. The search for the St. John Baobab. *Journal of the Virgin Island Historical Society* 6 (1): 8–12.
———. 1993. Bonsai baobabs. *Bonsai-Shunju (Journal of the Japan Bonsai Association)* 241: 32–37.
———. 1994. Africa's baobab tree: Why monkey names? *Journal of Ethnobiology* 14 (2): 26–34.
———. 1994a. Africa's Baobab tree: Why monkey names? *Journal of Ethnobiology* 14 (2): 26–34.
———. 1994b. *The search for the St. John baobab*, Virgin Islands Agriculture and Food Fair, Bulletin Number 8, 16–18. University of the Virgin Island.
———. 1994c. Jamaica's settlement vegetation, agroecology, and the origin of agriculture. *Caribbean Geography* 5 (1): 32–50.

———. 1995. The distribution, history and use of the African baobab in Barbados. In *Proceedings of the 31st annual meeting of the Caribbean Food Crops Society, July 7–13*. Zamarano: Honduras.

———.1991a. The Grove Place Baobab Tree. Virgin Islands Agriculture and Food Fair, Bulletin Number 5, pp 65-69. University of the Virgin Islands.

———. 1991b. Arawak, Spanish and African contributions to Jamaica's settlement vegetation. *Jamaica Journal* 24 (3): 17–23.

Rashford, John 1995. The Distribution, History and Use of the African Baobab in Barbados. Proceedings of the 31st Annual Meeting of the Caribbean Food Crops Society, July 7-13. Zamarano: Honduras.

———. 1997a. *An explanation for tamarind and baobab trees growing together in Africa and the Caribbean: The case of St. Croix*, Virgin Islands Agriculture and Food Fair, Bulletin Number 11, 22–26. University of the Virgin Islands.

Rashford, John 1997b. Africa's Baobab Tree in Jamaica: A Further Comment. Jamaica Journal, 26(2):51-58.

———. 2002. Evidence of seasonal hunger in the oral traditions of Jamaicans: An ethnobiological perspective. In *Ethnobiological and biocultural diversity*, ed. John R. Stepp, Felice S. Wyndham, and Rebecca K. Zarger. Athens: University of Georgia Press.

Rashford, John 2012a. Candomble's cosmic tree and Brazil's Ficus species. In African Ethnobotany in the Americas, R. Voeks and J. Rashford, eds. New York: Springer.

Rashford, John 2012 b. The meaning of "God tree" in Jamaica: An ethnobiological perspective. Unpublished.

———. 2015. The uses of the baobab flower (*Adansonia digitata* L.). *Ethnobotany Research and Applications* 14: 211–229. ISSN 1547-3465. http://journals.sfu.ca/era/index.php/era/article/view/1134. Accessed 23 Feb 2019.

———. 2018. The use of baobab leaves (*Adansonia digitata* L.) for food in Africa: A review. *Economic Botany* 72 (4): 478–495.

——— 2019. *Human beings and baobab and tamarind trees in the Caribbean: Ethnographic, historical and evolutionary perspectives on an ancient African memetic complex*. Elsa Goveia Memorial Lecture, Department of History and Archaeology, University of the West Indies, Mona, Jamaica, March 12, 2019. Publication forthcoming.

———. 2022a. Evidence of seasonal hunger in the oral traditions of Jamaicans: An ethnobiological perspective. In *Ethnobiological and biocultural diversity*, ed. John R. Stepp, Felice S. Wyndham, and Rebecca K. Zarger. Athens: University of Georgia Press.

———. 2022b. Nyame's altar: A cultural history of the 'God tree' in Jamaica. In *The cultural value of trees: Folk value and biocultural conservation*, ed. Jeffrey Wall, 75–87. New York: Routledge.

Rasmussen, Mattias Borg, and Ben Orlove. 2017. Anthropologists exploring water in social and cultural Life. *American Anthropologist* 119 (4): 1548–1433.

Ray, J. 1693. *Historia plantarum*. London: Samuel Smith and Benjamin Walford.

Readal, Maryann. 2022. *Baobab tree – The African "tree of life"*. The Herb Society of America Blog. https://herbsocietyblog.wordpress.com/2022/09/19/baobab-tree-the-african-tree-of-life/.

Reader, John. 1988. *Man on Earth*. Austin: University of Texas Press.

Reddy, A.S., K.B. Anjaria, and Rama Rao Vendrapati. 2002. Baobab: An exotic tree with a promise? *Asian Agri-History* 6: 343–350.

Relethford, John. 1990. *The human species: An introduction to biological anthropology*. Mountain View: Mayfield Publishing Company.

Reno, Stephen J. 1977. Religious symbolism: A plea for a comparative approach. *Folklore* 88 (1): 76–85.

RHDEL (Random House Dictionary of the English Language). 1969. New York: Random House.

RHWCD (Random House Webster's College Dictionary). 2000. New York: Random House.

Richards, A.I. 1939. *Land, labour and diet in Northern Rhodesia*. London: Oxford University Press.

Richardson, James B. 1972. The pre-Columbian distribution of the bottle gourd (*Lagenaria siceraria*): A Re-evaluation. *Economic Botany* 26 (3): 265–273.

Riches, David. 1996. Hunting and gathering societies. In *Encyclopedia of social and cultural anthropology*, ed. Alan Barnard and Jonathan Spencer, 288–291. New York: Routledge.
Richmond, Brian G., David R. Begun, and David S. Strait. 2002. Origin of human bipedalism: The knuckle-walking hypothesis revisited. *Yearbook of Physical Anthropology* 44: 70–105.
Ridley, H. 1930. *The dispersal of plants throughout the world*. Kent/Ashford: Reeve and Ashford/ Lovell Reeve and Co.
Riedel, Frank, Sebastian Erhardt, Chrispen Chauke, Annette Kossler, Elisha M. Shemang, and Pavel Tarasov. 2012. Evidence for a permanent lake in Sua Pan (Kalahari, Botswana) during the early centuries of the last millennium indicated by distribution of Baobab trees (*Adansonia digitata*) on "Kubu Island". *Quaternary International* 253: 67–73.
Riley, Bernard W., and David Brokensha. 1988. *The Mbeere in Kenya (Volume 11): Botanical identities and uses*. New York: University Press of America.
Rindos, David. 1980. Symbiosis, instability, and the origins and spread of agriculture: A new model. *Current Anthropology* 17: 593–612.
Rindos, David. 1984. *The origins of agriculture: An evolutionary perspective*. New York: Academic.
Rival, Laura. 1998. *The social life of trees*. New York: Oxford.
Robben, Antonius C.G.M. 2017. *Death, mourning, and burial: A cross-cultural reader*, ed. Antonius Robben, 2nd ed. Hoboken: Wiley-Blackwell
Roberts, R.G., and M.I. Bird. 2012. Evolutionary anthropology – Homo 'incendius'. *Nature* 485 (7400): 586–587.
Robson, Arthur J., and Hillard S. Kaplan. 2003. The evolution of human life expectancy and intelligence in hunter-gatherer economies. *The American Economic Review* 93 (1): 150–169.
Rodman, P.S., and H.M. McHenry. 1980. Bioenergetics and the origin of hominid bipedalism. *American Journal of Physical Anthropology* 52 (1): 103–106.
Romero, Claudia, Dalali B.K. Dovie, James Gambiza, Emmanuel Luoga, Susanne Schmitt, and Isla Grundy. 2014. Effects of commercial bark harvesting on *Adansonia digitata* (Baobab) in the Save-Odzi Valley, Zimbabwe, with considerations for its management. In *Bark: Use, management, and commerce in Africa*, ed. Anthony B. Cunningham, Bruce M. Campbell, and Martin K. Luckert. New York: The New York Botanical Garden Press.
Roodt, V. 1995. *The Shell field guide to the common trees of the Okavango delta and Moremi game reserve*. Gaborone: Shell Oil Botswana.
Root, Alan, and Joan Root. 1973. Baobab: Portrait of a tree. BBC-TV.
Rose, M.D. 1976. Bipedal behavior of olive baboons (Papio Anubis) and its relevance to an understanding of the evolution of human bipedalism. *American Journal of Physical Anthropology* 44 (247): 262.
Ross, Eric S. 1995. Touba: A spiritual metropolis in the modern world. *Canadian Journal of African Studies* 29 (2): 222–259.
———. 2008. Palaver trees reconsidered in the Senegalese landscape. In *African sacred groves: Ecological dynamics and social change*, ed. Michael J. Sheridan and Celia Nyamweru. Athens: Ohio University Press.
———. 2012. *Historic baobab trees of Senegal*. https://ericrossacademic.wordpress.com/2012/01/25/historic-baobab-trees-of-senegal/. Retrieved 2 June 2014.
Rowlett, Ralph M. 1999. Did the use of fire for cooking lead to a diet change that resulted in the expansion of brain size in Homo erectus from that of Australopithecus africanus? *Science* 283 (541): 2004–2005.
Russell, C. 1978. Fire control and human evolution. *Biology and Human Affairs* 43 (2): 14–20.
Sahlins, M. 1968. Notes on the original affluent society. In *Man the hunter*, ed. R.B. Lee and I. DeVore, 85–89. New York: Aldine Publishing Company.
———. 1972. *Stone age economics*. Chicago: Aldine.
Sahn, D.E. 1989. *Seasonal variability in third world agriculture*. Baltimore: Johns Hopkins University Press.
Sai, F.T. 1969. The problem of food and nutrition of West Africa. *World Review of Nutrition and Dietetics* 10: 77–99.

Samson, David R., Alyssa N. Crittenden, Ibrahim A. Mabulla, Audax Z.P. Mabulla, and Charles L. Nunn. 2017a. Hadza sleep biology: Evidence for flexible sleep-wake patterns in hunter-gatherers. *American Journal of Physical Anthropology* 162: 573–582. https://doi.org/10.1002/ajpa.23160.

Samson, David R., Alyssa N. Crittenden, Ibrahim A. Mabulla, and Audax Z.P. Mabulla. 2017b. The evolution of human sleep: Technological and cultural innovation associated with sleep-wake regulation among Hadza hunter-gatherers. *Journal of Human Evolution* 113: 91–102.

Sanchez, Aida Cuni. 2010. *Predicting suitable areas for cultivation and conservation of the baobab tree and investigating superior sources of planting material*. PhD thesis, Faculty of Engineering and the Environment, University of Southampton, Southampton, UK.

Sanchez, Aida Cuni, S. De Smedt, N. Haq, and R. Samson. 2011. Comparative study on baobab fruit morphological variation between Western and South-Eastern Africa: Opportunities for domestication. Genetic Resources and Crop Evolution 58: 1143–1156.

———. 2011. The baobab tree in Malawi. *Fruits* 66: 405–416.

Sarich, V.M., and A.C. Wilson. 1967. Immunological time scale for hominid evolution. *Science* 158: 1200–1203.

Sassaman, Kenneth E. 1992. Lithic technology and the hunter-gatherer sexual division of labor. *North American Archaeologist* 13 (3): 249–262.

Sastry, Vaidya P. Satyanarayana. 2000. Baobab: The wishing tree. *Asian Agri-History* 4: 315–318.

Sauer, C.O. 1947. Early relations of man to plants. *Geographical Review* 37: 1–25.

———. 1952. *Agricultural origins and dispersals*. New York: American Geographical Society.

———. 1956. The agency of man on the earth. In *Man's role in changing the face of the earth*, ed. Sol Tax, vol. 1, 46–69. Chicago: University of Chicago Press.

Sayers, Ken, and C. Owen Lovejoy. 2008. The chimpanzee has no clothes: A critical examination of pan troglodytes in models of human evolution. *Current Anthropology* 49 (1): 87–114.

Schaller, G.B. 1972. *The Serengeti lion*. Chicago: University of Chicago Press.

Schoeninger, M.J., H.T. Bunn, S.S. Murray, and J.A. Marlett. 2001a. Composition of tubers used by Hadza foragers of Tanzania. *Journal of Food Composition and Analysis* 14: 15–25.

Schoeninger, M.J., H.T. Bunn, S. Murray, T. Pickering, and J. Moore. 2001b. Meat-eating by the Fourth African Ape. In *Meat-eating and human evolution*, ed. Craig B. Stanford and Henry T. Bunn, 179–195. Oxford: Oxford University Press.

Schumann, Katharina, Rüdiger Wittig, Adjima Thiombiano, Ute Becker, and Karen Hahn. 2010. Impact of land-use type and bark- and leaf-harvesting on population structure and fruit production of the baobab tree (*Adansonia digitata* L.) in a semi-arid savanna, West Africa. *Forest Ecology and Management* 260 (11): 2035–2044.

———. 2012. Uses, management, and population status of the baobab in Eastern Burkina Faso. *Agroforestry Systems* 85: 263–278.

Sclater, W.L., and R.E. Moreau. 1933. Taxonomic and field notes on some birds of North-Eastern Tanganyika Territory. —Part IV. *Ibis* 75 (2): 187–219. https://doi.org/10.1111/j.1474-919X.1933.tb03632.x.

SCUC (Southampton Centre for Underutilised Crops). 2006. *baobab manual, field manual for extension workers and farmers*. University of Southampton, Southampton.

Scudder, Thayer. 1962. *The ecology of the Gwembe Tonga*. Manchester: Manchester University Press.

Scupin, Raymond, and Christopher R. DeCorse. 2008. *Anthropology: A global perspective*. Upper Saddle River: Pearson.

Seale, Shelley. 2012. *Hunting with the Hadzabe, perceptive travel website*. https://perceptivetravel.com/blog/2012/08/19/hunting-with-the-hadzabe/.

Sébire, A. 1899. *Les plantes utiles du Sénégal: plantes indigenes, plantes exotiques*. Paris: Bailière.

Ségalen, L., J.A. Lee-Thorp, and T. Cerling. 2007. Timing of C4 grass expansion across sub-Saharan Africa. *Journal of Human Evolution* 53 (5): 549–559.

Sharma, P.P., and R.J. Savant. 2012. Some less-known plants parts as supplementary foods. *International Multidisciplinary Research Journal* 2 (12): 12–13.

Shepstone, H.J. 1919. Water tree of the Sudan. *Scientific American* 121 (11): 260–272.
Sherry, D.S., and F.W. Marlowe. 2007. Anthropometric data indicate nutritional homogeneity in Hadza foragers of Tanzania. *American Journal of Human Biology* 19: 107–118.
Shipman, P. 1984. Ancestors: Scavenger hunt. *Natural History* 93 (4): 20–27.
Shipman, P. 1986. Scavenging or hunting in early hominids. *American Anthropologist* 88: 27–43.
Shipman, P. 2010. The animal connection and human evolution. *Current Anthropology* 51 (4): 519–538.
Shreeve, James. 1996. Sunset on the Savanna. Discover Magazine, July 1996, pp.116–125. https://www.discovermagazine.com/the-sciences/sunset-on-the-savanna.
Shostak, M. 1981. *Nisa, the life and words of a !Kung woman*. New York: Vintage Books.
Sidibé, M., and J.T. Williams. 2002. *Baobab: Adansonia digitata L.* Southampton: International Center for Underutilised Crops.
Sidibe, M., J.T. Williams, A. Hughes, N. Haw, and R.W. Smith. 2002. *Baobab. Adansonia digitata L.* Southampton: International Centre for Underutilised Crops.
Silberbauer, George B. 1972. The G/wi Bushmen. In *Hunters and gatherers today*, ed. M.G. Bicchieri, 271–325. New York: Holt, Rinehart and Winston.
Silf, Margaret. 2008. Sacred space for transformation: 'The cost of living our vocation is to be fixed and rooted in the cold soil.' *America: The Jesuit Review*, January 7–14, 9.
Simitu, Parnwell, and Zenroku Oginosako. 2002. Socio-economic survey of *Adansonia digitata* and *Tamarindus indica* in Kitui. In *Utilization and commercialization of dryland indigenous fruit tree species to improve livelihoods in East and Central Africa*. Proceedings of a Regional workshop, KEFRI Kitui, Kenya, 20–24 June 2005, compiled by Parnwell Simitu. World Agroforestry Centre – Eastern and Central Africa Regional Programme (ICRAF ECA), ECA Working Paper No. 7.
Simoons, Frederick J. 1998. *Plants of life, plants of death*. Madison: The University of Wisconsin Press.
Sinclair, A.R.E., M.D. Leakey, and M. Norton-Griffiths. 1986. Migration and hominid bipedalism. *Nature* 324: 307–308.
Skaanes, Thea. 2015. Notes on Hadza cosmology. *Before Farming: The Archaeology and Anthropology of Hunter-Gatherers* 1 (2): 247–267.
Slow Food Foundation. 2018. *Mambuyu*. https://www.fondazioneslowfood.com/en/ark-of-taste-slow-food/mambuyu-baobab-seed-drink/. Accessed 18 June 2020.
Smith, Adam. 1776. *An inquiry into the nature and causes of the wealth of nations*. London: Printed for W. Strahan; and T. Cadell.
Smith, John. 1882. *A dictionary of popular names of the plants which furnish the natural and acquired wants of man, in all matters of domestic and general economy: Their history, products, & uses*. London: Macmillan and Co.
Smith, Robertson. 1894. *Lectures on the religion of the Semites*, ed. J.S. Black. London.
Smitu, Parnwell. 2005. Utilization and commercialization of dryland indigenous fruit tree species to improve livelihoods in East and Central Africa. In *Proceedings of a regional workshop*, ed. Parnwell Smitu. KEFRI Kitui, Kenya, 20–24 June 2005. World Agroforestry Centre – Eastern and Central Africa Regional Programme (ICRAF ECA), ECA Working Paper No. 7.
Soni, Neha. 2015. Important fertility symbols and rites (Northern India). *IOSR Journal of Humanities and Social Science* 20 (4): 51–53.
Sousa, C., D. Biro, and T. Matsuzawa. 2009. Leaf-tool use for drinking water by wild chimpanzees (Pan troglodytes): Acquisition patterns and handedness. *Animal Cognition* 12: 5115–5125. https://doi.org/10.1007/s10071-009-0278-0.
Sousa, Cláudia, Dora Biro, and Tetsuro Matsuzawa. 2009. Leaf-tool use for drinking water by wild chimpanzees (*Pan troglodytes*): acquisition patterns and handedness. *Animal Cognition* 12 (Suppl 1): S115–S125. https://doi.org/10.1007/s10071-009-0278-0. Epub 2009 Aug 21. PMID: 19697068.
Southey, Olivia. 2014. *Baobab birth*. http://www.baobab-birth.com/about.html.
Sowunmi, M.A. 1985. The beginnings of agriculture in West Africa: Botanical evidence. *Current Anthropology* 26 (1): 127–129.

Speth, J.D. 1987. Early hominid subsistence strategies in seasonal habitats. *Journal of Archaeological Science* 14: 13–29.

Spielmann, Katherine A., and James F. Eder. 1994. Hunters and farmers: Then and now. *Annual Review of Anthropology* 23: 303–323.

Spjut, Richard W. 1994. A systematic treatment of fruit types. *Memoirs of the New York Botanical Garden* 70 (12): 1–181.

Spottiswoode, Keith S., K. Begg, and Colleen M. Begg. 2016. Reciprocal signaling in honeyguide-human mutualism. *Science* 353 (6297): 387–389. https://doi.org/10.1126/science.aaf4885.

Spottiswoode, Claire N., Katherine Faust Stryjewski, J.F.R. Suhel Quadera, and M.D. Sorenson Colebrook-Robjent. 2011. Ancient host specificity within a single species of brood parasitic bird. *Proceedings of the National Academy of Sciences* 108 (43): 17738–17742. https://doi.org/10.1073/pnas.1109630108.

Stanford, Craig B. 2003. *Upright: The evolutionary key to becoming human*. Boston: Houghton Mifflin.

Stapleton, C.C. 1955. *Common Transvaal trees*, Department of Forestry Bulletin No. 37. Pretoria: Government Printer.

Starin, Dawn. 2009. What will happen when the baobab goes global? *The New York Times op-ed*, May 26. http://www.nytimes.com/2009/05/26/opinion/26starin.html.

Stavrou, C., and M.S.L. Mills. 2013. *Observations of birds of the Soyo area, northwest Angola*. www.birdsangola.org/.../Stavrou%20&%20Mills.%202013.%20Malimbus.

Stephenson, James. 2000. *The language of the land: Living among a Stone-Age people in Africa*. New York: St. Martin's Press.

Sterner, Lars Johan. 2012. *On issues of timing controlled and habitual fire use: Testing the strengths of the short chronologies with focus on Western Eurasia*, Course Paper (Master Research and Thesis, ARCH 1044WY). Leiden: University of Leiden, Faculty of Archaeology.

Steward, Julian H., and Robert M. Adams. 1955. *Irrigation civilizations: A comparative study*. Social Science Section, Department of Cultural Affairs, Pan American Union.

Story, R. 1958. *Some plants used by the bushman in obtaining food and water*, Botanical Survey Memoir No. 30. Pretoria: Department of Agriculture, Division of Botany.

———. 1964. Plant lore of the Bushman. In *Ecological studies in Southern Africa*, ed. D.H.S. Davis, 87–99. The Hague: Junk Publishers.

Stow, G.W. 1905. *The Native races of South Africa*. London: Swan Sonnenschein.

Sturtevant, E.L. 1972. *Sturtevant's edible plants of the world*, ed. U.P. Hedrick. New York: Dover Publications, Inc.

Sugiyama, Michelle Scalise. 2011. The forager oral tradition and the evolution of prolonged juvenility. *Frontiers in Psychology*. https://doi.org/10.3389/fpsyg.2011.00133.

Swanepoel, C.M. 1993. Baobab phenology and growth in the Zambezi Valley, Zimbabwe. *African Journal of Ecology* 31 (1): 84–86.

Swanepoel, C.M., and S.M. Swanepoel. 1986. Baobab damage by elephant in the middle Zambezi Valley, Zimbabwe. *African Journal of Ecology* 242: 129–132. https://doi.org/10.1111/j.1365-2028.1986.tb00352.x.

———. 2008. Baobab damage by elephant in the middle Zambezi, Valley, Zimbabwe. *African Journal of Ecology* 24 (2): 129–132. https://doi.org/10.1111/j.1365-2028.1986.tb00352.x.

Sweeney, C. 1969. *Jebels by moonlight*. London: Chatto and Windus.

———. 1973. *Background of baobabs*. London: Constable.

———. 1974. *Naturalist in the Sudan*. New York: Taplinger.

Symes, C.T., and M.R. Perrin. 2003. Seasonal occurrence and local movements of the grey-headed (brown-necked) parrot Poicephalus fuscicollis suahelicus in southern Africa. *African Journal of Ecology* 41 (4): 299–305.

———. 2004. Breeding biology of the greyheaded parrot (*Poicephalus fuscocollis suahelicus*) in the wild. *Emu – Austral Ornithology* 104 (1): 45–47.

———. 2008. The habitat and associated bird assemblages of the Grey-headed Parrot *Poicephalus fuscocollis suahelicus* in Limpopo Province, South Africa. *Ostrich* 79 (1): 1–15.

TAHDEL. 2000. *The American Heritage Dictionary of the English Language*. 4th ed. Boston: Houghton Mifflin.

Tamari, Tal. 1991. The development of caste systems in West Africa. *The Journal of African History* 32: 221–250.

Tanaka, Soichi. 1969. Natural environments of Mangola. In *Kyoto University African Studies*, ed. Tada Umesao, vol. 3, 27–54.

Tanaka, J. 1976. Subsistence ecology of central Kalahari San. In *Kalahari Hunter-Gatherers*, ed. R.B. Lee and I. DeVore, 98–119. Cambridge: Harvard University Press.

Tanner, Nancy and Adrienne Zihlman 1976. Women in Evolution. Part I: Innovation and Selection in Human Origins. Signs, Journal of Women in Culture and Society 1(3):585–608.

Tanner, Nancy M. 1981. *On becoming human*. New York: Cambridge University Press.

Tanner, N., and A. Zihlman. 1978. Women in evolution. Part I: Innovation and selection in human origins signs. *Journal of Women in Culture and Society.* 1: 585–608. https://doi.org/10.1086/493245.

Tarboton, Warwick. 2011. *Roberts guide to the nests and eggs of Southern African Birds*. Johannesburg: Jacana Media.

Teel, W. 1985. *A pocket directory of trees and seeds in Kenya*. Nairobi: Kango.

Teleki, G. 1975. Primate subsistence patterns: Collector-predators and gatherer-hunters. *Journal of Human Evolution* 4: 125–184.

Ten Raa, Eric. 1971. The nature of genesis as a social myth. *Paideuma, Bd.* 17: 137–144. Published by Frobenius Institute.

TFD (The Free Dictionary). 2004. *Farlex, Inc.* https://www.thefreedictionary.com/. Accessed 16 May 2021.

Thevet, A. 1558 (1878). *Les Singularitez de la France Antarctique*. Paris: Maisonneuve.

Thomas, E.M. 1959. *The harmless people*. New York: Knopf.

Thompson, Nicholas, and Scott Thompson 2000. The baobab and the mango tree lessons about development – African and Asian contrasts. Bloomsbury Academic.

Thorpe, Susannah K.S., R.L. Holder, and Robin Huw Crompton. 2007. Origin of human bipedalism as an adaptation for locomotion on flexible branches. *Science* 316 (5829): 1328–1331.

Thorpe, Susannah K.S., Juliet M. McClymont, and Robin H. Crompton. 2014. The arboreal origins of human bipedalism. *Antiquity* 88: 906–926.

Tidemann, Sonia, and Andrew Gosler. 2010. *Ethno-ornithology: Birds, indigenous peoples, culture and society*. London: Earthscan.

Tiger, Lionel, and Robin Fox. 1969. *Men in groups*. New York: Vintage Books.

Tobias, P.V. 1971. *The brain in hominid evolution*. New York: Columbia University Press.

Tomita, Kozo. 1966. The sources of food for the Hadzapi Tribe: The life of a hunting tribe in East Africa. *Kyoto University African Studies* 1: 157–171.

Toutain, B. 1978. *Inventaire floristique du Sahel de Haute-Volta et du nord du pays Gourmantche. Ecologies des plantes, nom vernaculaire, interet fourrager*. Maisons-Alfort: IEMVT.

Trivers, R.L. 1971. The evolution of reciprocal altruism. *Quarterly Review of Biology* 46: 35–57.

Tsy, Jean-Michel Leong Pock, Roselyne Lumaret, Diana Mayne, Abdallahi Ould Mohamed Vall, Syahia I.M. Abutaba, Maurice Sagna, Soaharin'ny Ony Rakotondralambo Raoseta, and Pascal Danthu. 2009. Chloroplast DNA phylogeography suggests a West African center of origin for the baobab, *Adansonia digitata* L. (Bombacoideae, Malvaceae). *Molecular Ecology* 18: 1707–1715.

Turnbull, Colin M. 1962. *The forest people*. New York: Simon and Schuster.

———. 1965. *Wayward servants: The two worlds of the African pygmies*. Garden City: Natural History Press.

Tuttle, Russell H. 1981. Evolution of hominid bipedalism and prehensile capabilities. Philosophical Transactions of the Royal Society, London. B29289–94. http://doi.org/10.1098/rstb.1981.0016.

———. 1968. The importance of flux in two hunting societies. In *Man the hunter*, ed. Richard B. Lee and Irven DeVore, 132–137. Chicago: Aldine Publishing Company.

Turnbull-Kemp, P. 1967. *The leopard*. San Francisco: Tri-Ocean Books.

Turner, Pamela S. 2002. We step back in time with the Hadza. *The Christian Science Monitor*, August 13.

Tylor, E.B. 1871. *Primitive culture: Researches into the development of mythology, philosophy, religion, language, art and custom*. London: J. Murray.
Taylor, Stuart. 2000. The behavioral ecology of the Brown-headed Parrot in Southern Africa. *American Federation of Aviculture* 27 (6): 40–42.
Tuttle, R.H. 1981. Evolution of hominid bipedalism and prehensile capabilities. *Philosophical Transactions of the Royal Society* 292 (1057): 89–94. https://doi.org/10.1098/rstb.1981.0016.
Ungar, Peter S., F.E. Grine, and M.F. Teaford. 2006. Diet in early Homo: A review of the evidence and a new model of adaptive versatility. *Annual Review of Anthropology* 35: 209–228.
Uphof, J.C.T. 1968. *Dictionary of economic plants*. New York: Hefner Service Agency Inc.
Utley, Francis Lee. 2005. Tree of life. In *Encyclopedia Americana*. Scholastic Library Publishing, Inc.
UWED (Encarta World English Dictionary). 1999. New York: St. Martin's Press.
Vaid, K.M. 1964. Concluding chapter of a "Kalpa-vriksha". *Indian Forester* 90: 1963–2964.
Vaid, K.M. 1978. Where is the mythical 'wishing tree'? *Science Today*, April 1978, 35–44.
———. 1978a. Where is the mythical 'wishing tree'? *Science Today*, April, 35–44.
———. 1978b. Temple and the tree. *Wildlife Newsletter of Indian Forest College*, October, 58–59.
Vaillant, Janet G. 2002. Homage to Léopold Sédar Senghor: 1906-2001. *Project Muse* 33 (4): 17–24.
Van Balen, J.H., C.J.H. Booy, J.A. Van Franeker, and E.R. Osieck. 1982. Studies on hole-nesting birds in natural nest sites. *Ardea* 55 (1–2): 1–24.
van der Pijl, L. 1936. Fledermäuse und blumen. *Flora* 131: 1–40.
———. 1969. Evolutionary action of tropical animals on the reproduction of plants. *Biological Journal of the Linnean Society* 1 (1–2): 85–96. https://doi.org/10.1111/j.1095-8312.1969.tb01814.x.
Van Der Post, Laurens, and Jane Taylor. 1984. *Testament to the bushmen*. New York: Viking.
van Gennep, Arnold. 1960. *The rites of passage*. Chicago: University of Chicago Press.
van Schaik, Carel P., Robert O. Deaner, and Michelle Y. Merrill. 1999. The conditions for tool use in primates: Implications for the evolution of material Culture. *Journal of Human Evolution* 36: 719–741.
Van Valkenburgh, B. 1985. Locomotor diversity within past present guilds of large predatory mammals. *Paleobiology* 11: 406–428.
Van Zyl, L., and A.V. Ferreira. 2004. Physical and chemical carcass composition of springbok (*Antidorcas marsupialis*), blesbok (*Damaliscus dorcas phillipsi*) and impala (*Aepyceros melampus*). *Small Ruminant Research* 53: 103–109.
Vavilov, N.I. 1926. Studies on the origins of cultivated plants. *Bulletin of Applied Botany and Plant Breeding* 16: 11–245.
Velton, R. 2000. *Mali*. Chalfont St. Peter: Bradt Publications.
Venter, Sarah M. 2012. *The ecology of baobabs (Adansonia digitata L.) in relation to sustainable utilization in northern Venda, South Africa*. Dissertation, University of the Witwatersrand, Johannesburg, South Africa.
Venter, F., and J.A. Venter. 1996. *Making the most of Indigenous trees*. Victoria: Briza Publications.
Venter, Sarah M., and T.F. Witkowski. 2010. Baobab (*Adansonia digitata* L.) density, size-class distribution and population trends between four land-use types in northern Venda, South Africa. *Forest Ecology and Management* 259 (2010): 294–300.
———. 2011. Baobab (*Adansonia digitata* L.) fruit production in communal and conservation land-use types in Southern Africa. *Forest Ecology and Management* 261: 630–639.
Venter, S.M., and T.F. Witkowski. 2013. Where are the young Baobabs? Factors affecting regeneration of Adansonia digitata L, in a communally managed region of southern Africa. *Journal of Arid Environments* 92: 1–13.
Venter, S.M., K.L. Glennon, E.T.F. Witkowski, D. Baum, V. Cron, R. Tivakudze, and N. Karimi. 2017. Baobabs (*Adansonia digitata* L.) are self-incompatible and 'male' trees can produce fruit if hand-pollinated. *South African Journal of Botany* 109: 263–268.
Veslingius, J. 1638. De Plantis Aegyptiis Observationes et Notae ad Prosp. Alpinum. In *De Plantis Aegypti Liber*, ed. P. Alpino, 2nd ed. Venice: Franciscum de Franciscis Senensem.

Vincent, Anne S. 1985a. Plant foods in savannah environments: A preliminary report of tubers eaten by the Hadza of Northern Tanzania. *World Ethnoarchaeology* 17 (2): 131–148.
———. 1985b. *Wild tubers as a harvestable Resource in the East African SAVANNAS*, Ecological and ethnographic studies. Vol. 2. Berkeley: University of California.
Virani, Munir. 2011. *Simon Thomsett on the African Crowned Eagle – Part 1. African Raptors: The online home of African Raptor interests.* http://www.africanraptors.org/simon-thomsett-on-the-african-crowned-eagle-part-1/.
Voeks, R. 1997. *Sacred leaves of Brazilian Candomblé: African magic, medicine, and religion in Brazil.* Austin: University of Texas Press.
von Haartman, Lars. 1957. Adaptation in hole-nesting birds. *Evolution* 11 (3): 339–347.
Von Hagen, Victor Wolfgang. 1943. *The Aztec and Maya papermakers.* New York: J. J. Augustin Publishers.
von Maydell, H.J. 1986. *Trees and shrubs of the sahel: Their characteristics and uses.* Rossdorf: TZ-Verlagsgesellschaft.
Vrba, E.S. 1993. The pulse that produced us. *Natural History* 102: 47–51.
———. 1995. On the connections between paleoclimate and evolution. In *Paleoclimate and evolution*, ed. E.S. Vrba et al., 24–25. New Haven: Yale University Press.
Vrba, E.S., G.H. Denton, T.C. Partridge, and L.H. Burkle. 1995. *Paleoclimate and evolution.* New Haven: Yale University Press.
Walker, S.S. 1990. Everyday and esoteric reality in the Afro-Brazilian Candomblé. *History of Religions* 30 (2): 103–128.
Ward, C.V., W.H. Kimbel, and D.C. Johanson. 2011. Complete fourth metatarsal and arches in the foot of *Australopithecus afarensis. Science* 331: 750–753.
Washburn, S.L. 1960. Tools and human evolution. *Scientific American* 203: 63–75.
———. 1963. *Classification and human evolution.* New York: Wenner-Gren Foundation.
———. 1968. Speculations on the problem of man's coming to the ground. In *Changing perspectives on man*, ed. B. Rothblatt, 193–206. Chicago: University of Chicago Press.
Washburn, S.L., and R.L. Ciochon. 1974. Canine teeth: Notes on controversies on the study of human evolution. *American Anthropologist* 76: 765–784.
Washburn, S.L., and C.K. Lancaster. 1968. The evolution of hunting. In *Man the hunter*, ed. R.B. Lee and I. DeVore, 293–303. Chicago: Aldine.
Watson, Rupert. 2007. *The African baobab.* Cape Town: Struik Publishers.
Watt, G. 1885. *A dictionary of the economic products of India. Vol 1, Abaca to Buxus.* Calcutta: Allen.
Watt, John Mitchell. 1972. Magic and witchcraft in relation to plants and folk medicine. In *Plants in the development of modern medicine*, ed. Tony Swain, 67–102. Cambridge: Harvard University Press.
WD (Webster's Dictionary). 1828. *American dictionary of the English language* (Online edition). http://webstersdictionary1828.com/. Accessed 18 June 2020.
Weaver Watch. 2016. *Weavers breeding in Baobabs.* Cape Town, South Africa. http://weavers.adu.org.za/spcat.php?spc=22. Accessed 01/02/2016.
———. 2017. *Red-billed Quelea Quelea quelea.* http://weavers.adu.org.za. Accessed 18 June 2020.
———. 2020. *Weavers breeding in baobabs.* https://weavers.adu.org.za. Retrieved 6/3/2020.
Weiner, J. 1973. Dressing percentage, gross body composition and caloric value of the roe deer. *Acta Theriologica* 18: 209–222.
Weiss, E.A. 1979. Some indigenous plants used domestically by East African coastal fishermen. *Economic Botany* 33 (1): 35–51.
Wekesa, Linus, Esther Wang'ombe, Michael Mukolwe, Isamu Yamada, and Yuichi Sato. 2005. *Marketing study of ecological resource products.* Report prepared for ISFP. ISFP, Nairobi, Kenya.
Wells, Spencer. 2011. *Pandora's seed: Why the hunter-gatherer holds the key to our survival.* New York: Random House.
Welman, John. 1966. The baobab. *Blackwood's Magazine* 1804 (299): 150–159.
Werema, Chacha. 2017. Preliminary comparison of birds inhabiting exotic Acacia and native bushland habitats in semi-arid East-Central Tanzania. *Scopus* 37 (2): 32–37.

Werema, Chacha, and Cuthbert L. Nahonyo. 2015. Preliminary observations of birds of Songo Songo Island, Tanzania. *Scopus* 35: 39–43.
Wescott, R.W. 1967a. The exhibitionistic origin of human bipedalism. *Man* 2: 630.
———. 1967b. Hominid uprightness and primate display. *American Anthropologist* 69: 738.
Wheeler, P.E. 1984. The evolution of bipedality and loss of functional body hair in hominids. *Journal of Human Evolution* 13 (1): 91–98.
———. 1991. The thermoregulatory advantages of hominid bipedalism in open equatorial environments: The contribution of increased convective heat loss and cutaneous evaporative cooling. *Journal of Human Evolution* 21: 107–115.
———. 1992. The thermoregulatory advantages of large body size for hominids foraging in savannah environments. *Journal of Human Evolution* 23 (4): 351–362.
Whistler, Arthur W. 2000. *Plants in Samoan culture: The ethnobotany of Samoa*. Honolulu: Isle Botanica.
———. 2009. *Plants of the Canoe people: An ethnobotanical Voyage through Polynesia*. Lawai, Kauai: National Tropical Botanical Garden.
White, T., B. Asfaw, Y. Beyene, Y. Haile-Selassie, C. Lovejoy, G. Suwa, and G. WoldeGabriel. 2009. *Ardipithecus ramidus* and the paleobiology of early hominids. *Science* 326 (5949): 64. 75–86.
Whiten, A., J. Goodall, W.C. McGrew, T. Nishida, V. Reynolds, Y. Sugiyama, C.E.G. Tutin, R.W. Wrangham, and C. Boesch. 1999. Cultures in chimpanzees. *Nature* 399: 682–685.
Wickens, G.E. 1978. The uses of the baobab (Adansonia digitata L.) in Africa. In *Proceedings of the IX Plenary Meetings of Association pour l'Etude de la Taxonomize de la Flore d'Afrique Tropicale (AETFAT)*, 27–34. 18–23 March 1978. Las Palmas de Gran Canaria, Canary Islands, Spain.
———. 1979. The use of the baobab (*Adansonia digitata* L.) in Africa. Proc. IX Plenary Meeting of A.E.T.F.A.T., Las Palmas de Gran Canaria, 18–23 March 1978: 27–34.
———. 1982. Baobab: Africa's upside-down tree. *Kew Bulletin* 37 (2): 173–209.
Wickens, G.E., and P. Lowe. 2008. *The baobabs: Pachycauls of Africa, Madagascar and Australia*. Dordrecht: Springer.
Wiehle, Martin, Kathleen Prinz, Katja Kehlenbeck, Sven Goenster, Seifeldin Ali Mohamed, Reiner Finkeldey, Andreas Buerkert, and Jens Gebauer. 2014. The African baobab (*Adansonia digitata*, Malvaceae): Genetic resources in neglected populations of the Nuba Mountains, Sudan. *American Journal of Botany* 101 (9): 1498–1507.
Wiessner, Polly W. 2014. Embers of society: Firelight talk among the Ju/'hoansi Bushmen. *Proceedings of the National Academy of Sciences of the United States of America* 111 (39): 14027–14035.
Williams, Sharlotte Neely. 1971. The limitations of the male/female activity distinction among primates: An extension of Judith K. Brown's "A note on the division of labor by sex". *American Anthropologist* 73 (3): 805–806.
Williams, J.G., and N. Arlott. 1986. *A field guide to the birds of East Africa*. London: Collins.
Williams, Vivienne L., Anthony B. Cunningham, Alan C. Kemp, and Robin K. Bruyns. 2014. Risks to birds traded for African traditional medicine: A quantitative assessment. *PLoS One* 9 (8): e105397. https://journals.plos.org/plosone/article?id=10.1371/journal.pone.0105397. Accessed 18 June 2020.
Williamson, Jessie. 1955. *Useful plants of Nyasaland*. Zambia: Government Printer.
Wilmsen, E. 1989. *Land filled with flies: A political economy of the Kalahari*. Chicago: University of Chicago Press.
———. 1993. On the search for (truth) and authority: A reply to Lee and Guenther. *Current Anthropology* 34 (5): 715–721.
Wilmsen, E.N., and J.R. Denbow. 1990. Paradigmatic history of San-speaking peoples and current attempts at revision. *Current Anthropology* 31 (5): 489–524.
Wilson, Don E. 2020. *Bat*. Encyclopedia Britannica (Online). Retrieved 17 Mar 2022.
Wilson, R.T., and M.P. Wilson. 1986. Best building by the Hamerkop *Scopus umbretta*. *Ostrich* 57: 224–232.

Winder, Isabelle C., Geoffrey C.P. King, Maud H. Devès, and Geoffrey N. Bailey. 2013. Complex topography and human evolution: The missing link. *Antiquity* 87: 333–349.

———. 2014. Human bipedalism and the importance of terrestriality. *Antiquity* 88: 906–926.

Winterhalder, Bruce. 2001. The behavioural ecology of hunter-gatherers. In *Hunter-gatherers: An interdisciplinary perspective*, 12–38. Cambridge: Cambridge University Press.

Wise, Christopher. 2006. *Comparative Literature Studies* 43 (1/2): 19–38.

Wittfogel, Karl. 1957. *Oriental despotism; A comparative study of total power*. New York: Random house.

WNIDEL (Webster's New International Dictionary of the English Language). 1949. Springfield: G. & C. Merriam Co.

WNRUD (Webster's 11 New Riverside University Dictionary). 1984. Boston: Houghton Mifflin Publishing.

Wolf, Eric R. 1982. *Europe and the people without history*. Berkeley: University of California Press.

Wood, Brian M. 2006. Prestige or provisioning? A test of foraging goals among the Hadza. *Current Anthropology* 47 (2): 383–397.

Wood, B., and M. Collard. 1999. The changing face of genus *Homo*. *Evolutionary Anthropology* 8 (6): 195–207.

Wood, Brian M., Herman Pontzer, David A. Raichlen, and Frank W. Marlowe. 2014. Mutualism and manipulation in Hadza – Honeyguide interactions. *Evolution and Human Behavior* 35 (2014): 540–546.

Wood, Brian M., Jacob A. Harris, David A. Raichlen, Herman Pontzer, Katherine Sayre, Amelia Sancilio, Colette Berbesque, Alyssa N. Crittenden, Audax Mabulla, Richard McElreath, Elizabeth Cashdan, and James Holland Jones. 2021. Gendered movement ecology and landscape use in Hadza hunter-gatherers. *Nature Human Behavior* 5: 436–446.

Woodburn, James. 1959. *Hadza conceptions of health and disease*. Paper presented at the one day symposium on Attitudes to Health and Disease among some East African tribes. Kampala, Uganda, East Africa, 1959, 89–94.

———. 1962. The future of the Tindiga. *Tanganyika Notes and Records* 59: 268–273.

———. 1964. *The social organization of the Hadza of North Tanganyika*. PhD dissertation, Cambridge University, Cambridge, England.

Woodburn, James. 1968. An introduction to Hadza ecology. In *Man the hunter*, ed. Richard B. Lee and Irven DeVore, 49–55. Chicago: Aldine Atherton.

———. 1968a. An introduction to Hadza ecology. In *Man the hunter*, ed. Richard B. Lee and Irven DeVore, 49–55. Chicago: Aldine Atherton.

———. 1968b. Stability and flexibility in Hadza residential groupings. In *Man the hunter*, ed. Richard B. Lee and Irven DeVore, 103–110. Chicago: Aldine Atherton.

———. 1970. *The material culture of the Nomadic Hadza*. London: The British Museum.

———. 1972. Ecology, nomadic movement and the composition of the local group among hunters and gatherers: An East African example and its implications. In *Man, settlement and urbanism*, ed. Peter J. Ucko, Ruth Tringham, and G.W. Dimbleby, 193–206. Cambridge, MA: Schenkman Publishing Company.

———. 1982a. Egalitarian societies. *Man, New Series* 17 (3): 431–451.

———. 1982b. Social dimensions of death in four African hunting and gathering societies. In *Death and the regeneration of life*, 187–210. Cambridge: Cambridge University Press.

———. 1999. The Hadza of Tanzania. In *Cambridge encyclopaedia of hunter-gatherers*, ed. R.B. Lee and R. Daly, 31–37. Cambridge: Cambridge University Press.

———. 2013. Foreword. In *Hadzabe: By the light of a million fires*, ed. Daudi Peterson, Richard Baalow, and Jon Cox, xiv–xv. Dar-es-Salaam: Mkuki na Nyouta Publishers Ltd.

Woodburn, J.C., and Sean Hudson. 1966. *The Hadza: The food quest of a hunting and gathering tribe of Tanzania (16mm film)*. London: London School of Economics.

Wrangham, Richard. 1963. *Classification and human evolution*. Chicago: Aldine.

———. 2009. *Catching fire: How cooking made us human*. New York: Basic Books.

Wrangham, Richard, and Rachel Carmody. 2010. Human adaptation to the control of fire. *Evolutionary Anthropology* 19: 187–199.

Wrangham, Richard, and Zarin Machanda. 2013. Honey exploitation by chimpanzees and hunter-gatherers indicates an ancient use of fire by humans. In *The 82nd annual meeting of the American Association of Physical Anthropologists*.

Wrangham, R., J.H. Jones, G. Laden, D. Pilbeam, and N.L. Conklin-Brittain. 1999. The raw and the stolen: Cooking and the ecology of human origins. *Current Anthropology* 40: 567–594. https://doi.org/10.1086/300083.

Wrangham, Richard, Dorothy Cheney, Robert Seyfarth, and Esteban Sarmiento. 2009. Shallow-water habitats as sources of fallback foods for hominins. *American Journal of Physical Anthropology* 140: 630–642. https://doi.org/10.1002/ajpa.21122.

Wright, I.M., and O. Kerfoot. 1966. The African baobab: Object of awe. *The Journal of Natural History* 75 (5): 50–53.

Yellen, John E., and Richard B. Lee. 1976. The Dobe-/Du/da environment. In *Kalahari hunter-gatherers*, ed. R.B. Lee and I. DeVore, 27–46. Cambridge, MA: Harvard University Press.

Yu, Douglas. 2016. *Baobab foods predicts baobab will explode in the US beverage industry*. William Reed Business Media Ltd. https://www.beveragedaily.com/Article/2016/02/16/Baobab-Foods-predicts-baobab-will-explode-in-the-US-beverage-industry#.

Zazamalala Foundation. 2019. *Joy and tragedy under the baobab*. https://www.facebook.com/264123037853082/posts/voor-nederlands-zie-hieronderjoy-and-tragedy-under-the-baobabmany-birds-use-our-/332394051025980/.

Zihlman, A. 1978. Women in evolution Part II: Subsistence and social organization among early hominids. *Signs: Journal of Women in Culture and Society* 4: 4–20. https://doi.org/10.1086/493566.

Zihlman, Adrienne. 1991. Did the Australopithecines have a division of labor? In *The archaeology of gender: Proceedings of the Twenty-second Annual Conference of the Archaeological Association of the University of Calgary*, ed. Dale Walde and Noreen D. Willows, 64–70.

Zihlman, Adrienne L. 1978a. Motherhood in Transition: From Ape to Human. In: Family Formation and First Child, W. Miller and L. Newman (eds.), pp. 35-50. Chapel Hill: Carolina Population Center Publications.

Zihlman, Adrienne L. 1978b. Women and Evolution. Subsistence and Social Organization among Early Hominids. Journal of Women in Culture and Society 4: 4-20.

Zihlman, A. L. and Tanner, N. 1978c. Gathering and the Hominid Adaptation. In: L. Tiger and H. Fowler (eds) Female Hierarchies, pp. 163-194. Chicago: Beresford Book Service.

Zihlman, A.L., J.E. Cronin, D.L. Cramer, and V.M. Sarich. 1978. Pygmy chimpanzee as a possible prototype for the common ancestor of humans, chimpanzees and gorillas. *Nature* 275: 744–746. PMID 703839 DOI: 10.1038/275744A0.

Zihlman, A., and N. Tanner. 1978. Gathering and the hominid adaptation. In *Female hierarchies*, ed. L. Tiger and H. Fowler, 163–194. Chicago: Beresford Book.

Zimmerman, D.A., D.A. Turner, and D.J. Pearson. 1996. *Birds of Kenya and Northern Tanzania*. Princeton: Princeton University Press.

———. 1999. *Birds of Kenya and Northern Tanzania*. Princeton (New Edition): Princeton University Press.

Zwarts, Leo, Rob G. Bijlsma, Jan van der Kamp, Marten Sikkema, and Eddy Wymenga. 2015. Moreau's paradox reversed, or why insectivorous birds reach high densities in savanna trees. *Ardea* 103 (2): 123–144. https://doi.org/10.5253/arde.v103i2.a2.

Index

J. Rashford, *Baobab*, https://doi.org/10.1007/978-3-031-26470-2